# Research Methods and Design Beyond a Single Discipline

*Research Methods and Design Beyond a Single Discipline* presents and discusses 15 research methods after an overview of each, an illustration of the research process, and an examination of related topics. The focus of this book is on data collection techniques (e.g., interviewing, observation, questionnaires) and research design.

Research methods are named and categorized by data collection techniques in this book rather than by the quantitative and qualitative dichotomy used for data analysis techniques. Each of the 15 research methods included in the book is depicted regarding its functions, features, applications, and dos and don'ts, with examples and case studies from various disciplines. Research methodology development in research design starts with selection of data collection techniques, which takes into consideration research questions of a study and research method features. Data analysis techniques and selection are described as part of the research methodology development to help readers choose the most appropriate methodology for their own research, and put research design into practice.

This book is essential for masters and doctoral students, and researchers who wish to learn about research methods and design comprehensively and systematically, as well as instructors delivering research methods courses across the social and behavioral sciences.

**Heting Chu** is a Professor at Palmer School of Library and Information Science, Long Island University, USA.

# Research Methods and Design Beyond a Single Discipline

## From Principles to Practice

**Heting Chu**

Routledge
Taylor & Francis Group
LONDON AND NEW YORK

Designed cover image: vasabii / Getty Images

First published 2024
by Routledge
4 Park Square, Milton Park, Abingdon, Oxon OX14 4RN

and by Routledge
605 Third Avenue, New York, NY 10158

*Routledge is an imprint of the Taylor & Francis Group, an informa business*



*British Library Cataloguing-in-Publication Data*
A catalogue record for this book is available from the British Library

ISBN: 978-0-367-89883-0 (hbk)
ISBN: 978-0-367-89885-4 (pbk)
ISBN: 978-1-003-02171-1 (ebk)

DOI: 10.4324/9781003021711

Typeset in Optima
by KnowledgeWorks Global Ltd.

# Contents

# Figures and tables

## Figures



## Tables

# Preface

I feel delighted to have this opportunity to share with you my learning, thinking, understanding, and using research methods since the 1980s. My teaching of this subject in the past decades also contributes considerably to the formation of this book as my interaction with students, especially those in our Ph.D. program, sparks interesting ideas and thoughts. I often have been having the urge to put down what is in my mind in the manuscript writing process. Once a section or chapter is completed, I eagerly look forward to working on the next one. I have to admit that I do not experience such kind of excitement in my major writing projects frequently.

The organization of this book, as shown in the Table of Contents, is unique in two ways. First, it mainly focuses on data collection methods and research design because data collection must be done before data analysis and numerous excellent books on data analysis, especially statistics or quantitative data analysis, are already in existence. In addition, research design, the other theme of this book, also starts with selection of data collection techniques. Second, research methods in this book are named after and categorized by data collection technique (e.g., interview, observation, questionnaire). Detailed rationale for these two features of this book is provided in Chapter 1.

The entire book consists of 14 chapters in four parts. Part I offers an introduction and overview of this book. Chapter 1 explains what constitutes a research method and why research methods should be categorized by and named after data collection technique. After a walkthrough of major research methods utilized in the social sciences, it presents a taxonomy of research methods by applicability. A note on the orientation of this book, *Research Methods and Design Beyond a Single Discipline*, is made at the end of Chapter 1.

Chapter 2 illustrates the general research process that comprises problem identification and statement, literature search and review, research question formulation, research method selection, data collection and analysis, results presentation and discussion, as well as conclusion drawing. Essentials in levels of measurement and sampling are depicted along with a discussion of validity, reliability, and ethics in research.

Chapters 1 and 2 should be read in their entirety before those on individual research methods and the last two chapters of this book. All the research methods included in this book are grouped into stand-alone methods, super-methods, meta-methods, and contingent methods.

Part II concentrates on stand-alone methods that are further divided into six chapters. Chapter 3 is for questionnaire, Chapter 4 for interview, Chapter 5 for observation, Chapter 6 for theoretical approach, Chapter 7 for metric-based methods, and Chapter 8 is devoted to four less often used stand-alone methods: focus groups, transaction log analysis, research diaries, and historical method.

Part III contains four chapters, each describing one type of research methods. Chapter 9 is about a super-method of content analysis. Chapters 10 and 11 present meta-methods – experiment and ethnography, respectively. Chapter 12 depicts two contingent methods: think-aloud protocols and Delphi study.

Part IV, the last part of this book, is devoted to method selection and research design in two separate chapters. Chapter 13 explores how different types techniques are chosen for analyzing the quantitative and/or qualitative data collected. Chapter 14 aims to demonstrate how to design research based on research questions and functions/features of various techniques for data collection. The last chapter concludes with some final words on research methods and design.

All the topics and concepts included in this book are presented clearly and logically. Although some well-known monographs have decent coverage of one or two major research methods (e.g., interview, questionnaire, observation), this book provides a panoramic and coherent perspective of 15 research methods in one place as well as with abundance of tips and advice on method selection and research design. Furthermore, all the topics are described and presented in such a way that it is easy for the reader to comprehend.

Case studies are offered for each research method discussed in this book to further illustrate and demonstrate how each is utilized in real research. I make great efforts in choosing case studies that are representative in their use of individual research methods. In general, two case studies are done for major research methods (e.g., questionnaire, interview, observation), while only one is done for less frequently utilized research methods (e.g., focus groups, research diaries). Case studies are assessed primarily based on how a given method is used in the study. However, other points relevant to scholarly writings (e.g., research question formulation) are also made where appropriate in order to help set up a high standard for reporting research. Case studies are performed more critically in this book than in most evaluations in the hope to point out the weaknesses and limitations of those studies so that similar mistakes will not be repeated in actual research. My intention however is not to be negative about those studies. Instead, each study examined in this book is exemplary in using a particular method in the first place. Furthermore, each case study publication is in the public domain (see Appendix: List of Case Study Publications) and can be readily accessed via Google Scholar by searching its title unless indicated otherwise.

Links to online sources, including all the case study publications, might have changed or become unavailable when you read this book even though I have done my best to check them before submitting my manuscript for publishing. If such incidents occur, do try to search again using a related search tool such as search engine or database to locate the publication, website, or any other items.

This book project started right before the outbreak of the COVID-19 pandemic around the world. During the time span of over three years, I received enormous support and assistance from so many individuals. The entire book project might not have come to fruition if it were not with the unceasing encouragement and generous support provided by Hannah Shakespeare, then Senior Editor of Research Methods at Routledge. My heart-felt thanks also go to (1) Alexis O'Brien, Marlena Sullivan, and Alexandra Simmons who were my Development Editors in chronological order for their professional assistance; (2) Lucy Kennedy and Eleanor Taylor who functioned as Senior Editor of Research Methods for their patience, invaluable advice, and guidance in the latter part of this book project after Hannah left for a new position within Taylor and

Francis; and (3) Matthew Bickerton, Senior Editorial Assistant of Research Methods and Psychology, for guiding me through this book project according to Taylor and Francis' policies. The librarians, especially Louis Pisha, at my affiliation Long Island University always provided timely help in locating what I was looking for in spite of the COVID-19 pandemic.

I also would like to thank my husband who served as the first reader for each of my book chapter regardless of the challenging and serious nature of the book subject. The arrival of my granddaughter in November 2022 has been a continuous source of happiness and laughter even though she "took away" some of my book writing time.

Heting Chu
Glen Cove, New York

# Part I

# Introduction and Overview

# 1 Research Methods

## An Overview

**Chapter Overview Questions**

- What is research?
- What are research methods? What constitutes a research method?
- How should research methods be named and categorized?
- Why cannot research methods be categorized as quantitative and qualitative?
- What major research methods are there in the scholarly world?
- How can research methods be grouped by applicability?
- Why does this book approach research methods and design beyond a single discipline?

### 1.1 Research and research methods

People need to conduct research, specially planned and structured investigation, for a variety of purposes (e.g., solving problems, understanding reality, learning from the past) using a carefully developed methodology. There are two major kinds of research depending on if study findings are expected to have real-life applications: basic research and applied research. Basic research, also known as pure or theoretical research, refers to studies carried out for achieving a fuller understanding of a phenomenon or an issue without immediate consideration of how the findings will be applied. Example questions of basic research include: What is information? What is the origin of human beings? Answers to such questions are sought but the immediate goal of researchers is to understand and explain certain things, rather than apply their research results in real-life situations. In comparison, applied research or empirical research is conducted to solve practical problems, discover new knowledge, and so on, so that study findings can be utilized immediately in the real world. What can be done to improve the usability of e-commerce sites? What measures can be taken to enhance people's standards of living? These sample questions represent what applied research does.

When people conduct basic or applied research, they must choose and develop one or more techniques for collecting and analyzing data for their scholarly endeavors. A method is a particular form of procedure for accomplishing or approaching something, especially a systematic or established one. A research method, simply put, is a systematic procedure for conducting research. This procedure involves planning, selecting, and developing specific techniques for both data collection and data analysis.

DOI: 10.4324/9781003021711-2

Data collection techniques commonly seen in the scholarly world include experiment, interview, observation, and questionnaire. In contrast, data analysis techniques are too many to be completely enumerated in this context. They however can be roughly grouped into two broad categories: quantitative data analysis techniques and qualitative ones. Generally speaking, data collection techniques are chosen based on research questions, while data analysis techniques are selected according to the type of data collected. If quantitative data (i.e., numbers) are collected, they should be analyzed quantitatively using statistical techniques such as frequency, mean, and correlation analysis. It is in this sense that quantitative data analysis is typically called statistical analysis. Qualitative data analysis techniques (e.g., comparison, qualitative coding) would be adopted if qualitative data (i.e., texts) are collected. There are overall many more techniques for quantitative data analysis than that for qualitative data analysis although the latter are by no means easier or simpler.

The focus of this book is placed on how to select and develop data collection techniques according to one's research problem and subsequently decide whether the data collected should be analyzed quantitatively, qualitatively, or using both quantitative and qualitative techniques. Coverage of specific quantitative data analysis techniques or statistical procedures is beyond the scope of this book partly because numerous excellent monographs on statistics and qualitative data analysis are readily available. Nevertheless, this book will make references to some such books and resources in Chapter 13, Techniques for Data Analysis and Their Selection, when covering how to select techniques for analyzing quantitative and qualitative data.

## 1.2 Naming and categorizing research methods

The concept of research method has been given many naming variations ever since its inception. Sample names for research method include research strategy, research framework, research design, and research methodology. Those naming variations imply that different scholars have different understandings and interpretations of this concept.

In classifying research methods used in studies published in core journals in library and information science, Järvelin and Vakkari (1990) came up with research strategy (e.g., empirical, conceptual), data collection method (e.g., interview, observation, questionnaire), type of analysis (e.g., qualitative, quantitative), and type of investigation (e.g., empirical, descriptive, comparative, conceptual). In the authors' viewpoint, these four terms represent the different components of research methods. The same classification scheme was followed in some later studies on research methods. But elements in the four specified facets of research methods are sometimes not mutually exclusive. For instance, "citation analysis" is listed both as a research strategy and data collection method. The same can be said about "experiment" or "IR (information retrieval) experiment" under research strategy and data collection method, respectively. Other identical pairs in Järvelin and Vakkari's classification scheme encompass "qualitative method" in research strategy and type of analysis, "empirical" or "conceptual" in both research strategy and type of investigation. Such redundancies in the classification confirm that components (e.g., research strategy) other than data collection and analysis techniques may not be part of research methods.

Of the terms some scholars suggest for research method, research strategy and research design emphasize the approaches scholars take in their investigations. Research framework, on the other hand, deals with the overall structure of a study instead of actual techniques for data collection and analysis. It should be noted that "method" and "technique" are essentially synonymous. But in this book, for clarity purpose, "method" refers to all data collection and analysis techniques employed in a single study, while "technique" is a term associated with specific procedures for data collection or analysis. Research methodology in comparison has a broad connotation, covering a set of methods applied in a study or a system of methods used in a particular area of study or activity. Researchers sometimes do use "method" and "methodology" interchangeably without a clear distinction between the two. In addition, not all naming variations of research methods (excluding data collection and analysis techniques) are considered in every study on research methods. For instance, research paradigm (e.g., naturalism, phenomenology, positivism) is more meaningful at the conceptual level than directly linking to any specific data collection or analysis technique. The same can be said about research strategy, research framework, or any other naming variations of research method with the exception of data collection and analysis techniques.

Research methods have also been categorized using other criteria in the past. Common categorization criteria include data collection technique (e.g., interview, observation, questionnaire), data analysis technique (e.g., quantitative, qualitative), and temporal dimension (e.g., cross-sectional, longitudinal). Because research methods, as stated earlier, consist of data collection techniques and data analysis techniques, any criterion that is not part of research methods cannot and should not be chosen as a criterion for categorization purpose. Otherwise, categorization results would not be collectively exhaustive and mutually exclusive – the two fundamental principles any categorization effort must observe.

The one criterion for categorizing and naming research methods therefore has to be chosen between data collection technique and data analysis technique. Of the two, which one is more appropriate for naming and categorizing research methods? The answer is data collection technique because of the following three reasons: (1) Data collection techniques (e.g., experiment, interview, questionnaire) are both unique and indicative. (2) Data must be first collected before being analyzed. It is the type of data collected that dictates the kind of technique to be used for data analysis. (3) The three broad types of data analysis techniques (i.e., quantitative, qualitative, integrated[1]) are neither adequate nor indicative for categorizing research methods. If specific data analysis techniques such as central tendency or ANOVA are used for categorization purpose, the indicative issue of naming and categorizing research methods remains unsolved. For these three reasons, research methods should be named after and categorized by data collection technique. Further elaboration on the reasons why research methods cannot be named and categorized by data analysis technique is provided in the next section.

## 1.3 Labeling research methods by data collection technique: A further discussion

As one of the two components that make up research methods, data analysis techniques are often broadly categorized as qualitative or quantitative. Qualitative and quantitative data analysis techniques can each be further classified into more specific ones based on the nature of data collected (e.g., frequency distribution or correlation analysis for quantitative data, qualitative coding, or content analysis for qualitative data). Individual data analysis techniques, as indicated earlier, will not be discussed extensively in this book given its focus and scope. Instead this book will explore how to select proper data analysis techniques given a study's research problem and what is already collected.

One major reason why research methods should not be labeled and categorized as qualitative and quantitative is because almost all the data collection techniques (e.g., questionnaire, observation) can be used to gather both qualitative and quantitative data despite the fact that each may have an orientation toward one of the two. Take questionnaire as an example, it can contain open-ended questions that aim to gather qualitative data even though questionnaire is generally adopted for collecting quantitative data. Likewise, observation leans more toward qualitative data collection (e.g., how social media users interact with one another) but can also be employed to gather quantitative data (e.g., number of contacts each social media user has). Experiment can also serve as an example in this context. It can be used to collect both qualitative data (e.g., observation notes for a usability test of search engines) and quantitative data (e.g., usability performance measures of search engines) in its design although experiment is often considered a method for quantitative research in publications.

It becomes apparent by now that whether qualitative or quantitative techniques should be used for data analysis in a study depends entirely on the type of data (i.e., qualitative or quantitative) collected. When qualitative as well as quantitative data are collected, for example, via questionnaire, both qualitative and quantitative techniques must be employed for analyzing them, respectively. If we place questionnaire into the quantitative category only, how about the qualitative data gathered via open-ended questions that must subsequently be analyzed qualitatively? It is thus unsuitable to categorize and label research methods by data analysis technique. Existing examples of such inappropriate practices include calling questionnaire a quantitative method and observation a qualitative one. In other words, research methods should be named and categorized by data collection technique. There is no data collection technique that can be 100% quantitative or qualitative. Nor shall any research method be called quantitative or qualitative.

## 1.4 A quick walkthrough of major research methods

When explaining the etymology of the term "method", Glazier (2017, 50) specifies that method is the pursuit after some end or a procedure employed to attain a certain end. The end in this context is the data to be collected and the method is the means. Experiment, interview, observation, and questionnaire are some means researchers take to achieve the end of data collection. Scholars, for example, use questionnaires for data collection in questionnaire research. Investigators conduct experiments to gather data in experimental research. That is, the data collection technique used in a study is often part of the research name. Table 1.1 lists major research methods used in the scholarly world on the basis of an extensive review of publications on research methods as well as a content analysis and categorization effort of the data collection techniques reported in over 1,900 research publications from three major journals in information studies (Chu and Ke 2017).

Table 1.1 records more than 14 individual research methods that are named after corresponding data collection techniques. Below is a brief overview of them before each method is depicted and discussed from Chapters 3 to 12 with ample examples as well as case studies. The overview of these research methods is presented alphabetically by their name for the sake of simplicity. No assumptions should be made about each method's significance in the presentation order.

### *1.4.1 Bibliometrics and other metric-based methods*

Bibliometrics is a method used for collecting publication and reference/citation data from written publications. References and citations are two sides of the same coin. If this concept is looked at from the perspective of the citing author, it is called "reference"

*Table 1.1* Major Research Methods in the Scholarly World

| |
|---|
| Bibliometrics and other metric-based methods |
| Content analysis (including discourse analysis and secondary analysis) |
| Delphi study |
| Ethnography/field study |
| Experiment |
| Focus groups |
| Historical method |
| Interview |
| Observation |
| Questionnaire (including index, inventory, scale, and test) |
| Research diary/journal |
| Theoretical approach (e.g., conceptual analysis, modeling, and theory building) |
| Think-aloud protocol |
| Transaction log analysis |
| Other methods (e.g., action research, card sorting) |

when the citing author makes a reference to a publication. When the same activity is viewed from the cited author's perspective, we would say that the cited author receives a "citation". References and citations are thus synonymous. But citations will be used for presenting both citations and references throughout this book for brevity purpose.

Bibliometrics is often applied in research with an evaluative intention. In this book, bibliometrics also includes citation analysis, informetrics, and scientometrics because they all share the same type of data – publication and citation data. It is the research focus that actually differentiates one from the other. When the research focus is on citations, it is called citation analysis. When the research focus is on information, it is called informetrics. If the research emphasis is on sciences, it is called scientometrics. In addition to publication and citation data, bylines, acknowledgments, and other similar components of publications can also serve as source data.

Webometrics, by and large, is bibliometrics in the web environment where webpages and websites are generally regarded as "publications", inlinks (i.e., links a webpage/site receives), and outlinks (i.e., links a webpage/site makes to others) as citations and references, respectively. Moreover, the web has its own "citation" or link indexes in the form of commercial search engines, and so it is ready for researchers to exploit (Thelwall 2008). Webometrics became a technique for collecting data from the web via search engines in the late 1990s and early 2000s. Link analysis or cybermetrics are often used in place of webometrics. A study may be considered using the link analysis method if it collects only link data. The internet, of which the web is a part, was sometimes called cyberspace in the late 1990s. In that sense, cybermetrics is a synonym of webometrics to a great extent.

Other metric data (e.g., article downloads, views, mentions) are increasingly available from academic social media (e.g., academia.edu, Mendeley), digital libraries, and other similar platforms. Non-academic social media sites (e.g., Facebook, Twitter[2]) also gradually play an important role in today's society. All those new forms of metric data serve as a catalyst for the emergence of altmetrics, short for alternative metrics. Altmetrics in recent years has become a research method for collecting such data as opposed to the traditional bibliometric or webometric data. Altmetrics may soon replace webometrics as a research method for data originating from the internet, especially social media sites, if the current trend continues.

### *1.4.2 Content analysis*

Content analysis refers to collecting data by systematic examination of texts or other passages in the contexts of their use (Krippendorff 2013, 10). It can be used for both data collection and data analysis. When it is employed as a data collection technique, researchers "analyze" the chosen data source in order to gather what is needed for a scholarly project. Two explanations should be offered regarding the word "analysis" at this point. First, analysis in this context includes abstraction, comparison, synthesis, and other analytical techniques applied in the data collection process. For this reason, comparative research is not listed as a research method separately from content analysis. Second, analysis in this and other research method names (e.g., citation analysis, transaction log analysis) means that analysis must be performed for data collection purpose.

Content analysis can be further divided into two types: manifest and latent. Manifest content analysis denotes what to be analyzed is physically present, observable, and countable. Manifest content analysis is easier in comparison to latent content analysis. Analyzing what research methods are used in a scholarly publication is an example of manifest content analysis. Latent content analysis means what to be analyzed is hidden (underlying), conceptual, unobservable, and uncountable. An example would be to find out what research paradigm or theoretical origin (e.g., phenomenology, positivism, post-positivism, relativism) is applied in one study. Latent content analysis, when contrasted with the manifest counterpart, is no doubt harder.

In this book, content analysis also includes discourse analysis and secondary analysis since they all can be employed for achieving the objective of data collection by examining some kind of data source. It is the data source that essentially determines whether a research method is content analysis, discourse analysis, or secondary analysis. A research method is called discourse analysis if the target source is discourse data (e.g., interview transcripts). Similarly, a research method would be secondary analysis if the target source is secondary data (i.e., data collected previously for other purposes). Census data are a typical example of secondary data used in many studies while they are always collected for a census initially. Meta-analysis is a method for gathering data, usually statistical in nature, from individual studies to summarize their results. Content analysis is widely used in research even though it may not be explicitly specified as a data collection technique in study reports. Systematic reviews and scoping reviews belong to the broad category of content analysis as well.

### *1.4.3 Delphi study*

The Delphi method is generally used for collecting data with a questionnaire from a group of experts to address a research problem in order to reach consensus, make forecasts, or achieve similar objectives via several rounds of exchanges between the researcher and participants. Unlike some other research methods (e.g., content analysis, theoretical approach), a Delphi study can be easily identified in a study because of its distinctive features like several rounds of data collection using a questionnaire. Strictly speaking, the Delphi study is a method contingent upon questionnaire and cannot be used alone. Further explanation of contingent research methods will be provided in Section 1.5.

### *1.4.4 Ethnography/Field study*

Ethnography and field study share many characteristics in data collection because both rely on multiple techniques (e.g., observation, interview) for gathering data in a natural setting where research participants live or work. They are hence grouped as one type of research method in this book. The only difference between the two methods lies in their participants. Ethnography normally chooses people of the same race or nationality who share a distinctive culture as research participants, while field study does not have a similar orientation. In addition, there is no fixed, specific set of data collection techniques for either ethnography or field study. Both ethnography and field study can select individual data collection techniques according to their research goals.

### *1.4.5 Experiment*

Experiment is a long-established method for data collection by following a procedure to test or try out what is studied in either a laboratory or real setting. The former is called laboratory experiment while the latter field experiment. Experiment is by far the most often used method in psychology and biomedical science. Experiments used in other fields (e.g., information studies) nevertheless often differ from what Campbell and Stanley (1966) depicted in their seminal book. For example, experiments in information studies are conducted mainly for evaluating new procedures (e.g., key-phrase extraction from texts), algorithms (e.g., search result ranking), or systems (e.g., digital libraries). Such experiments rarely take the classic design of experimental versus control groups with pretest, treatment, and posttest. Rather, experiments are normally performed on the target in a laboratory or simulated environment with only some of the experiment essentials (e.g., experimental group, control group, pretest). Such experiments would belong to the quasi- or pre-experiment category Campbell and Stanley (1966) presented.

Experiment is also a method frequently employed in conjunction with other data collection techniques. Typical implementations could include the use of questionnaire or interview for pretest/posttest, observation for collecting behavioral data, and/or think-aloud protocol for gathering cognitive data during the experiment. Experiment is therefore a meta-method, which will be fully discussed in Section 1.5.

### *1.4.6 Focus groups*

Focus groups, as a research method, are for collecting data by carrying out a discussion on a research problem between a moderator and a group of study participants. Several groups are usually invited to participate in one study, which explains why the plural form of "groups" is chosen in naming the method. Focus groups are sometimes regarded as focus group interviews or group interviews in the literature. However, the emphasis of focus groups is not placed on question-and-answer as in interviews. It should be on discussion and interaction among group members instead. Although focus groups were primarily used in commercial market research in the past, this method is increasingly used in academic social research (Bloor et al. 2001, vi) and other domains.

### *1.4.7 Historical method*

The historical method played a dominant role in many fields in the scholarly world until the 1980s. This method is used for data collection by examining, synthesizing, summarizing, and interpreting existing published and unpublished materials related to a historical research problem. The historical method bears some similarity to content analysis because they both analyze the content of materials for data collection purpose. On the other hand, these two methods are different in that the historical method deals exclusively with problems of historical nature.

### *1.4.8 Interview*

Interview is a technique for data collection by asking individual participants questions relating to a research problem. There are three kinds of interviews depending on how interview questions are prepared and if probe or follow-up questions are allowed. If the researcher has a prepared list of questions for the interview, it is called structured interview. Structured interview, to a great extent, is analogous to a questionnaire that is administered verbally. If the researcher does not use a prepared list of questions and formulates specific interview questions on site, it is called unstructured interview. If the researcher asks some questions from a prepared list but also raises some probe or follow-up questions, it is called semi-structured interview. Among the three different kinds of interview, semi-structure interview is perhaps the most adopted one thanks to its flexibility in application. In addition, interview is one of the methods that researchers often use when they want to collect data, particularly nonverbal language (e.g., facial expressions, body posture, gestures, eye movements), from study participants in real time.

### *1.4.9 Observation*

Observation is a method for gathering data via watching and making notes of what is studied carefully and attentively. All five senses of the observer (i.e., the researcher) need to be used in data collection (Baker 2006). Observation can be made in a lab or the field. Other common subcategories of observation include obtrusive versus unobtrusive, and participant versus nonparticipant.

Observation is also regularly done as part of meta-methods like experiment and field study. In the case of classic experiments, for example, observation is a must-have technique for collecting behavior data before and after the treatment is administered. Some form of observation is also needed in non-classic experiments in order to obtain data for comparing the before-after and old-new differences. Field study often utilizes observation as one of the multiple data collection techniques chosen for the research. Consequently, observation is generally discussed in the context of field study (e.g., Yin 2016). In addition, observation is made when applying such data collection techniques as interview and focus groups even though observation then may not be singled out as a separate method in those studies. For instance, the interviewer would closely observe any nonverbal language of the interviewee during the interview process.

### *1.4.10 Questionnaire*

Questionnaire, also known as survey, is a technique for data collection by using a preformed list of questions. A questionnaire can comprise both closed-ended and open-ended questions although in most cases the former make up the majority or the only type of questions in the data collection instrument. Certain indexes, inventories, scales (e.g., Emotional Intelligence Scale), and tests (e.g., Personality Test) can also be regarded as some form of questionnaire and are therefore placed in this category of research method in the current book. Furthermore, questionnaire is frequently used for pretest and posttest in experiment. It is also a prerequisite technique for carrying out the Delphi study – a contingent research method briefly mentioned earlier in this chapter. In those two cases, questionnaire may not be considered a separate data collection technique in research reports.

### *1.4.11 Research diary/journal*

Research diary or journal is used to gather data about events, activities, thoughts, reflections, or other aspects relating to an individual who keeps the diary over a period of time. Scholars rely on the research diary for gathering data that can only be obtained from participants themselves in a synchronous mode. They resort to the research dairy for data collection in order to study what participants would be doing at a given time, on a daily basis, or at certain time intervals (e.g., every three hours, bi-weekly). Research diary can be structured (e.g., using a form), semi-structured, or unstructured (i.e., open-ended with no particular format requirement). It enables researchers to collect data that would not be gathered otherwise. Research diary can be used alone or together with other data collection techniques in a study.

### *1.4.12 Theoretical approach*

Theoretical approach, as a research method, is a technique for gathering data through conceptual analysis, theoretical examination, or other similar activities. Although few studies name this research method using the same term, it is frequently applied by researchers in an interdisciplinary field (Chu 2015). It also bears some similarity to content analysis, especially latent content analysis. The distinctive feature of this approach lies in its focus on "theoretical" in research. Investigators would perform conceptual or theoretical analysis according to existing theories, models, and frameworks or with an aim to develop a theory, model, or framework in a respective domain. It has to be pointed out that the theoretical approach is not simply a review because the latter only does evaluation without any further conceptual analysis, modeling, or theory building.

### *1.4.13 Think-aloud protocol*

Think-aloud protocol is a research method that attempts to collect data about participants' cognitive activities via the verbal reports of their thoughts (i.e., think-alouds) while taking part in an experiment or performing some tasks (Ericsson and Simon 1980). This method originates in cognitive psychology but has been adopted in recent decades for research in other fields (e.g., information studies). Like the Delphi method, think-aloud protocol cannot be used alone. It is instead contingent on experiments of a certain kind. More will be discussed in Section 1.5 with regard to contingent methods.

### *1.4.14 Transaction log analysis*

Transaction log analysis, as a research method, gains momentum when computerized systems are used for information processing, storage, and access. Transaction logs become the by-product of computerized information systems. This method allows researchers to gather data by analyzing transaction logs that are automatically captured at either the server or client side. Transaction logs that interest researchers include search logs (e.g., search terms entered, sites visited), logs from e-commerce sites (e.g., pages bookmarked, length at a page, items placed in the shopping cart), and information system usage data (e.g., abstracts viewed, articles downloaded). Captured screens can be roughly grouped with transaction logs as the screen displays interaction data between the user and system. Transaction log analysis has been increasingly used as a data collection technique in recent decades thanks to the development of data-capturing applications and the emergence of data science which explores data of gigantic size.

### *1.4.15 Other methods*

In addition to the 14 research methods depicted above, there are additional data collection techniques scholars utilize although they belong to the less-often-used category. Because of their low usage frequencies, they are amalgamated into one category of "Other methods" in this book.

- Action research can be considered a close cousin of ethnography or field study in that it also uses multiple data collection techniques. Nevertheless, action research is always designed to solve an immediate problem or to produce guidelines for best practice. It relies on techniques such as participant observation, interview, and focus groups for data collection. Action research would actuate changes or actions in the target setting or environment (Berg 2009, 258).
- Card sorting is a technique for data collection by asking participants to put cards of assorted entities (e.g., concepts, website headings) into groups. Through this technique, the researcher can gather data about how participants categorize the given entities.
- Drawing, as a data collection technique, asks participants to sketch what they see, feel, think, and/or do. This technique is often used together with other research methods to help gather visual data.
- Photo survey serves as the researcher's visual notebook during data collection. Photo survey researchers would use a camera to take photos to record the scenes and surroundings of their study targets. The photo survey is always complemented by observation and other techniques to make up the complete research methodology. As smartphones with a camera become an indispensable device to most people, it would only be easier and more feasible for researchers to utilize this technique for data collection along with other methods (e.g., observation).

There are apparently more research methods (e.g., eye-tracking) other than what is already presented. Those methods however appear too marginal to be included in this book. Readers interested in such data collection techniques should be able to locate articles or even monographs devoted to them without much difficulty.

Before proceeding to the next section, it seems necessary to briefly discuss what cannot be considered research methods for the rationale given below.

- Case study – Although some prior publications (including monographs) list case study as one research method, case study only implies that a small number of cases or sampling units, typically one or several, as opposed to a sizeable sample or the entire population, are studied in the research. A case study alone does not have any indication about which research method or data collection technique it may use. In reality a case study can choose any research method that suits its research objective.
- Grounded theory – According to Bawden (2012, 156), "... grounded theory is, notoriously, not a research method in itself, but rather a general strategy". Grounded theory

means that a study is conducted without adopting an existing framework or theory and intended to ground its work on the data collected and analyzed. What methods are appropriate for a given study is determined by the problem it explores rather than if the study takes the grounded theory approach. The grounded theory approach itself does not suggest any particular research method or data collection technique for a study.

- Library research – Also known as desk research, library research refers to a step or component (i.e., literature search) in the research process. Library research itself does not imply any research method or data collection technique, while all researchers are supposed to conduct literature search in order to learn more about what has been done in the past on the chosen research topic.
- Longitudinal/Cross-sectional study – Longitudinal study is one kind of research whose temporal coverage expands over a period of time (e.g., January to December in 2019). Its counterpart is the cross-sectional study which collects data for a given topic at one point of time (e.g., one point in 2019) as opposed to over a period of time. In other words, it purely deals with the temporal scope or coverage of a study and provides no natural linkage to any research method. A longitudinal study can theoretically choose any technique for data collection. The same holds true for its counterpart of cross-sectional study.

The afore-described list of non-research methods can no doubt be expanded although it is not the intention of this book in the first place. Simply put, any term that does not represent a technique for data collection is not a research method when we name research methods by data collection technique. This book aims to provide an interdisciplinary, systematic coverage of major research methods in the scholarly world and how to design a research project by choosing one or more of them according to one's research objective.

## 1.5 Taxonomy of research methods by applicability

Table 1.2 presents a taxonomy of research methods by applicability to facilitate a better understanding of those individual methods. For example, some methods (e.g., questionnaire) can be used alone, whereas others (e.g., think-aloud protocol) have to be employed in conjunction with another method. This taxonomy is also used for organizing and presenting individual methods in this book. Specifically, stand-alone methods are introduced and illustrated first, followed by super-methods, meta-methods, and contingent methods.

Super-methods in this taxonomy refer to those methods that are employed in almost all research to some degree. Among the research methods covered in this book, content analysis can be regarded as a super-method in that it is used in almost all research with varied degrees. Some studies may exclusively focus on analyzing the content of data sources for data collection purpose. Other investigations perform content analysis as a supplemental method. Likewise, techniques such as comparison also belong to the category of super-methods because comparison is often made in the process of data collection. It is however not listed as a separate technique since comparison, like summarization, abstraction, deduction, and induction, is one of the major techniques adopted in performing content analysis but rarely listed as data collection technique.

The second category in the taxonomy of research methods is about stand-alone methods, which means that each method can be used alone in a study if needed. Compared with other categories in Table 1.2, this one is the largest in size. Some of the research methods (e.g., bibliometrics, interview, questionnaire) are widely applied and thus one

*Table 1.2* Taxonomy of Research Methods by Applicability and Corresponding Chapters

| *Applicability* | *Method* | *Chapter* |
|---|---|---|
| Super-method | Content analysis | 9 |
| Stand-alone method | Bibliometrics and other metric-based methods, focus groups, historical method, interview, observation, questionnaire, research diary, theoretical approach, transaction log analysis | 3–8 |
| Meta-method | Experiment, ethnography/field study | 10–11 |
| Contingent method | Delphi study, think-aloud protocol | 12 |

chapter is devoted to covering each of them in this book. Other stand-alone methods (e.g., focus groups, research diary) are used less frequently, and therefore each of those data collection techniques is discussed in one section of Chapter 8.

Meta-methods denote those involving multiple techniques for data collection in research. In the classic sense, experiment is a pertinent example of meta-method. When experiment is chosen as the research method for a study, several individual techniques such as interview, observation, questionnaire, and/or think-aloud protocol need to be used in various combinations according to the experimental study's purpose. In fact, only when multiple methods are employed simultaneously in a study, can the experiment be conducted. Ethnography or its synonymous field study is another example of meta-methods. Focus groups, interview, observation, and research diary are among the common tools that ethnographers utilize. That is, each ethnography study must use several techniques suitable for its data collection.

Contingent methods, as a group, possess the attribute that they have to be used together with another method in order to be functional. Research participants, for instance, must conduct test searches if think-aloud protocol is chosen for gathering data about their cognitive activities in the search process. Such test searches are one type of experiment and provide the basis for the participants to think aloud while searching. Otherwise, the participants would not be able to verbalize anything meaningful in relation to their search activities. The same is true with Delphi study in which a questionnaire must be developed and utilized for data collection purpose.

## 1.6 A note on the orientation of this book

The research methods enumerated in Table 1.1 actually originate from different disciplines (e.g., experiment from psychology, ethnography/field study from anthropology and sociology), which leads to the discipline-specific approach in many monographs on research methods. Experiment, for example, dominates psychology as a research method, while field study prevails in sociology and anthropology. Questionnaire and interview on the other hand are natural choices for many other disciplines in the scholarly world. Typically, experiment is presented extensively in a text for psychology. Field study might be the only research method described in a book devoted to sociology.

The usage of these methods however is not limited to the discipline for which the method is developed or based on how frequently the method is applied in one discipline. Rather, a particular method can be used in any discipline as long as it can help researchers to address their research problems. If we only cover in a book several methods that are most likely used in a specific discipline (e.g., education, communications, political science), readers in that discipline will not have the opportunities to learn other research methods that might be valuable to them. This discipline-specific approach obviously is not helpful and beneficial to students and junior scholars who are building their toolbox of research methods when our fields of study are becoming increasingly multidisciplinary or interdisciplinary. It must be pointed out that discipline and field are treated as synonyms in this book and thus used interchangeably.

More and more fields feature multidisciplinarity or interdisciplinarity. This implies that one discipline has to include or draw on two or more branches of knowledge and methods in its research and development. Such methodologies comprise research methods originated from other disciplines for addressing issues and problems one particular field faces. An excellent and highly relevant example of interdisciplinarity can be found in the seminal work of Machlup and Mansfield (1983, 6) who examined the composition of the study of information by creating a list of contributing disciplines: "We begin our list with destinations that contain the letters of *inform*[3]: information theory, information science, and informatics. We continue our list with a similar primitive scheme of classification: first the disciplines with names ending in *ics*, then those ending in *logy*, followed by those called *science, analysis, theory*, or *research*: bibliometrics, cybernetics, linguistics, phonetics, psycholinguistics, robotics, scientometrics, semantics, semiotics, systemics; cognitive psychology, lexicology, neurophysiology, psychobiology; brain science, cognitive science, cognitive neuroscience, computer science, computing science, communication sciences, library science, management science, speech science, systems science; systems analysis; automata theory, communication theory, control theory, decision theory, game theory, general system theory; artificial-intelligence research, generic-information research, living systems research, operations research, pattern-recognition research, telecommunications research". Data science is another interdisciplinary field whose components would entail a long list of discrete contributing disciplines and domains.

Both information studies and data science are apparently more interdisciplinary than the established fields such as education or linguistics. Nevertheless, multidisciplinarity or interdisciplinarity becomes a characteristic of many disciplines as these fields develop and advance. For instance, linguistics has close ties with psychology and sociology because any language after all is created and used by people (i.e., relating to psychology) living in a particular culture (i.e., relating to sociology). People in linguistics therefore need to learn about research methods often utilized in psychology and sociology too besides those in their own discipline. Overall, more and more research projects rely on two or more methods originating from various disciplines to fully address their research problems (e.g., Chu 2015).

In addition, the fundamentals of research methods remain basically the same regardless of to which field these methods are applied. There are no different techniques or procedures for a same data collection technique to be used in different disciplines. Take questionnaire as an example, there will not be much difference regarding how the questionnaire is developed and utilized in education or political science. The guidelines for how to create factual questions, opinion questions, and filter questions, for example, stay the same whether the questionnaire is applied in education or political science research. What actually differs in these two fields in this context is only the question contents in the two questionnaires because the contents must be specific to the discipline or, more accurately, to the research project under consideration.

What is presented above underscores the very reason why this book attempts to approach the subject of research methods and design beyond a single discipline. All major research methods adopted in the scholarly world are to be described and discussed without having an orientation toward one individual discipline. The fundamentals and characteristics of every research method are to be portrayed and presented along with some advice on the dos and don'ts in development and application. Examples and case studies provided throughout this book are purposely taken from different disciplines. Students and scholars can then design and carry out their projects by choosing appropriate methods based on their research problems as well as method features. Chapter 14, Research Design via Method Selection, will further demonstrate the benefits of this novel and unconventional approach.

**Chapter Summary**

- Research is the conduct of a specially planned and structured investigation using a carefully developed methodology. Research is usually categorized as basic research and applied research depending on if study findings are to be implemented in the real world.
- Research methods are systematic procedures for conducting research and they are composed of data collection and analysis techniques.
- Research methods should be named after and categorized by data collection techniques such as experiment, interview, and questionnaire.
- Data analysis techniques can be broadly labeled as quantitative and qualitative. But research methods should never, ever be categorized as such because each research method, named after a data collection technique, can be used to collect both quantitative and qualitative data.
- An overview of more than one dozen major research methods is provided in this chapter. Those methods include bibliometrics, content analysis, Delphi study, ethnography/field study, experiment, focus groups, historical method, interview, observation, questionnaire, research diary/journal, theoretical approach, think-aloud protocol, and transaction log analysis.
- Research methods can be grouped by applicability as stand-alone methods, super-methods, meta-methods, and contingent methods.
- This book approaches research methods and design beyond a single discipline because no research method is discipline-specific. Scholars and students from different fields and domains need to adopt this approach in order to evolve and thrive in the scholarly world.

## Notes

1 "Mixed" rather than "integrated" is the term widely used in the scholarly world as a third type of data analysis technique which involves both quantitative and qualitative measures. The present author, however, argues that data analysis techniques should be carefully chosen and integrated according to the data collected in research as opposed to simply mixing some individual techniques together. The same holds true for choosing more than one technique for data collection.
2 Twitter was rebranded as X in 2023. However, it is still referred to as Twitter throughout this book.
3 Italics are in the original text of the book.

## References


Baker, Lynda M. 2006. "Observation: A Complex Research Method." *Library Trends* 55 (1): 171–89. https://doi.org/10.1353/lib.2006.0045.

Bawden, David. 2012. "On the Gaining of Understanding Syntheses, Themes and Information Analysis." *Library and Information Research* 36 (112): 147–62. https://doi.org/10.29173/lirg483.

Berg, Bruce L. 2009. *Qualitative Research Methods for the Social Sciences*. 7th ed. Boston: Allyn and Bacon.

Bloor, Michael, Jane Frankland, Michelle Thomas, and Kate Robson. 2001. *Focus Groups in Social Research*. Thousand Oaks: Sage.

Campbell, Donald T., and Julian C. Stanley. 1966. *Experimental and Quasi-Experimental Designs for Research*. Chicago: Rand McNally.

Chu, Heting. 2015. "Research Methods in Library and Information Science: A Content Analysis." *Library & Information Science Research* 37 (1): 36–41. https://doi.org/10.1016/j.lisr.2014.09.003.

Chu, Heting, and Qing Ke. 2017. "Research Methods: What's in the Name?" *Library and Information Science Research* 39 (4): 284–99. https://doi.org/10.1016/j.lisr.2017.11.001.

Ericsson, K. Anders, and Herbert A. Simon. 1980. "Verbal Reports as Data." *Psychological Review* 87 (3): 215–51.

Glazier, Jack D. 2017. "Domain Assumptions of Research." In *Research Methods in Library and Information Science*, edited by Lynn Silipigni Connaway and Marie L. Radford, 46–60. 6th ed. Denver: Libraries Unlimited.

Järvelin, Kalervo, and Pertti Vakkari. 1990. "Content Analysis of Research Articles in Library and Information Science." *Library & Information Science Research* 12 (4): 395–421.

Krippendorff, Klaus. 2013. *Content Analysis: An Introduction to Its Methodology*. 3rd ed. Los Angeles: Sage.

Machlup, Fritz, and Una Mansfield, eds. 1983. *The Study of Information: Interdisciplinary Messages*. New York: John Wiley & Sons.

Thelwall, Mike. 2008. "Bibliometrics to Webometrics." *Journal of Information Science* 34 (4): 605–21. https://doi.org/10.29085/9781856049986.018.

Yin, Robert K. 2016. *Qualitative Research from Start to Finish*. 2nd ed. New York: The Guilford Press.

# 2 The Research Process and Research Design

**Chapter Overview Questions**

- What are the features of a research process?
- What steps are there in the research process?
- What are the major sources for research problem identification?
- What are the key points in literature search and review?
- How should research questions be formulated?
- What levels of measurement are there, and why should we consider them in data collection?
- What are the major points in selecting a sampling method in data collection?
- What are validity and reliability in research?
- What are the major concerns in research ethics?

Research is both a noun and a verb. As a noun, research refers to scholarly publications that report study findings and outcomes. As a verb, research means investigations that aim to address problems and issues by following a set of established practices. In this chapter, research is regarded as a verb and involves a series of activities taking place over a period of time. Research "is a cyclical, self-correcting process consisting of seven major interrelated stages: definition of the research problem, statement of the hypothesis, research design, measurement, data collection, data analysis, and generalization" (Frankfort-Nachmias, Nachmias, and DeWaard 2015, xi). Indeed, research is a process rather than one single activity. It also should be pointed out that the term "self-correcting" in Frankfort-Nachmias, Nachmias, and DeWaard's statement above does not imply that research is error free and able to correct itself. Rather, it is the researchers who should review, critique, and revise their research procedures in order to correct any mistakes or errors they are likely to make. In addition, the research process is spiral (Berg 2009, 26) and recursive (Gorman and Clayton 2005, 37). Some publications may occasionally use a linear flow chart with arrows to illustrate these two features.

The research process is spiral in that all the steps or stages involved are nonlinear yet progressive. The recursive characteristic of the research process is reflected by the fact that a study often needs to redo one research step or more after it commences to ensure its successful completion. Few researchers can start a project and proceed through the entire process without returning to a previous step for review or revision. Even though each research project is unique in terms of topic and content, there are common steps

DOI: 10.4324/9781003021711-3

to be taken in the research process. Those steps, seven in total, are to be described and discussed one by one in this chapter.

- Problem identification and statement
- Literature search and review
- From research problem to research questions
- Development of research methodology
- Data collection and analysis
- Result presentation and discussion
- Conclusion drawing

Do keep in mind that the linear presentation of the research steps in this chapter is dictated by the two-dimensional space for presentation. It does not imply that those steps can and should be taken linearly. As indicated earlier, the research process instead is spiral, recursive, nonlinear, and progressive. Researchers should review, readjust, and revise any research step whenever necessary throughout the research process.

## 2.1 Problem identification and statement

Any research project starts with problem identification. Do note that the word "problem" is used generically in this context, covering any topic that needs to be studied. It is termed "problem" instead of "topic" in "problem identification" and "problem statement" to be in line with the research terminology we commonly use. In no way does the word "problem" imply that the only purpose of research is to solve problems. Rather, research is conducted for a variety of purposes as we have briefly discussed in the very beginning of Chapter 1.

Researchers must identify which problem needs to be researched and then justify why it is so or explain the importance of studying the problem. A research problem is a question or an issue that requires a response in the form of a structured scientific inquiry (Frankfort-Nachmias, Nachmias, and DeWaard 2015, 46). Nevertheless, not all problems are research problems, and there are three prerequisites for a problem to meet before it can be considered a research problem.

The first prerequisite is that the problem is in need to be researched. This means that, for example, no one has studied it before or a study needs to be replicated if already done before. The second prerequisite is that the problem is researchable. This implies that the problem can be addressed using research methods. There are numerous problems that are proven unfathomable in reality. Examples of such problems include how to acquire research skills without learning and practice, or how to develop a universal measure for guaranteeing happiness. The third prerequisite is that researchers have proper expertise and resources needed for addressing the problem. For instance, a research problem in psychology can be studied only if the researcher has knowledge in that subject domain. Meanwhile adequate resources, mostly in the form of time and money, should be available to fund a research project because scholars cannot conduct research without such support.

With the three prerequisites in mind, researchers usually rely on one or more of the following sources for problem identification.

- Request for proposals (RFPs) – RFPs, normally issued by grant-making agencies (e.g., US National Science Foundation) or similar institutions, constitute one of the major sources for problem identification. The topics listed in the RFPs are always current or important.

- Research agendas and call for papers (CFPs) – Research agendas issued by professional associations and CFPs made by professional conferences and scholarly journals represent additional sources for current research topics. Compared to RFPs, research agendas and CFPs encompass more research problems although no funding is connected with this type of research problem identification source. After scholars identify a problem from such sources, they have to seek research funds from agencies or institutions that provide them.
- Research reports – Research reports in journals, conference proceedings, and other scholarly publications are often another type of sources for problem identification. Researchers typically state at the end of their reports what they plan to do in the future as part of the "conclusions" section. Occasionally, a separate section is designated with the heading of "future research and recommendations" to point out what needs to and should be done beyond the present research. This source is particularly useful for identifying research problems if the study report is relevant to one's research interest.
- Conversations with colleagues – This kind of source for problem identification is categorized as informal scholarly communication (aka informal communication) channel via conversations, talks, and the like as opposed to the formal scientific communication (aka formal communication) channel via written materials (e.g., reading research reports). Yet informal communication plays a more significant role in discoveries and cutting-edge research because it normally takes one year or even longer for a research report to get published even in the age of digital publishing. In contrast, informal communication can take place instantaneously or synchronously. Colleagues usually have shared research interests and thus can communicate with each other about research ideas more efficiently and effectively. Such conversions are also more likely to spark new ideas for research. Those new ideas could well become new research problems.
- Day-to-day operations and services – Many disciplines, due to their nature and orientation, focus on applied research problems (e.g., what is an optimal approach to teaching and learning) instead of basic research problems (e.g., how to define what creativity is). Therefore, many research problems can be identified through day-to-day operations and services in those disciplines.
- One's own experience – One's own experience can become a valuable source for problem identification especially if that person is mindful and observant and has been a researcher for years or even decades. Researchers would do better in identifying problems from their own experience precisely because such experience enables them to spot problems more readily.
- An open and inquisitive mind – Of all the sources enumerated till this point, an open and inquisitive mind is undoubtedly the most important one. We all know that opportunities only favor those who are prepared. The same holds true for problem identification. Charles F. Kettering, an American inventor, engineer, and businessman, once stated: "There will always be a frontier where there is an open mind and a willing hand" (Kettering 1945, 12). The researcher with an open and inquisitive mind shall be able to identify problems that need to be explored along with the other sources discussed above for problem identification.

After identifying a research problem, the researcher should state the problem by following the established practice. A problem statement is an overall statement of the research problem in consideration. Kettering summarized succinctly the importance of making a proper problem statement: "A problem well stated is a problem half solved" (McLean 2005, 53). Structurally, a problem statement consists of three parts:

1) Subject (e.g., this study, the present research)
2) Verb (e.g., examine, compare, investigate)
3) Object (i.e., a phrase describing what is to be studied)

Below is a sample problem statement taken from a research report by Negriff (2019, 1): "The current study examined the risk content of Facebook posts among a sample maltreated and comparison youth and compared the rates of risk content produced by the participants versus their Facebook friends." In this problem statement, "The current study examined … and compared …" is the opening part of the problem statement, while the rest describes the problem Negriff explored in the research.

Will you be able to identify the subject, verb, and object in Negriff's problem statement above based on the discussion so far? If the same problem is stated before the actual study is carried out which is usually the case, we can begin the problem statement with, for example, "The current study intends to examine … and compare …", followed by the problem to be researched.

The problem statement not only serves as a goal for the researcher to achieve but also functions as the basis from which research questions are to be derived. The latter will be discussed in Section 2.3. In other words, research questions of a study are never formulated arbitrarily. They instead must be derived naturally and logically from the problem statement. Although the problem statement may get modified during the research process, it is still essential for the researcher to draft one after a research problem is identified because subsequent steps are built on or determined by the problem statement.

## 2.2 Literature search and review

Literature review is a step the researcher must take before the study can proceed further. The purposes of literature review are multifold and can be recapitulated as follows (Boote and Beile 2005; Mauch and Birch 2003, 119–22):

- Gain good knowledge of the field as to what facts, who the eminent scholars, where the parameters, and which important ideas, theories, questions/hypotheses.
- Generate more questions/hypotheses for further studies.
- Become familiar with the methodologies common to the field and their usefulness, appropriateness in various settings.
- Get a reinforcement of an early hope that the conceived research is really needed.
- Help narrow a research problem already identified.

The objectives of literature review, in brief, are to find out (1) what has been done, (2) how well prior related research is conducted, and (3) what remains to be done. Learning about (1) helps the researcher avoid doing studies that others already completed. Learning about (2) enables the researcher to evaluate how previous studies are carried out while noting down if there is anything (e.g., data collection techniques) in them that can be adapted for the project under consideration. Learning about (3) assists the researcher to become clearer about the void in previous research and thus finalize the identified research problem.

### 2.2.1 Literature search

Before the literature review can be conducted, the researcher must first locate publications that are relevant to the stated research problem. It should be pointed out that "literature" in both "literature search" and "literature review" refers to research publications instead of any publication or literary work. This implies that only academic databases (for articles and other nonbook publications), online public access catalogs (OPACs) or simply library catalogs (for books and other library collection items), and similar information systems (e.g., institutional repositories) should be chosen for literature search. All other sources that are not vetted by fellow scholars should not be utilized for literature search because such sources (e.g., search engines) do not go through the rigorous peer review process and thus may contain invalid data or information. The peer review process, though not 100% objective or error free, is the established and accepted norm for controlling the quality of scholarly writings.

On the other hand, Google Scholar (https://scholar.google.com) and other similar sources do comprise scholarly publications, most of which are also peer reviewed and a portion of them might even have a link to the full-text document if it is in the public domain. When a publication is in the public domain, it means that we can obtain and use it free for non-commercial purpose. However, such sources do not provide a comprehensive coverage of any discipline if compared with academic databases and the like. They are good for preliminary searches to find out which databases and other sources contain what one tries to locate. Therefore, Google Scholar and similar sources are primarily used as tools for selecting other sources for a given topic. If no dedicated sources such as databases are available, they then become the default sources for literature search. When selecting other sources for literature search, do ask for help from librarians, mentors, colleagues, and friends who are familiar with those sources if necessary. The rationale behind the above suggestion is because choosing proper sources for literature search is a prerequisite for quality literature review.

Literature search for a research project, unlike everyday or trivial searching, needs planning and practice after appropriate sources are selected. Get familiar with the chosen sources to learn about their content and temporal coverage as well as search functions by going over the "about", "help", "search tips", or other similar items the selected sources provide. Although reading such items may take some time (usually no more than 15 minutes), it would greatly facilitate the subsequent searching. Otherwise, the researcher might, for example, search in an inappropriate database or use a search technique that the database does not support.

This book will not detail search techniques and tips because of the following two reasons: (1) There are already many publications on how to conduct effective searches. Moreover, almost all academic libraries provide workshops and tutorials on searching, and researchers are encouraged to take advantage of both if in need of search support. (2) It is beyond the scope of this book on research methods to describe specific search techniques and strategies. However, one thing all researchers should strive to achieve when conducting a literature search is to locate as many relevant documents as possible in order to gain a comprehensive view of the research landscape of the chosen topic. The keyword in the search process is exhaustiveness. All techniques (e.g., choosing multiple sources for search, using multiple synonyms of a search term) that would help increase the number of search results should be utilized. Other approaches for achieving the same objective include checking footnotes and tracing references (White 1994). Checking footnotes means to look at the footnotes of a publication for locating additional related documents. This approach is especially helpful for finding more documents in disciplines (e.g., history) that use footnotes extensively. Tracing references, like checking footnotes, is for the researcher to look at the references of a document after reading it to see what other relevant documents are cited therein. All scholarly documents have references, and publications from social science disciplines usually have longer reference lists when compared with that in the sciences. Because all the references are cited by the citing document author, they must be relevant to the citing document in terms of subject content. Both checking footnotes and tracing references are therefore proven techniques for locating more related documents in literature search. Novice researchers should make use of them whenever possible.

As noted above, literature search for a research project must be on target as well as exhaustive. But when shall the researcher stop searching for additional documents? The question becomes intensely pertinent if the technique of tracing references is employed. In general, researchers can stop searching when they see that the same document is retrieved again. Researchers can also consult their mentors and colleagues in the search process.

### *2.2.2 Literature review*

Literature search with an aim of exhaustiveness usually yields more documents than what we can or should evaluate in the literature review. Therefore, researchers ought to categorize all the retrieved documents into three groups (i.e., directly relevant, indirectly relevant, and not relevant) based on their research problem and associated subproblems. Use scan, skim, or other quick reading techniques to complete this categorization process. For instance, the title, abstract, and keywords of each retrieved journal article usually provide brief yet rich information for you to quickly learn the aboutness of individual documents.

Retrieved documents in the directly relevant category are right on what the study intends to examine. If a study plans to explore factors affecting mobile phone usage, for example, all publications on this topic are directly relevant to the study.

The indirectly relevant category of search results refers to publications that are not exactly related to what the study intends to find out but might be of some relevance from a different perspective. Again, take the hypothetical study on factors affecting mobile phone usage as an example. Research publications on mobile devices (e.g., tablets) other than mobile phones are examples of indirectly relevant documents in that tablets are also mobile or handheld devices. Research on mobile devices is likely to shed light on studies about factors affecting mobile phone usage.

Documents that are not relevant to the chosen research problem usually account for a portion of what is retrieved. No matter how well the literature search is performed, search results always contain noise or irrelevant documents. This type of results should never be included in a literature review.

Apparently, the literature review should evaluate all the directly relevant documents retrieved although this category of documents is often small in quantity. That is the very reason why the intended research is needed. In contrast, more retrieved documents fall into the category of indirectly relevant. Whether to include this type of retrieved documents in the literature review is dependent on several considerations:

- Include if not many directly relevant documents are found.
- Include if the indirectly relevant documents are authored by well-known researchers of the field or from top journals.
- Include if the indirectly relevant documents provide different insights for the intended study.

Beginning researchers might ask how many documents should be reviewed for a single study. The answer is shaped by multiple factors (e.g., scale, topic) although there are no established guidelines for determining a number. If the study, for instance, is cutting-edge and focuses on one specific problem, reviewing around 15 documents should serve the literature review purpose well. If one conducts dissertation research, reviewing about 50–100 documents would be a common practice.

After determining which retrieved documents should be included in the literature review, researchers should read each of them in light of the study in progress. While each researcher may use different reading techniques to get to know the documents before reviewing them, do the following in the review process:

First, scan all the documents to get an overview of each. Research publications in general have the sections listed in Table 2.1. All the components with a bullet point (e.g., background) can be a separate section if needed. Some authors may combine results and discussion into one section. In addition, some authors tend to use terms different from those listed in the table as section headings. For example, "related studies" could be chosen as a heading in place of "literature review". Knowing the structure of a research publication would facilitate the scanning process because some sections (e.g., literature review) can be safely skipped.

*Table 2.1* Common Sections of a Research Publication

| *Section Heading* | *Section Content* |
|---|---|
| Introduction | State what the research problem is and explain why the problem needs to be researched |
| • Background | Provide any necessary information that would help the reader understand the stated research problem |
| Literature Review | Examine and evaluate prior related research publications |
| Methodology | Describe data collection and analysis techniques selected for the study as well as explain why those techniques are chosen |
| • Data collection | Describe data sources, data collection instruments (e.g., questionnaire, interview guide) if any, and data collection procedures |
| • Data analysis | Describe techniques used for data analysis |
| Results | Present results of the study |
| Discussion | Discuss the study's results and findings in relation to the study's research problem and questions |
| Conclusions | Draw conclusions based on the study's findings and in light of the significance of the study |
| • Limitations | Point out any limitations the current study may have |
| • Recommendations for further research | Outline how the limitations of the current study will be overcome in the future and suggest related topics for further research |

Second, group all the documents selected for literature review by subject content. Avoid reviewing related documents by author's last name or chronologically by publication year. Rather, organize the review by the subject content of each document. That is the very purpose for this grouping exercise and those selected documents will be reviewed by group in separate subsections. Using the same example of the hypothetical study on factors affecting mobile phone usage, we can group the selected documents by factors (e.g., demographic, socioeconomic). That is, all documents about demographic factors affecting mobile device usage will be evaluated in one subsection, and so on. If we already have draft research questions, they can be used for helping group the selected documents as well although, as briefly mentioned earlier, research questions are usually formulated after literature review is completed in the research process. A same document might be reviewed in two or more groups if it covers contents related to each of the groups.

Third, read the documents in each group in light of the undertaken study by making notes to aid the creation of the literature review. Reading at this point must be intensive. Hart (1998, 53) rightly points out the following when discussing how to read for review:

> Reading with the purpose of reviewing is very different from reading for pleasure. Reading to review has the goal of producing a product: an analytical evaluation of the research on your topic. This means that you are expected to unravel the reasoning that informs the research and arguments that you find in the literature. Therefore, our first concern must be with the question of what it means to read analytically. Reading analytically is not something that can be done in one reading or in a short time period.

Indeed, reading for review requires us to read critically and in light of the identified research problem. Meanwhile, reading one document several times is more a rule than an exception. During this kind of reading, the researcher should note the following: (1) important findings and conclusions, (2) methodological strengths and weaknesses, and (3) how are the reviewed studies relating to the research in progress.

Fourth, develop a literature map for all the selected documents to further help organize the literature review by subject content. Figure 2.1 presents a literature map example for an article titled "Research methods: What's in the name?" (Chu and Ke 2017). This literature review is structured according to the three research questions as well as what the selected documents encompass as a whole. Each end box on the literature map contains a specific heading and related individual documents to be reviewed underneath. For example, two articles by "Peritz 1977" and "Hildreth and Aytac 2007", respectively, are reviewed under the heading "Methodology". As mentioned earlier, one same document may appear two or more times in different parts of the literature review if it is related to two or more aspects of the intended research project. Take "Hildreth and Aytac 2007" as an example, it is reviewed in all three literature review subsections of Chu and Ke (2017).

Fifth, write the literature review by organizing and expanding the notes taken for each document read. Avoid writing one paragraph for each document in the review. Instead integrate all the studies selected for review according to the document groups created earlier in the review process and shown in the literature map before presenting them logically and coherently based on their contents. Within each section, note differences among studies and also look for obvious gaps or areas needing further research.

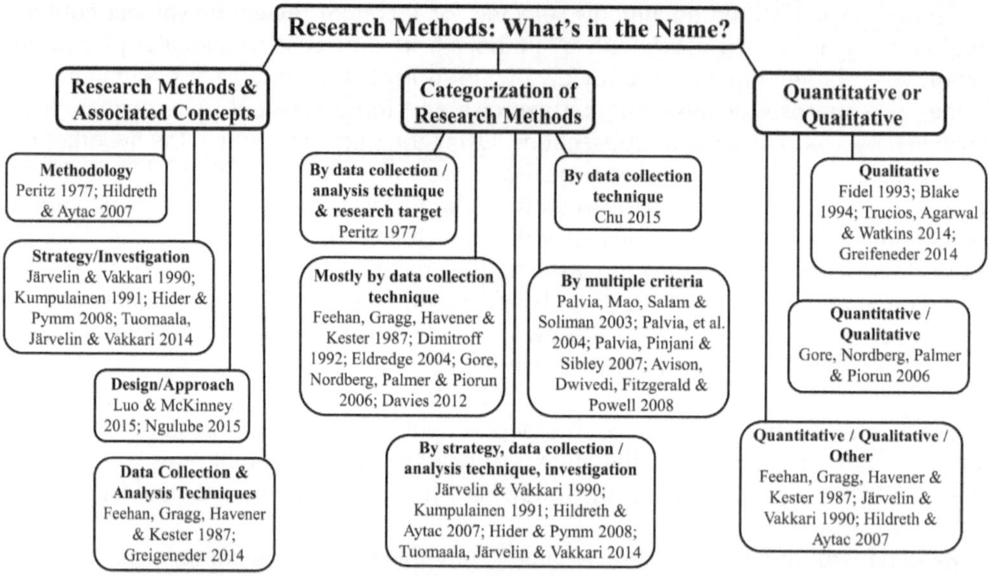


*Figure 2.1* A Literature Map[1]

Sixth, summarize the gaps and limitations identified in previous studies in the context of the ongoing research at the end of literature review. Those gaps and limitations also help justify why the present study is needed. Candace Hastings (2009) gave a talk on how to prepare a literature review which is well worth watching.

In general researchers would do the literature search and scan retrieved documents before data collection. The actual literature review, however, is usually not written until data analysis is completed even though literature review is presented right after literature search in the book. In addition to what is already discussed in this section, pay attention to the following dos and don'ts when preparing the literature review:

- Dos
  - Do evaluate the selected publications through analysis and synthesis.
  - Do organize and present the literature review as a coherent writing.
  - Do paraphrase what is taken from the publication if you do not plan to quote.
  - Do list the page number if a direct quote is taken from the publication or a reference is made to a book.
- Don'ts
  - Don't just summarize the selected publications without any evaluation of them.
  - Don't organize and present the literature review by author's last name or publication year, nor write one paragraph per publication in the literature review.
  - Don't cite just based on what is reviewed by others (i.e., cite from citations) unless the original document is read because the cited document is already interpreted and evaluated once by the citing author. Distortion is likely to occur when citing from citations and without reading the cited document first-hand.
  - Don't over-quote in the literature review.

## 2.3 From problem statement to research questions

This section has two parts: One deals with how to transform a problem statement to research questions, whereas the other is devoted to hypotheses.

### 2.3.1 *Transforming a problem statement to research questions*

Research questions could be the same as or more specific than the problem statement in terms of extent. Therefore, one or more questions can be derived from the problem statement depending on the number of subproblems contained within. As indicated earlier, the problem statement of a study forms the basis for formulating research questions. Decomposing a research problem denotes that the researcher looks closely at the problem statement to see how many different subproblems it contains. After that, one research question should be formulated for each subproblem identified in the problem statement.

In addition to the problem statement, related publications and the researcher's own knowledge about the problem can also help to formulate research questions. According to Kerlinger (1986, 16–18), a good research question should meet three criteria: (1) express what needs to be studied, (2) state clearly and unambiguously in question form, and (3) imply possibilities of empirical testing. Criterion (1) and (2) are both fairly self-explanatory. Criterion (3) nevertheless asks for some explanation. This criterion means that variables in the research question should be operationally defined if needed and the study can be conducted using scientific methods. Scientific methods refer to those methods that are established and enable researchers to carry out studies systematically and structurally. Major scientific methods will be discussed from Chapters 3 through 12 and shall be further elaborated in Chapter 13, Techniques for Data Analysis and Their Selection, as well as in Chapter 14, Research Design via Method Selection. The operational definition of a variable, however, necessitates some further elucidation at this point.

A variable is an entity (e.g., a person, an attribute, a factor) that can take two or more values. Gender, age, race, and income are all examples of variables. A variable can be defined conceptually or operationally. The conceptual definition for a variable can be easily obtained from a dictionary. But it does not specify how a variable is measured or what values a variable can take in a study. For some variables, their operational definitions are already established. For instance, male and female are the two common values the gender variable can take. For other variables, in contrast, their empirical attributes (i.e., values) cannot be readily or directly measured. Happiness, satisfaction, anger are a few examples of such variables that should be defined operationally in studies. Researchers would choose or devise a way to specify the "operations" to be taken for measuring the values of a target variable. This also explains the meaning of the word "operational" in operational definition. Take "intelligence" as an example, its conceptual definition can be readily found from a major dictionary (e.g., Oxford English Dictionary). But how can "intelligence" be measured empirically? Consequently, various IQ tests have been developed for measuring intelligence over the decades. The researcher may decide to choose, for example, the Wechsler Adult Intelligence Scale as an operational definition for the "intelligence" variable in the study. The Wechsler Adult Intelligence Scale therefore defines the operations needed for measuring the variable.

The quality of a publication serves as another example of a variable in need of an operational definition because there is no readily available measure to determine it even though we may know what a quality publication is when reading one. Relevant operational definitions for such a variable include the number of citations a publication receives or the impact factor (IF) score of the journal that publishes the target document. Journal impact factor or JIF in short is the ratio between the number of citations a journal receives in one year and the total number of articles published in the journal during the two preceding years (see Section 7.3.2.1 for details). Calculating either the citation frequency of a publication or the IF score of a journal that publishes the document is thus the operation taken for measuring the quality of a publication. In sum, an operational definition of a variable defines what to do in order to measure the target variable. How would you operationally define the variable of "happiness" based on what we have discussed so far?

As indicated earlier, only some variables need to be defined operationally while others do not. It might seem obvious to make a distinction between the two. In reality, however, novice researchers may erroneously try to provide an operational definition for variables such as country and income when measures for those variables are long established (e.g., country names for the country variable, dollar amounts for measuring income). Therefore, novice researchers are advised to ask themselves the following question when not sure if a variable should be defined operationally: Are there already established values or empirical attributes the variable can take? If yes (e.g., country), no operational definition needs to be provided for the target variable. Otherwise, define the variable operationally by specifying what operations should be taken to make the variable measurable.

Transforming a problem statement to research questions involves several steps. First, determine if there are subproblems in the problem statement. If the problem statement contains no subproblem, only one research question is to be formulated to cover the entire problem. If there are two or more subproblems in the problem statement, formulate one research question for each of the subproblems identified. Never attempt to cover two or more subproblems in one research question because, as we will discuss in Section 2.3.2, you would run into difficulties if trying to make one hypothesis for the research question that contains two or more subproblems. Therefore, how many research questions should be created for a study depends entirely on the number of subproblems the problem statement has. Second, phrase each research question according to the three criteria that Kerlinger (1986, 16–18) specified and we have elucidated above.

A quick explanation of the relationship concept is necessary at this point. Relationship, in research terminology, refers to the association between or among variables. If there are two variables involved in an association, it is a bivariate relationship. If there are three or more variables involved, we call it a multivariate relationship. Do note that we stop counting when the number of variables reaches three in exploring relationships. We would simply call them collectively as multivariate. Whenever a relationship is examined, we need to understand two more concepts: independent variables and dependent variables. An independent variable is a variable that determines, affects, or causes changes in other variable(s). A dependent variable is a variable that is dependent on or affected by the independent variable. When we try to find out what determines college graduates' starting salary, we would examine variables such as GPA (grade point average), subject area (e.g., anthropology, computer science, economics, sociology), college reputation, and experience. GPA, subject area, college reputation, and experience are all examples of independent variables in this case while starting salary is the dependent variable. This hypothetical study explores relationship between the independent variables (e.g., GPA, subject area, college reputation, and experience) and the dependent variable (i.e., starting salary). Gender, race, and other demographic variables are common candidates of independent variables in research although they should not be considered in deciding college graduates' starting salary and other similar matters. We will further discuss variables and relationships in the rest of this book where appropriate, including in Chapter 13, Techniques for Data Analysis and Their Selection.

In Section 2.1, we quoted a problem statement from Negriff (2019, 1): "The current study examined the risk content of Facebook posts among a sample maltreated and comparison youth and compared the rates of risk content produced by the participants versus their Facebook friends". How many subproblems are there in Negriff's problem statement? The answer is two: (1) examine the risk content of Facebook posts among a sample maltreated and comparison youth, and (2) compare the rates of risk content produced by the participants versus their Facebook friends. Accordingly, two research questions can be derived from the problem statement although Negriff should have done so in the report:

- Research Question 1: What risk content do a sample of maltreated and comparison youths post on Facebook?
- Research Question 2: What are the rates of risk content produced by the research participants versus their Facebook friends?

Up to this point, we have satisfied the first two criteria Kerlinger (1986, 16–18) posited for formulating research questions. The third criterion of implying possibilities of empirical testing involves operational definition of variables and application of research methods. By looking at the two research questions, we can see that "risk content" and "maltreated youths" are the two variables that need to be defined operationally because there are no established measures for both variables. According to Negriff (2019, 1), risk content is operationally defined as those Facebook posts with references to alcohol, marijuana, hard drugs, partying, and sexual content while maltreated youths are those who received maltreatment (i.e., physical abuse, sexual abuse, emotional abuse, or neglect). Even though it is too early at present for us to consider what research methods should be used in Negriff's study just based on the research questions formulated, we do notice that content analysis Negriff has selected sounds a proper research method for the investigation. Comparison youths in the study are those participants without maltreatment and included in the research for comparison purpose.

Now let us take another example to explain how to transform a problem statement into research questions. Assume that a problem statement reads the following: The study intends to explore factors affecting the adoption of handheld devices. The following research questions can be formulated:

- Research Question 1: Is age a factor affecting the adoption of handheld devices?
- Research Question 2: Is gender a factor affecting the adoption of handheld devices?
- Research Question 3: Are years of education a factor affecting the adoption of handheld devices?
- Research Question 4: Is prior ownership of small electronic devices a factor affecting the adoption of handheld devices?
- ...
- Research Question n: Are there other factors that affect the adoption of handheld devices?

Because the sample problem statement contains several subproblems, multiple research questions are formulated for each of the subproblems identified. These research questions, in accordance with the three criteria Kerlinger (1986, 16–18) suggested, are meant to be specific and precise. Meanwhile variables are defined operationally where necessary. For example, the variable of education level is operationalized by years. That is, education level is to be measured by years of education one receives. An alternative operational definition for the variable of education level can be words indicating the various education levels (e.g., high, medium, and low). But this definition is not preferred for reasons we shall explain when discussing levels of measurement in Section 2.5.1. Another variable appearing in each research question that requires an operational definition is "adoption". What is regarded as adoption of handheld devices? Purchasing and owning one (e.g., a mobile phone) or borrowing one (e.g., a tablet) from a library is defined as adoption of the device. The last research question is created to cover any factors that are not already specified on the above list of questions. Dean (2010) offers an informative video lecture on writing problem statements and research questions. This video is highly recommended to readers of this book.

### *2.3.2 Hypotheses*

Once the research questions are formulated, it is time to consider if a hypothesis can be stated for any of the research questions based on existing theories, prior research as well as one's own knowledge and experience. A hypothesis is a tentative and educated answer to a given research question, which implies that a hypothesis can be formulated only if there is a tentative and educated answer to a given research question. If no tentative and educated answer is available for a given research question according to existent theories, prior research as well as one's own knowledge and experience, then no hypothesis can be made. In other words, it is the research question that ultimately determines whether a hypothesis can be formulated. In the previous example about factors affecting handheld device adoption, we are certain that, for instance, age and years of education would affect the adoption of handheld devices based on past research as well as our knowledge and experience. We can thus formulate one hypothesis for Research Question 1 and Research Question 3, respectively, as follows.

- Age is a factor that negatively affects the adoption of handheld devices.
- Years of education are a factor that positively affects the adoption of handheld devices.

As for Research Question 2, Research Question 4, and Research Question n, we cannot provide a tentative and educated answer based on past research as well as our knowledge and experience. Hence, no hypothesis can be made in those cases. In addition, Research Question n is an open-ended question (i.e., a question without a fixed answer) and we would not anticipate a fixed answer. Research questions of this kind in fact are never appropriate for having a hypothesis.

The above example demonstrates that (1) there is one and only one hypothesis for each research question, and (2) not every research question is eligible for having a hypothesis. These two points also determine that research questions must be formulated before hypotheses can be made. If there is no research question, no basis exists for forming a hypothesis because a hypothesis is after all a tentative and educated answer to the research question. Any attempt in making a hypothesis before formulating a research question may ruin the research project. That is, the prerequisites for making a hypothesis are two-fold: (1) a research question is already formulated, and (2) a tentative, educated answer to the research question is available based on existing theories, prior research as well as one's own knowledge and experience. Consider if you are able to formulate any hypothesis for the two research questions for Negriff's study (2019), and why.

In Section 2.3.1, we recommend that never attempt to cover in one research question two or more subproblems from a problem statement. Otherwise, we will run into difficulties when trying to make a hypothesis for such a research question. For example, we would not be able to formulate a meaningful hypothesis if we combine Research Question 1 (i.e., "Is age a factor affecting the adoption of handheld devices?") and Research Question 3 (i.e., "Are years of education a factor affecting the adoption of handheld devices?") we have discussed earlier into one research question because the hypothesis for each of them is different. One reads "Age is a factor that negatively affects the adoption of handheld devices" and the other states "Years of education are a factor that positively affects the adoption of handheld devices". This illustrates why we cannot "kill two birds with one stone" when it comes to research question creation and subsequent hypothesis formulation. Some readers may wonder what to do if one research question contains more than one subproblem from the problem statement. The answer is to reformulate your research question so that each would address one and only one subproblem from the problem statement.

Frankfort-Nachmias, Nachmias, and DeWaard (2015, 55–56) maintain that hypotheses must be clear (i.e., defining all involved variables conceptually and operationally), specific (i.e., pointing out the relationship expected to be found in terms of direction), testable (i.e., having proper techniques for the testing), and value-free (i.e., without any personal opinions and biases). The hypothesis depicted and presented up to this point is the so-called research hypothesis. Research hypotheses should be transformed to null hypotheses and hypotheses in statistical terms before being tested. As this book plans to only cover data analysis briefly as part of research design and method selection, limited discussion on hypothesis testing will be offered. A quick note needs to be made on qualitative hypotheses. Qualitative hypotheses deal with qualitative data. The testing of qualitative hypotheses, like any other kinds of qualitative data analysis, involves only words and texts via reasoning, comparison, and other similar analytical techniques. This process is entirely different from that for quantitative hypothesis testing.

In summary, research questions are derived from the problem statement. Hypotheses can only be made when a tentative, educated answer to the research question is available based on existing theories, prior research as well as one's own knowledge and experience. Research questions determine (1) what kind of data should be gathered, and (2) what techniques are to be selected for collecting such kind of data in a study. The latter is a major theme of the present book and will be covered from Chapters 3 to 12. Moreover, research questions also serve as a guide in the research process and help the researcher to stay focused.

## 2.4 Development of research methodology

This step in the research process considers what techniques can be selected for data collection and analysis in a study. The selection of data collection techniques is done after literature review and research question formulation because literature review would also inform us of the techniques used in previous related studies, while formulated research questions would guide us to see what data should be gathered and what techniques can enable us to collect such data.

Nevertheless, data collection technique selection cannot be fully discussed at this point because individual techniques are yet to be presented in Chapters 3–12. Similarly, how to select techniques for analyzing the data collected will be depicted in Chapter 13,

Techniques for Data Analysis and Their Selection. In Chapter 14, the last chapter of this book, further description and discussion of methodology development will be provided to elaborate on research design via method selection.

## 2.5 Data collection and analysis

Once data collection techniques (e.g., interview, questionnaire) are chosen, the data collection process can begin. How to collect data using individual techniques will be presented as an integral part of each data collection techniques from Chapters 3 through 12. Regardless of which technique is chosen for gathering data, researchers need to consider which units of data are to be collected or data collection units (Yin 2016, 91). Data collection units could be individuals (e.g., Participant A, Participant B), groups (e.g., millennials, Facebook users), or organizations (e.g., companies, universities). They also could be tangible objects (e.g., paintings, buildings) or intangible units (e.g., attitudes, behaviors, opinions). The research problem and associated questions of a study determine the data collection unit. For example, we can collect in a study test scores from individual students (individuals), students in different classes (groups), or students at different universities (organizations). If our focus is on comparing how students in different classes do in a test, the data collection unit would be groups. The data collection unit basically defines the "whom" or the "what" to be collected in research.

Data collection units can also be used in combination or at different levels. For instance, researchers can gather data about the opinions (intangible units) of individual employees (individuals) regarding income equality in different companies (organizations). Here, three data collection units are involved in the study. The data collection unit would start with organizations, to individuals and then individuals' opinions about income equality. Typically, one single study may have more than one unit of data collection. Researchers should therefore plan accordingly at the time of data collection.

The geographical dimension is another aspect the researcher must consider when collecting data. It defines the geographical scope of what is to be collected. The geographic dimension includes, for example, election districts, school districts, towns, cities, counties, states/provinces, and countries. The larger the geographic dimension, the wider geographic scope of data collection.

Similarly, the time dimension constitutes one more point of consideration in data collection. There are two common types of temporal dimension: longitudinal or cross-sectional. The former implies that data are collected over a period of time (e.g., several weeks, one year) and data collection is done at several or more time points, while the latter limits its data collection to just one point of time. Longitudinal research is conducted to study changes over time and is usually specified in the title or other parts of the research report. In comparison, the time dimension is not specifically taken into consideration in cross-sectional studies and usually not mentioned as a result.

Of all longitudinal research, there are three kinds that deserve some description (Babbie 2016, 106–11). (1) Trend studies – research for studying a general population over time (e.g., people's alcohol assumption in the past three years); (2) cohort studies – research for studying a group with a common feature such as age and race over time (e.g., Generation Xers' adoption of wearable technology); (3) panel studies – research for studying the same group of people over time. The "Up" series of documentary films by Paul Almond and Michael Apted clearly illustrates the panel study idea (https://en.wikipedia.org/wiki/Up_(film_series)). The series followed the lives of 14 British children since 1964 when they were seven years old and at the interval of seven years. In 2019, the ninth film in the series "63 Up" was introduced. Panel studies are also known as the Delphi studies which are to be discussed in Section 12.2. It is the different type of participants that determines the kind of longitudinal research.

After deciding the units and dimensions of data collection and completing the data gathering (to be discussed from Chapters 3 to 12), researchers are ready to choose proper techniques to analyze collected data. The selection of data analysis techniques in the research process is essentially dependent on the type of data (i.e., quantitative or qualitative) collected, levels of measurement (see Section 2.5.1), and research objectives (e.g., exploring correlation between two variables, revealing underlying themes of a data set). Chapter 13 is dedicated to the discussion on the selection of data analysis techniques. Given the focus and scope of this book, how to perform data analyses will only be covered briefly because many excellent publications, especially those on statistics, are readily available in many disciplines.

Before moving to the next step in the research process – results presentation and discussion – it is necessary for us to get familiar with two fundamental topics relating to data collection: (1) levels of measurement, and (2) sampling.

### *2.5.1 Levels of measurement*

Levels of measurement refer to how to measure data collected for a study. There are four different levels of measurement: nominal, ordinal, interval, and ratio. They ultimately determine what techniques will be used for analyzing collected data. As variables take values in the form of data, levels of measurement can also be applied to categorize or label variables (e.g., nominal variables, interval variables) in addition to data measurement (e.g., nominal data, interval data).

#### *2.5.1.1 Nominal level*

The nominal level is for measuring data that can be classified into categories. Gender, race, geographical location, and student type are some variable examples that take data measured at the nominal level. For example, male and female are the two main data categories for gender. Data categories for race include African American, Asian, Hispanic, and White in the United States. All the categories for nominal variables must be established by following two principles: (1) collectively exhaustive (i.e., including all possible categories), and (2) mutually exclusive (i.e., no data case should be in more than one category). Because of the first principle, researchers should provide a category of "Other (please specify): ..." whenever enumerated categories for a variable that cannot exhaust all the possibilities in a data collection instrument (e.g., questionnaire). Take the variable of race as an example, African American, Asian, Hispanic, and White do not cover all possible races in the United States. Thus, create the "Other (please specify): ..." category in order to allow American Indian, Pacific Islander, and other races to identify themselves.

At the nominal level, different data categories represent different values. Therefore, only two types of analyses can be performed on nominal data: "equal to", and "not equal to". Any other data analysis is not applicable to nominal data. For instance, it does not make any sense if one tries to say female is greater than male even though male might be coded as 1 while female as 2 for data processing purpose. Keep in mind that 1 and 2 in this context are merely labels or symbols without any quantitative implications because gender is a variable measured at the nominal level.

#### 2.5.1.2 Ordinal level

The ordinal level includes all the features of the nominal level. Furthermore, it possesses an inherent order and data values measured at this level can be compared using the "greater than" or "less than" mechanism. Examples of ordinal data encompass social classes, consumer satisfaction, job titles, and public opinions about societal issues. Such data are still recorded by categories (e.g., lower, middle, upper as social class categories) and hence have the property of normal data. Nevertheless, ordinal data have an intrinsic order and can be compared to determine which value is greater or less than the rest. The middle class, for example, is at a position higher than the lower class but lower than the upper class on the spectrum of social classes.

Numbers are frequently assigned to rank ordinal data in data collection and data coding. Such numbers represent only ranked values and have no equal interval between any two adjacent categories. If we designate 1 for lower class, 2 for middle class, and 3 for upper class, we cannot say that 1 (lower class) plus 2 (middle class) will lead to 3 (upper class) because those numbers purely are labels or symbols for ranking. Data measured at the ordinal level hence can just be used for calculations of "equal to" or "not equal to" and "greater than" or "less than". The Likert scale and other similar scales are all originally designed to collect ordinal data (e.g., research participants' opinions or attitudes about an issue). Each point on the scale signifies a value greater than or less than the one adjacent to it. The Likert scale will be discussed further in the next subsection on interval level. Detailed description of how to use the Likert scale in data collection instruments is presented in Section 3.2 with respect to opinion/attitude question creation in questionnaire or other data collection instruments.

#### 2.5.1.3 Interval level

The interval level includes all the features for data measured at the nominal or ordinal levels with the additional characteristic that data values are defined in a standard unit of measurement (e.g., years for age, degrees for temperature). Age, years of education, IQ score, temperature, and standard test scores (e.g., SAT) are examples of interval variables. In comparison with ordinal data, we are able to know not only which data value at the interval level is greater or less than the other but also how much more or less through addition and subtraction. For instance, undergraduates receive four more years of education than high school graduates. This kind of difference between two data values, however, cannot be specified for data measured at the ordinal level.

Data measured on the Likert and other similar scales, as briefly mentioned in Section 2.5.1.2, belong to the ordinal level in nature. Yet they are often treated as interval data if we explicitly specify in the data collection instrument (e.g., questionnaire) that all the intervals between any two adjacent points on the scale are equal. Such specifications would become clearer if we visually present the scale with equal intervals as shown below. Further details about Likert scale usage in data collection are provided in Section 3.2.1.

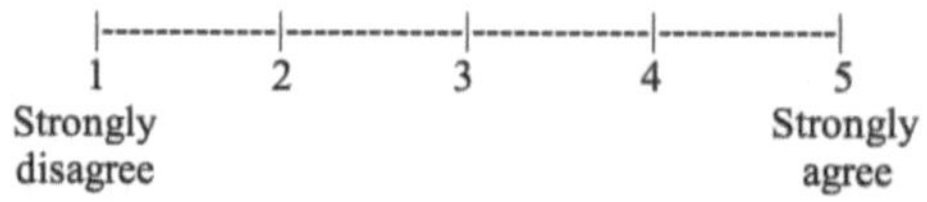


On the other hand, interval data start measurement at an arbitrary point instead of a natural zero point where the data property or value we attempt to observe is absent in its totality (Loether and McTavish 1980, 18). Age can serve as an example in this case. There is no natural zero point in the measurement of age because a baby's age cannot be zero as soon as the baby is born even though age at this point would be measured in minutes or even seconds. Further explanation about the natural zero point is to be offered in Section 2.5.1.4.

Some variables can be defined both at the ordinal and at the interval level. In such cases, we should always collect data measured at a higher level so that more techniques would be available for data analysis later. Take education level as an example, it can be measured either at the ordinal level (e.g., high, medium, low) or at the interval level (i.e., number of years one receives education) according to their corresponding characteristics. We hence should always measure education at the interval level whenever possible to take advantage of having more techniques for data analysis afterward.

Dichotomous or two-point variables, though taking only two values and measured at the nominal level (e.g., yes or no), can nonetheless be regarded as ordinal or interval instead of nominal in data analysis (Andrews et al. 1998, 14). Further discussion on levels of measurement and data analysis will be provided in Chapter 13, Techniques for Data Analysis and Their Selection.

#### *2.5.1.4 Ratio level*

The ratio level retains all the features for nominal, ordinal, and interval variables. In addition, its scale of measurement has a natural zero point (Loether and McTavish 1980, 18). A natural zero point is a theoretical defined point on a scale representing the absence of any of the properties being measured. This level of measurement is most commonly encountered in physical sciences although abundant variables in the social sciences are measured at the ratio level as well. Distance, length, time, number of students in a class, and budget size in dollar amount are some examples of ratio variables.

A simple example of ratio variable is the number of text messages one receives daily. If one receives no text messages in a day, this means that the property being measured (i.e., number of text messages received daily) is absent. In contrast, age, years of education, and other interval variables do not have a natural zero point. For instance, years of education are at most an interval variable and do not have a natural zero point even though it is measured in number of years because a person who does not receive any formal education can still learn by self-study or other means.

Temperature is another interval variable with the exception of temperature measured on the Kelvin scale where an absolute zero on Kelvin or 0K, equal to either −273°C or −459.7°F, indicates the total absence of any temperature or that "all thermal motion ceases in the classical description of thermodynamics" (Wikipedia Contributors 2023a). The same, however, is not true to temperature measured on both the Fahrenheit and Centigrade scale because a zero degree in either case does not imply the absence of temperature.

It becomes apparent that it is the natural zero point that sets the ratio level apart from the interval level. Because of this feature, we can do multiplication and division with ratio data. We can also apply the statement "A is X times as much as B" where A and B are two values of a variable to find out if a variable is measured at the ratio level. If such a statement makes sense (e.g., "Joe earns twice as much as John"), the variable is measured at the ratio level. Otherwise, the variable is measured at the interval level. For instance, we cannot say that "70F is twice as warm as 35F" because temperature in Fahrenheit is an interval variable.

#### 2.5.1.5 Levels of measurement: A summary

Regardless of at which level the data are measured, we can roughly group nominal and ordinal data as qualitative while interval and ratio data as quantitative. This categorization can help determine both the type of data (i.e., qualitative or quantitative) to be collected and the techniques needed for data analysis. Table 2.2 recapitulates the features[2] of and calculations allowed for different levels of measurement.

Calculations allowed at each level of measurement dictate what kinds of data analysis can be performed later on the data collected. Generally speaking, nominal and ordinal data are mainly analyzed qualitatively along with some basic descriptive statistics (e.g., frequency, percentage), whereas interval and ratio data are usually analyzed quantitatively. The latter two do share many statistical or quantitative analysis techniques even though they are categorized as two different levels of measurement. For this reason, many statistical analysis techniques group interval and ratio data together. Some qualitative techniques (e.g., qualitative coding, content analysis), on the other hand, can also be adopted for analyzing quantitative data to complement the number-only quantitative approaches. No matter whether qualitative and/or quantitative techniques are selected for data analysis, always measure data at the highest possible level at the time of data collection. The rationale behind this recommendation is that more data analysis techniques would be available as the level of measurement moves from nominal to ordinal, interval, or ratio. We can only, for example, compare nominal data as equal or non-equal. In comparison, a wide variety of techniques can be utilized for analyzing ratio data. As indicated earlier, education level is an example variable that can be measured either at the ordinal level (e.g., high, medium, low) or at the interval level (e.g., years of education). Choose years of education to measure the education level variable whenever possible because there are more analytical techniques for interval data than that for ordinal data. Chapter 13 is devoted to discussing how to choose proper techniques for analyzing data that are measured at different levels.

### 2.5.2 Sampling

More often than not, researchers ought to select a sample for an investigation instead of working with the entire population or universe mainly in consideration of time, resources, and feasibility.

*Table 2.2* Levels of Measurement: Features and Calculations Allowed

| *Level of Measurement* | *Features* | | | | *Calculations Allowed* |
|---|---|---|---|---|---|
| | *Collectively Exhaustive, Mutually Exclusive Categories* | *Categories Having an Ordering* | *Categories Having a Standard Unit of Measurement* | *Meaningful Zero Point on Scale* | |
| Nominal | √ | | | | =, ≠ |
| Ordinal | √ | √ | | | =, ≠, >, < |
| Interval | √ | √ | √ | | =, ≠, >, <, +, − |
| Ratio | √ | √ | √ | √ | =, ≠, >, <, +, −, ×, ÷ |

Time is always an issue in any research. Working with a subset of a population for a research project no doubt would reduce the time needed to complete the undertaking. The same holds true in terms of resources required in research. As for the feasibility issue, it must be addressed in that few investigations are able to include the entire population in the research process even if time and resources are not factors involved in deciding on the sample versus population matter. For example, there are always students who cannot participate in a study for a variety of reasons (e.g., sick leave) regardless of whether researchers have abundant time and resources to recruit participants from the student population. Moreover, research using a sample that is randomly chosen would yield similar results as that with the entire population according to the probability theory (Babbie 2016, 190–201).

A sample, by definition, is a subset of cases from a population while the aggregation of all cases relevant to a study constitutes the population. In a higher learning institution, for example, its student population consists of all the students registered in all the programs the institution offers in a semester. A subset of the students from that institution can serve as a sample for the population. If the institution has 25,000 students, a sample can theoretically be any size between 1 and 24,999 students.

Sampling is traditionally introduced when describing survey or questionnaire research in numerous publications on research methods. This practice gives the reader a false impression that sampling is only done in questionnaire research. To the contrary, sampling is utilized for data collection in many other kinds of research. For instance, content analysis as a data collection technique needs to select a sample of entities (e.g., documents, social media posts) to be analyzed while bibliometrics would choose a subset of all publication and citation records on a given topic area for examination and analysis. In other words, sampling is needed whenever it is not economical or feasible to use the entire population in a study. Sampling is by no means limited just to questionnaire research!

#### *2.5.2.1 Major considerations in sampling*

The first major consideration in selecting a sample is to define a population in research. A population comprises the complete set of cases or sampling units relevant to a study (e.g., all the students who enrolled at an academic institution in Spring 2023). Sampling units can be people (e.g., individuals, groups), institutions (e.g., companies, organizations, schools), objects (e.g., e-gadgets, tools), artifacts (e.g., events, publications, websites), or social interactions (e.g., relations, marriages, divorces).

The actual list of sampling units from which a sample is drawn is called a sampling frame. It may or may not be equal to the population size. Take a college class roster as an example, it would not be equal to the class population size if it is not updated after the add/drop period in the beginning of a semester. In addition, it is sometimes impossible to obtain or compile a complete list of sampling units in a population. If we, for instance, intend to find out all the residents in a large city such as New York, we would not be able to achieve our objective no matter how hard we try due to the mobility, privacy, and other factors.

The second major consideration in sampling is to determine a sample size or the number of sampling units to be drawn from the population. As briefly mentioned earlier, a sample could contain one sampling unit, all but one sampling unit, or any other number in between. Let us return to the example given above, a sample can be any size between 1 and 24,999 students if the student population or the total number of students at the institution is 25,000. That is, we can in theory choose a sample of one student, 24,999 students, or any number of students in between the two (e.g., 2, 3, … 24,998) when we conduct a study about the students from that institution. Obviously, it makes little sense to select a sample of one student only or 24,999 students even though both are legitimate sample size. We therefore should determine a sample size that would both represent the population well and meet the objective of having a sample.

There are, however, three common misconceptions about sample size. (1) A sample should be a regular proportion (e.g., 5%) of the population. (2) A sample should have a total of about 2,000 sampling units regardless of the population size. (3) Any increase in sample size will significantly increase the accuracy of estimation. The first two misconceptions are groundless, while the third one touches on some fact but is simply untrue. The reality is diminishing returns (i.e., estimation accuracy) from bigger samples, which is demonstrated via the example below.

In quantitative data analysis, a property of a sample (e.g., average number of courses taken in a semester) is called statistic while that of a population is labeled as parameter. The Sample Size Calculator (http://www.surveysystem.com/sscalc.htm) by Creative Research Systems (2021) is a free application for calculating sample size after the confidence interval, confidence level, and population size are specified. Both the confidence interval (aka margin of error) and confidence level are involved in inferential statistics. Let us try to approach these two concepts without getting too deep into statistics. The confidence interval is the plus-and-minus figure (e.g., 47%±3) usually reported in opinion poll results and refers to the range of values within which a true population parameter (e.g., mean, percentage) is estimated to lie (Babbie 2016, 524; Scheuren 2004, 63). The figure of 47%±3 implies that between 44% (47−3) and 50% (47+3) of the population would have picked that answer for a question asked. To put it very simply, a wider confidence interval leads to a less accurate estimation due to the longer range, while a narrower confidence interval results in a more accurate estimation. For instance, the actual distance between New York City and Washington, DC is about 277 miles. If we cannot do a search to find the exact distance, we instead estimate that the distance between the two cities is from 242 to 312 miles (i.e., 277±35). Another estimation has a narrower confidence interval: 272–282 miles (i.e., 277±5). Of the two estimates, the latter is more accurate with a shorter range.

The confidence level, usually set at either 95% or 99%, represents the estimated probability that a population parameter lies within a given confidence interval (Babbie 2016, 524). Assuming 25,000 as the population size, the Sample Size Calculator yields the following sample sizes:

- 378 when confidence level is set to 95% and confidence interval at ±5
- 1,023 when confidence level is set to 95% and confidence interval at ±3
- 6,939 when confidence level is set to 95% and confidence interval at ±1

The above example shows that increasing the sample size can reduce the margin of error (i.e., confidence interval) and enhance estimation accuracy but not drastically. Specially, reducing the margin of error by 2% (i.e., from ±5 to ±3) requires an increase in sample size by 645 sampling units; reducing the margin of error by 4% (i.e., from ±5 to ±1) requires an increase in sample size by 6,595 sampling units. Because a sample is used for estimating a population parameter in any study that involves sampling, completely removing the margin of error or achieving full estimation accuracy is impossible no matter how large the sample is. A larger sample would help improve estimation accuracy but not to a great extent as illustrated in the above example. Also, a sample too large in proportion of the population (e.g., 6,939 out of 25,000) would eliminate the value of having a sample in research.

Sample size determination theoretically should be done through a formula[3] as employed in the Sample Size Calculator. In reality, determining a sample size in research of many disciplines turns out to be more practical and simpler than the formula approach. The researcher basically decides a sample size by taking into consideration time, resources, and research topic. Institutions that frequently conduct research with nationwide samples (e.g., National Opinion Research Center at the University of Chicago, PEW Research Center) often select a sample around 1,500 participants. Scheuren (2004, 10) resonates with such practice in stating that "the well-known national polls frequently use samples of about 1,000 persons to get reasonable information about national attitudes and opinions". Research projects with a smaller scope commonly choose a sample ranging from several dozen to several hundred of sampling units. Few studies determine a sample size using the formula approach.

The third major consideration with respect to sampling is to select a sampling method. A sampling method specifies the procedure for selecting individual sampling units from a population. There are two broad categories of sampling methods: probability and nonprobability. Probability or random sampling ensures the representativeness of a sample, while nonprobability sampling, also known as qualitative sampling, is mainly done in qualitative research or for convenience, feasibility, and economic reasons. Compared with nonprobability sampling, choosing a random sample in research has basically two implications: representativeness and generalizability. Representativeness means that research with a randomly chosen sample produces results equivalent to those that would be obtained had the entire population been used, and it can only be achieved through a random sample. Generalizability, arising from the concept of representativeness, implies that researchers can generalize or extend findings from studies using a random sample to the population from which the sample is drawn. For instance, we can generalize research findings with a random sample of 300 students at a higher learning institution to all the 25,000 students (i.e., the population) there.

Because of those implications, we should select a random sample whenever possible in research. Furthermore, both probability and nonprobability sampling methods can be further categorized into specific procedures according to their features. The next two sections detail basic probability and nonprobability sampling methods.

#### *2.5.2.2 Basic probability sampling methods*

Basic probability sampling methods include simple random sampling, systematic sampling, stratified sampling, and cluster sampling. Each of them will be introduced along with comparisons to its counterparts in helping the reader choose the most appropriate one for a project at hand.

##### 2.5.2.2.1 SIMPLE RANDOM SAMPLING

Simple random sampling is a procedure that gives each sampling unit in the population an equal, nonzero probability of being selected. Techniques for drawing a simple random sample used to be the "lottery method" by applying a procedure in lottery playing (e.g., tossing a coin, mixing slips in a hat) or the random number approach based on either a table of random digits appended to textbooks or a computer algorithm for generating random numbers. Nevertheless, those traditional simple random sampling techniques have been replaced by random number generators such as Research Randomizer (https://www.randomizer.org).

Unlike the traditional techniques, random number generators make the sampling process simpler and more efficient. After a sample size (e.g., 10) and a population range (e.g., 1–50) are given, for example, Research Randomizer would yield a list of random numbers that can be sorted and displayed according to the researcher's specifications.

The simple random sampling method is a one-step procedure when contrasted with other probability sampling methods to be discussed. It indeed becomes simple and true to its name thanks to the advent of applications like Research Randomizer. It is also a robust technique for random sampling in that each sampling unit in the population has an equal, nonzero probability of getting selected. However, simple random sampling perhaps is not the most appropriate choice when the sample size is large (e.g., in the thousands). This is because drawing actual sampling units from a huge population could be tedious and prone to error.

#### 2.5.2.2.2 SYSTEMATIC SAMPLING

Systematic sampling refers to a procedure that selects every kth sampling unit of the population after the sampling start point is randomly chosen from the first kth sampling units. The value of k is determined by dividing the population size N with a chosen sample size n (i.e., k = $N/n$). If we plan to select a sample of 10 ($n$) sampling units from a population of 50 (N), for example, the value of k would be 5 (i.e., $N/n$ = 50/10). After k is determined, we also need to randomly choose a number from 1 to 5 to serve as the start point in the sampling process using the simple random sampling technique we have just discussed. If the number of 2 is randomly selected as the start point for this systematic sampling, the subsequent sampling units will be 7, 12, 17, 22, 27, 32, 37, 42, and 47.

Although it may easily be overlooked, selecting a sampling start point randomly between 1 and k is an essential element in systematic sampling and for ensuring that every sampling unit gets an equal chance of being selected. If we do not do so, one of the following two consequences is likely to occur: (1) running out of sampling units before the sampling process completes if the start point is greater than k (e.g., 6 when k is 5), or (2) the last k sampling units in the population would never get selected if the sampling is set to start with 1. Both consequences naturally should be avoided in systematic sampling. You may want to plug in some numbers to better understand the two scenarios depicted above if the start point for systematic sampling is not chosen randomly between 1 and k. In addition, the sampling frame in systematic sampling should be randomized using a spreadsheet application (e.g., Excel) or other similar applications if it is not already so. An alphabetical list of personal names, for instance, is not random because names beginning with certain letters (e.g., B or S) have more chance to be chosen than others (e.g., names starting with the letter X or Z).

In comparison with simple random sampling, systematic sampling could be simpler before the creation of online randomizers. It is also straightforward and having regularity, thus suitable for drawing large samples. On the other hand, systematic sampling has to rely on a simple random sampling technique to choose a start point.

#### 2.5.2.2.3 STRATIFIED SAMPLING

Stratified sampling aims to select a sample that is proportional to the population makeup according to a particularly attribute (e.g., race) and attempts to ensure that different groups of a population are adequately represented. In a community with residents of different races, for example, we certainly want to choose a sample that reflects the racial composition of its members when conducting a study on them. If we know that White accounts for 50% of the community population, African-American 20%, Hispanic 15%, Asian 5%, and other 10%, the sample should have a similar representation of each race.

Race is hence the criterion applied in the above example of drawing a stratified sample from a residential community. Each portion in stratified sampling is called a stratum (e.g., White, African-American). In addition, one and only one criterion can be used for determining the strata because the end product will not be a stratified sample if two or more criteria are adopted in the sampling process.

The stratified sampling procedure consists of three steps. First, divide all sampling units of a population into homogeneous strata based on one chosen criterion. Race, age, or gender are some common criteria applied to identify homogeneous strata when human participants are invited in research. Second, determine the percentage of each stratum in the population. Third, use either the simple random or systematic sampling technique to select individual sampling units from each stratum identified. Do observe the rules for doing the simple random or systematic sampling in this part of stratified sampling.

As shown, stratified sampling features not only a three-step procedure but also relying on another sampling technique (i.e., simple random or systematic sampling) to draw a sample. Although this feature may require extra efforts in sampling, stratified sampling is the only approach among all the probability sampling techniques that enables researchers to draw a sample that represents the population in proportion based on one criterion.

#### 2.5.2.2.4 CLUSTER SAMPLING

Cluster sampling is designed for research when no sampling frame is available or the population is too huge to use any other probability sampling method alone. This kind of sampling involves splitting the target population into successively smaller clusters or groups for sampling through multiple stages. If we intend to conduct a nationwide study on a topic, for instance, we would go through the stages listed in Figure 2.2 to obtain a random sample of individuals.

Even though there is no complete sampling frame for all the individuals living in the United States, a sampling frame is available at every sampling stage from counties to towns, ..., at last to members of chosen families. Taking counties as an example, all the 3,142 counties and county equivalents in the 50 states and the District of Columbia can be obtained from Wikipedia (Wikipedia Contributors 2023b). Securing sampling frames for the rest of sampling stages would likewise not be an issue at all.

The previous example illustrates what cluster sampling is about. In general, the cluster sampling process is divided into several stages where a sampling frame is available. Then at each stage, use a probability sampling technique, most often either simple random or systematic, to create a group of entities (e.g., counties, towns) for sampling at the next stage. Repeat this process till the target sampling units (e.g., family members) are sampled.

Cluster sampling progresses from a population without a sampling frame to a cluster or group of sampling units that has a sampling frame through multiple stages. Like stratified sampling, cluster sampling also needs to determine the stages according to one and only criterion. In the above example, clusters of sampling units are all selected according to area, which explains why cluster sampling is also known as area sampling.

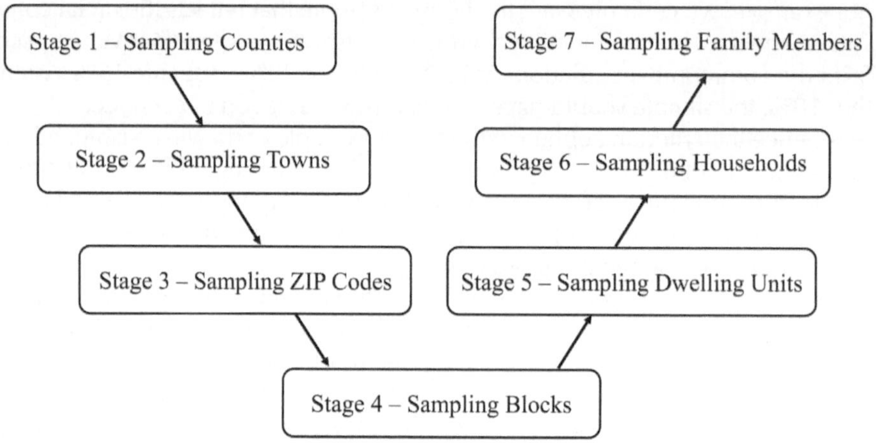


*Figure 2.2* Cluster Sampling through Multiple Stages

Cluster sampling is different from the other three probability sampling methods we have discussed so far in two points: (1) the multiple-stage process creates successively smaller clusters till the target sampling units can be obtained, and (2) apply a probability sampling method (i.e., either simple random or systematic) at each stage to select individual sampling units.

### *2.5.2.3 Basic nonprobability sampling methods*

In contrast to their probability counterparts, nonprobability sampling techniques are used when it is impractical or impossible to choose a probability sample. Nonprobability sampling is often done in qualitative research. Common sampling techniques in the nonprobability category encompass convenience, purposive, quota, and snowball sampling.

#### 2.5.2.3.1 CONVENIENCE SAMPLING

As its name suggests, convenience sampling selects sampling units that are conveniently available as long as they meet the basic requirement for serving as research participants. For example, if a study aims to explore what affects people's selection of social media sites (e.g., Facebook, Twitter), anyone with a social media account would be qualified to serve as a participant. No additional requirement would be made about potential participants of this research in the process of convenience sampling. Researchers may, although not recommended, even ask their family members, relatives, colleagues, and friends to take part in the study if they all have social media accounts.

Features of convenience sampling are distinctive both on the positive side as well as on the negative side. Less effort is needed for sampling if this approach is taken when compared with other sampling methods. However, the representativeness of samples obtained using this sampling technique is always out of the question. Researchers thus should think carefully and be fully aware that study results based on convenience sampling would not be generalizable. Other nonprobability sampling techniques, though facing the same issue of non-representativeness, are not as serious in contrast due to their respective characteristics that are to be presented shortly.

#### 2.5.2.3.2 PURPOSIVE SAMPLING

When a researcher chooses sampling units subjectively, it is called purposive sampling. Because such sampling is based on researchers' judgment, it is also known as judgmental sampling. Assuming we plan to conduct research on students and intend to invite students who visit the institution's library to participate in the study, we would need to use our judgment to determine if a person entering the library is a student. All other individuals (e.g., faculty, library staff) will not be invited to take part in the research. Ng'oma and colleagues (2019), for instance, chose a purposive sample of 22 women and explored their perceptions of perinatal depression and treatment needs in the Republic of Malawi. In the study, research participants must be pregnant women and mothers aged 18 years and above with six-week-old babies. These women were also screened positive for depression using the Edinburgh Postnatal Depression Scale.

While purposive sampling appears simpler and less time-consuming than all the probability approaches to sampling, it solely relies on researchers' judgment to select sampling units which might be prone to misjudgment. However, it outperforms convenience sampling in representativeness because of the judgment involved in the sampling process.

#### 2.5.2.3.3 QUOTA SAMPLING

Quota sampling is similar to stratified sampling we discuss in Section 2.5.2.2.3 except that the former is not a probability sampling method. Using this sampling technique, we can select sampling units that are as closely as possible a replica of the population by proportion, but not randomly. According to Perrin and Anderson (2019) from Pew Research Center, American adults' Facebook visit frequency is as follows: several times a day – 51%, about once a day – 23%, less often – 26%. If we want to conduct a study about the relationship between those Facebook users' visit frequency and their purposes of visiting the social media site, we can draw a quota sample of them. That is, we would make sure that 51% of our research participants coming from those who visit Facebook several times a day, 23% about once a day, and 26% less than once a day in order to explore if there exists any relationship between visit frequency and purposes. Thomson et al. (2010) investigate if sunbed use in children aged 11–17 in England increases their risk of developing malignant melanoma. The researchers use a quota sample of gender, aiming for an equal number of boys and girls in the study. That is, the sample for the research includes 50% of boys and 50% of girls.

Like the case of stratified sampling, common attributes used for establishing the quota include age, gender, and race. Furthermore, another nonprobability sampling technique (e.g., convenience sampling, purposive sampling) should be employed to draw individual sampling units from each quota. In contrast with stratified sampling, quota sampling shares all the features of stratified sampling but the randomness, which means that research results obtained via a quota sample cannot be generalized – a concept to be further discussed in Section 2.5.2.4.

#### 2.5.2.3.4 SNOWBALL SAMPLING

Snowball sampling is needed when researchers have no sampling frame or little knowledge about potential participants. All other sampling techniques we have discussed so far therefore cannot be used. Prospective research participants in such cases usually involve drug addicts, homeless people, victims of domestic violence, non-users of library, and more. They often are hard for researchers to identify or approach. It is for these reasons that snowball sampling is developed as a technique which starts with one or several individuals who meet the criteria for participation in research. These individuals would then be asked to make suggestions or nominations for additional qualified candidates until a desirable number of people are recruited for taking part in the study. Snowball sampling gets its name because the process of recruiting participants is identical to how a snowball is made.

A less often used name for this technique is chain referral sampling (Babbie 2016, 188). "The method yields a study sample through referrals made among people who share or know of others who possess some characteristics that are of research interest" (Biernacki and Waldorf 1981, 141). Unlike all other sampling methods depicted earlier, researchers using snowball sampling must actively and deliberately develop and control the sample's initiation, progress, and termination. Specific problem areas in this type of sampling include the following (Biernacki and Waldorf 1981):

- Finding participants and starting referral chains
- Verifying the eligibility of potential participants
- Engaging participants in the study
- Controlling the types of chains and number of cases in any chain
- Pacing and monitoring referral chains and data quality

Each of the problem areas could lead to difficulties in the sampling process. In describing those problem areas, the authors also illustrate the sampling procedures. Biernacki and Waldorf (1981) approach snowball sampling from the sociological research perspective with reference to large-scale projects on drug addiction. However, snowball sampling is applicable to other cases where participants can only be recruited via referral. For instance, Browne (2005) uses snowball sampling to recruit participants for her study of non-heterosexual women. Public libraries all over the world strive to outreach people who live in the community but do not go to libraries. Snowball sampling would be a suitable technique for getting in touch with those non-users of library to find out the reasons that prevent people from using libraries and then develop corresponding measures to encourage them to become library users.

#### *2.5.2.4 Generalizability and other considerations in selecting sampling methods*

The concept of generalizability denotes the extent to which research findings based on a sample can be generalized or applicable to the population from which the sample is drawn. It has profound implication in sampling. When a sample in a study is randomly drawn or obtained using a probability sampling method, its findings can be generalized. Otherwise, the research results are not generalizable.

Generalizability is also known as external validity while the validity concept will be introduced in Section 2.8. The extent of generalization is only as large as the size of the population from which the random sample is obtained. For example, when we draw a random sample of 300 students from an institution with a student population of 25,000, our findings can be generalized only to those 25,000 students. The same results based on the 300 students theoretically are not applicable to students in any other institutions regardless of how similar they all might be. Therefore, generalizability is one of the criteria

*Table 2.3* Summary of Basic Sampling Methods

| | *Sampling Method* | *Generalizability* | *Step* | *Other Features* |
|---|---|---|---|---|
| Random | Simple random | Y | 1 | Simplicity in sampling with randomizer sites, not good for large samples |
| | Systematic | Y | 2 | Straightforward, good for large samples, but must use simple random sampling to choose a start point |
| | Stratified | Y | 3 | Sample in proportion to the makeup of the population according to a particular attribute |
| | Cluster | Y | 4+ | Feasibility in sampling when there is no sampling frame, involving multiple sampling steps |
| Nonrandom | Convenience | N | 1 | Convenience in sampling, but at the cost of sampling representativeness |
| | Purposive | N | 1 | Convenience in drawing samples, but with subjectivity or possible misjudgment |
| | Quota | N | 3 | Similar to stratified sampling but without any generalizability |
| | Snowball | N | n | Sample without any sampling frame or decent pool of potential sampling units; sampling steps could be 2, 3, …, n |

we should consider when choosing between probability and nonprobability sampling methods.

Other considerations we need to give in selecting a particular sampling method comprise time, resources, and the distinctive features associated with each of the methods (e.g., availability of a sampling frame, steps needed to draw a sample). Table 2.3 summarizes the features of all basic sampling methods.

The decision on selecting a proper sampling method for research should be made by taking into consideration all relevant factors summarized in Table 2.3 in conjunction with time and resources needed for research. Despite the availability of many sampling options, do probability sampling if possible and feasible for the sake of generalizability. All other factors are secondary in comparison. Generally speaking, simple sampling or systematic sampling is most commonly used in quantitative research, while purposive sampling appears taking the lead in qualitative research.

A quick note on Sections 2.6 and 2.7 is due at this point. Both are depicted immediately below in order to provide a complete coverage of all the steps in a research process. Logically, both sections should be discussed after all the remaining 12 chapters of this book are done. That order of presentation, however, cannot be adopted in this book for two reasons. First, the focus of the present book, as indicated earlier, is on research methods and design. Covering in detail what to do in the last two steps of a research process is thus beyond the scope of this book. Second, how to accomplish these two steps largely depends on the results and objectives of individual studies. What can be briefly discussed below hence will merely be some general guidelines, not specific to individual studies, for completing the last two steps of a research project. Readers are encouraged to refer to these guidelines whenever needed in their research endeavors.

## 2.6 Result presentation and discussion

After data collection and analysis, the next step in the research process is for investigators to present and discuss their results. Because of the two reasons depicted above, it is inappropriate and ineffective to use examples to illustrate result presentation and discussion in this book. Instead, only some general guidelines are suggested for this research step.

- Decide whether results and discussion are presented in two separate sections, as required in some cases (e.g., journal submissions) or integrated into one section. The latter seems more logical and minimizes repetition when researchers can present one result and discuss it right afterward and follow this pattern until all results and discussion are done. If results and discussion are covered in two separate sections, certain degree of redundancy is inevitable in the discussion when researchers need to refer to the results presented earlier from time to time. Therefore, the general advice in this decision-making is to integrate result presentation and discussion in one section unless required otherwise.
- Organize results (i.e., analyzed data) according to research questions formulated earlier in the study. The ultimate purpose for all the steps up to this point is to answer the research questions. Now that the data are collected and analyzed, presenting the results and associated discussion based on research questions would help the reader better understand what answers to the research questions are obtained by the study.
- Describe your results as well as discuss and interpret them. In this step of research, result description alone would not do the job of answering research questions. Rather, it is through discussion that researchers explain what the results truly mean to the study. Researchers can also compare their current results with what has been done in the past to further confirm the importance of the study in consideration.
- Provide logic and structure in result presentation and discussion. For example, describe demographic results, if any, before reporting results about other aspects of the study.
- Create subheadings to highlight the theme of each subsection so that the reader would learn about what is to be presented by looking at the subsection headings. Short phrases derived from research questions are usually good candidates for subsection headings.
- Be selective of what you have in your presentation so that your research findings would not be buried in too many details and nuances. Your inclusion or exclusion decisions should be based on the following three points: (1) Report your findings faithfully and never only select those that support your research. (2) Report your findings within the given time (e.g., conference) or space (e.g., journal) limit. (3) Report your findings at a proper level of specificity and with minimal redundancy.
- Use tables or figures (e.g., graphs, diagrams) where appropriate to encourage synthesis, comparison, and visualization. Tables and figures are particularly effective for reporting results. However, avoid simply repeating what is already presented in tables and figures in narratives (e.g., 75% of research participants are female). Rather, use narratives to highlight, contrast, and discuss the results (e.g., there are twice more female participants than male counterparts in the study).

## 2.7 Conclusion drawing

The last step in the research process is to draw one or more conclusions of a study. Conclusions of a study represent the points the reader would take away after going over the study report. According to Yin (2016, 235), "... a conclusion is an overarching statement or series of statements that raises the interpretation of a study to a higher conceptual level or broader set of ideas. In one sense, the conclusion captures the broader significance of a study." Below are several guidelines for conclusion drawing even though a general procedure for conclusion drawing, as indicated before, is impossible due to the fact that what goes into the conclusions of each study is determined by the study itself.

- Draw conclusions as answers to your research questions based on your results or findings and in reference to their implications. Research questions raised in the study should be explicitly answered. These answers should be stated along with their implications. If a study finds out that misinformation is often disseminated via social media, for example, its report should not stop there. Instead, the researcher of the study should draw a conclusion by stating that social media users must take proper measures to evaluate the credibility and accuracy of information before using it.
- Avoid simply restating or paraphrasing your findings as conclusions. Findings are the results you obtained and interpreted in your study, while conclusions are statements that, based on your findings, indicate implications and significance of your study.
- Answer in your conclusions the "so what?' question that the reader would ask after going over your study.

In discussing how to write conclusions for qualitative research, Yin (2016, 235–43) summarizes five different ways of concluding: (1) calling for new research and making substantive propositions that are not methodological; (2) challenging conventional social stereotypes; (3) putting forward new concepts, theories, even discoveries about human social behavior; (4) generalizing to a broader set of situations; and (5) taking actions. Although Yin outlines those five ways of concluding from the perspective of qualitative research, particularly that of field study, the underlying points are applicable to other kinds of research as well. In addition, researchers often point out the limitations of their studies and suggest measures for overcoming them in the form of future research recommendations at the time of conclusion drawing.

Most books on research methods would devote one section or even one chapter to research report writing. Yet research report writing cannot be meaningful unless it is done in the context of a real research project. Furthermore, how well one can write also plays a major role in research report writing. For these reasons, the present book would not address the topic of research report writing. Rather it only provides, where appropriate, some general points in the form of suggestions. That being said, the guidance and advice Bem (2003) provides for empirical journal article writing is well worth reading even though Bem's intended audiences come from personality and social psychology.

## 2.8 Research validity and reliability

In addition to getting familiar with all the steps in a research process, we should also pay attention to validity and reliability – the two quintessential concepts in any study.

### 2.8.1 Validity

Validity, for any research project, comprises internal validity and external validity. Internal validity addresses the question if a research project studies what it is intended to do. We can verify a study's internal validity by checking its problem statement against the research report. For example, Negriff (2019, 1) states that "[T]he current study examined the risk content of Facebook posts among a sample maltreated and comparison youth and compared the rates of risk content produced by the participants versus their Facebook friends". Indeed, Negriff did exactly what she indicated to do according to her study report. If Negriff had looked at cyberbullying instead or had not recruited maltreated youths as research participants, we would say that the internal validity of Negriff's study is not assured.

External validity, on the other hand, refers to whether research findings of a study using a sample can be generalized to the population from which the sample is drawn. In that sense, external validity is synonymous with the concept of generalizability discussed in Section 2.5.2.4. If a study employs randomly sampling, its findings can be generalized to the population from which the sample is obtained. The study ensures its external validity. Otherwise, the study does not have external validity. In research with a social component, often times random sampling is not a viable option (e.g., cyberbullying victims). In those cases, it does not mean that we would discard the research. Rather, we just acknowledge not having external validity as a limitation in our study.

Numerous books of research methods reduce validity to accurate measurement or whether researchers measure what they intend to measure. Measurement, however, deals only with data collection while validity in fact is a concept relating to the entire research process.

### 2.8.2 Reliability

Reliability examines if similar results can be obtained when a study is repeated by following the same methodology. It is concerned with data collection and analysis in the research process. When a study, for example, uses questionnaire and interview for data collection, both the questionnaire and interview guide (i.e., instructions and a list of questions for conducting an interview) should be presented in the final research report as appendixes or available upon request. This means that other researchers would be able to collect data with the same questionnaire and interview guide when replicating a study.

In addition, researchers must provide detailed description of their data analysis procedures in the report as an assurance of reliability in a study. All techniques applied in data analysis likewise should be described "in sufficient detail to permit other researchers to replicate the analysis from the same body of data" (Babbie 2016, 505). For quantitative data analysis, reliability implies that other researchers would be able to, for instance, produce the same frequency distribution, obtain the same correlation coefficient, and attain the same factors and factor loadings. For qualitative data analysis, reliability means that other researchers can, for example, identify similar underlying concepts and themes from the interview transcripts although reliability is harder for researchers to achieve in qualitative data analysis due to the nature of qualitative research.

Similar to the case of validity, many books on research methods associate reliability with measurement consistency only. The consistency of measurement in data collection is one element in research. Many other elements or factors determine the reliability of a study. That is, the entire methodology of a study should be described with adequate details so that other researchers can replicate the research by using the same techniques for data collection and analysis. Furthermore, some authors of research methods discuss intercoder reliability in place of reliability. Intercoder reliability is a concept that researchers must consider when performing qualitative coding and differs from the reliability of a study. This book will explain intercoder reliability in Chapter 9, Content Analysis.

Additional discussion and elaboration on validity and reliability will be provided in the remaining chapters of this book where appropriate. Both concepts also serve as major criteria for evaluating research.

## 2.9 Research ethics

Researchers should be mindful of ethics regardless of their disciplines and research topics. Ethics involve moral principles that govern a person's behavior or the conducting of an activity. Accordingly, research ethics refer to moral principles that govern a person's research behavior or the conducting of research. Different from the code of ethics for one discipline (e.g., psychology) or one profession (e.g., nurses) which has broad ramifications, ethical guidelines in research are generally of two kinds, one regarding researchers and the other about protecting participants.

### *2.9.1 Ethical guidelines concerning researchers*

Ethical guidelines concerning researchers are those that direct investigators to seek truth and be honest in research. Researchers should never plagiarize (i.e., taking someone else's work or ideas and passing them off as one's own), falsify data (i.e., fabricating data that are not collected via an established method such as experiment and questionnaire for a given project), purposely exclude data of contrary results, and other unethical activities. The entire research community would not tolerate any such activity, and once discovered, the person involved will be severely disciplined (e.g., retraction of the research report published, termination of employment).

### *2.9.2 Ethical guidelines for protecting participants*

In addition to those ethical guidelines regarding researchers, there is another set of ethical guidelines developed to protect participants or do no harm to participants in research projects. The National Research Council (2003, 26–28) specifies the following six types of harms that researchers should avoid doing.

- Physical harm from research can include death, injury, pain, suffering, or discomfort.
- Psychological harm from research can include negative self-perception, emotional suffering (e.g., anxiety or shame), or aberrations in thought or behavior (e.g., agreeing to a hateful statement under pressure from the research environment).
- Social harm can involve negative effects on relationships or interactions with other people.
- Economic harm usually involves financial loss, which can result from study participation (e.g., the need to pay for transportation or child care in order to participate), from disclosure of study findings or participation (e.g., loss of health insurance or

employment), or as a side effect of other harms (e.g., having to pay court costs in a lawsuit that results from a breach of confidentiality).

- Legal harm can include arrest, conviction, incarceration, and civil lawsuits.
- Dignitary harm can result when individuals are treated as means to an end and not as people deserving respect for their own values and preferences.

Of the six types of harms, physical harm usually results from biomedical research. In order to eliminate such harm, researchers must follow strict safety procedures in their research implementation. Psychological, social, and legal harms mainly associate with how well researchers keep the confidentiality of their study and maintain the anonymity of their participants. In no circumstance shall researchers reveal the identity of participants who always ought to be labeled Participant 1, 2, ... n or Participant A, B, ... Z. Do not attempt to identify participants with their first names because first names, along with other related information, can be easily identified by friends, colleagues, and the like. Similarly, institutions involved in research should also be identified using a label (e.g., Institution A) or with a broad geographic identifier such as a college in the American Midwest.

Researchers should also obtain participants' informed consent before data collection. In doing so, researchers need to briefly describe their intended study, state its importance, articulate potential risks of harm to participants, and ask for permission to collect and use data gathered from participants. When proper actions (e.g., keeping confidentiality) are taken in the research process, the risks of harm could be completely eliminated. Researchers as a result can state in the informed consent form that there would be virtually no potential risks to participants in the research (Berg 2009, 87–90). In addition, researchers should try to compensate participants for costs (e.g., transportation, child care) incurred from taking part in the study.

Another ethical concern that researchers must consider is covertness or unobtrusiveness in studies using data collection techniques such as observation. The general guideline is that data can be collected covertly or unobtrusively only when it is necessary and justified. Necessary and justified are the two keywords that should be highlighted in this type of decision-making while by taking into consideration the actual research purpose, environment, and target participants. If covert research is to be conducted after careful and thorough consideration of ethics, investigators still need to observe all other ethical tenets (e.g., maintaining confidentiality) in the research process.

Furthermore, an institutional review board (IRB) is set up in almost all research and higher education institutions to protect the rights and welfare of human subjects recruited to participate in research activities. The IRB equivalence in the United Kingdom is the institutional ethics committee (EC). The IRB originally was created for federally funded research that involves human subjects (e.g., medical research). It is now applicable to all research that uses human participants. The IRB is charged to review and approve any research to be conducted by members in the parent institution. Researchers are required to apply for IRB approval before embarking on their studies.

Research ethics should be upheld in all studies, from research design to research reporting, in every step of the research process regardless of the size of a project, the nature of a study or the number of participants. This book will not specifically repeat the topic of research ethics in the subsequent chapters unless some highlighting is necessary. However, readers are expected to follow research ethical norms in all their research undertakings.

## Chapter Summary

- The research process is spiral, recursive, nonlinear, and progressive.
- Major steps in the research process include problem identification and statement, literature search and review, formulation of research questions, research method development, data collection and analysis, result presentation and discussion, and conclusion drawing.
- Major sources for research problem identification involve RFPs, CFPs, research reports, conversations with colleagues, day-to-day operations and services, one's own experience, and an open, inquisitive mind.
- Several key points in literature search and review need special attention. Literature in this step refers to scholarly publications. Literature search must be conducted thoroughly in carefully selected databases and other scholarly sources. Literature review should not only summarize what has been done in a target topic area but also evaluate how related studies have been done and point out what remains to be done.
- Research questions should be derived naturally and logically from the problem statement, each addressing one aspect of the stated problem if it contains two or more subproblems.
- Nominal, ordinal, interval, and ratio are the four levels of measurement used in data collection. They are essential in data collection and also determine what techniques are available for analyzing the data so measured.
- Select a probability sampling method if possible to ensure the generalizability of research findings. The choice decision on individual sampling methods (e.g., simple random sampling, purposive sampling) is based on their respective features.
- Validity consists of internal validity and external validity. Internal validity addresses the question if a research project studies what it is intended to do. External validity or generalizability refers to whether research findings of a study using a sample are applicable to the population from which the sample is drawn. Reliability examines if similar results can be obtained when a study is repeated by following the same methodology.
- Ethical concerns regarding researchers include plagiarism, data falsification, purposeful exclusion of contrary results, and the like. Ethical concerns about protecting participants focus on how to do no physical, psychological, social, economic, legal, and dignitary harms to them in the research process. Keeping confidentiality and maintaining privacy of participants are the two most essential measures researchers must take in any study.

## Notes

1 The literature map is based on publications reviewed in Chu and Ke (2017).
2 The features are adapted from Figure 2.1 in Loether and McTavish (1980, 19).
3 The actual formula is presented in many textbooks (e.g., Frankfort-Nachmias, Nachmias, and DeWaard 2015, 160-61) although it is applicable only after the researcher estimates the population standard deviation as well as an acceptable level of estimation accuracy. This makes the formula approach literately infeasible in research. This book decides not to elaborate on it any further partly because of the infeasibility of this approach.

## References


Andrews, Frank M., Laura Klem, Patrick M. O'Malley, Willard L. Rogers, Kathleen B. Welch, and Terrence N. Davidson. 1998. *Selecting Statistical Techniques for Social Science Data: A Guide for SAS Users*. 3rd ed. Cary: SAS Institute.

Babbie, Ear R. 2016. *The Practice of Social Research*. 14th ed. Boston: Cengage Learning.

Bem, Daryl J. 2003. "Writing the Empirical Journal Article." In *The Compleat Academic: A Practical Guide for the Beginning Social Scientist*, edited by John M. Darley, Mark P. Zanna, and Henry L. Roediger III. 2nd ed. American Psychological Association. Accessed July 1, 2023. https://dbem.org/writing-for-psychological-journals/.

Berg, Bruce L. 2009. *Qualitative Research Methods for the Social Sciences*. 7th ed. Boston: Allyn and Bacon.

Biernacki, Patrick, and Dan Waldorf. 1981. "Snowball Sampling: Problems and Techniques of Chain Referral Sampling." *Sociological Methods & Research* 10 (2): 141–63. https://doi.org/10.1177/004912418101000205.

Boote, David N., and Penny Beile. 2005. "Scholars Before Researchers: On the Centrality of the Dissertation Literature Review in Research Preparation." *Educational Researcher* 34 (6): 3–15. https://doi.org/10.3102/0013189x034006003.

Browne, Kath. 2005. "Snowball Sampling: Using Social Networks to Research Non-heterosexual Women." *International Journal of Social Research Methodology* 8 (1): 47–60. https://doi.org/10.1080/1364557032000081663.

Chu, Heting, and Qing Ke. 2017. "Research Methods: What's in the Name?" *Library & Information Science Research* 39 (4): 284–99. https://doi.org/10.1016/j.lisr.2017.11.001.

Creative Research Systems. 2021. "Sample Size Calculator." Accessed June 30, 2023. http://www.surveysystem.com/sscalc.htm.

Dean, Gary J. 2010. "Writing Problem Statements and Research Questions." In *Course Lecture for Practical Research in Adult and Community Education*. Indiana: Indiana University of Pennsylvania. Accessed June 29, 2023. https://www.youtube.com/watch?v=Z1yJEOtLsb8.

Frankfort-Nachmias, Chava, David Nachmias, and Jack DeWaard. 2015. *Research Methods in the Social Sciences*. 8th ed. New York: Worth Publishers.

Gorman, Gary E., and Peter Clayton. 2005. *Qualitative Research for the Information Professional: A Practical Handbook*. 2nd ed. New York: Neal-Schuman.

Hart, Chris. 1998. *Doing a Literature Review: Releasing the Social Science Research Imagination*. Thousand Oaks: Sage.

Hastings, Candace. September 25, 2009. *The Literature Review*. College Station: Lecture, Texas A&M University. Accessed June 29, 2023. https://www.youtube.com/watch?v=9la5ytz9MmM.

Kerlinger, Fred N. 1986. *Foundations of Behavioral Research*. 3rd ed. New York: Holt, Rinehart & Winston.

Kettering, Charles F. 1945. "Research Opens the Door." *Scientific American* 172 (2): 7–12. https://doi.org/10.1038/scientificamerican0145-7.

Loether, Herman J., and Donald G. McTavish. 1980. *Descriptive and Inferential Statistics: An Introduction*. Boston: Allyn and Bacon.

Mauch, James E., and Jack W. Birch. 2003. *Guide to the Successful Thesis and Dissertation: A Handbook for Students and Faculty*. 5th ed. New York: Marcel Dekker.

McLean, A. Peter. 2005. "A Problem Well Stated Is a Problem Half Solved." In *Family Assessment: Integrating Multiple Clinical Perspectives*, edited by Manfred Cierpka, Volker Thomas and Douglas H. Sprenkle, 53–80. Cambridge: Hogrefe & Huber.

National Research Council. 2003. *Protecting Participants and Facilitating Social and Behavioral Sciences Research*. Washington: National Academic Press.

Negriff, Sonya. 2019. "A Pilot Study Examining Risk Behavior in Facebook Posts for Maltreated and Comparison Youth Using Content Analysis." *Child Abuse and Neglect* 96: 1–9. https://doi.org/10.1016/j.chiabu.2019.104091.

Ng'oma, Mwawi, Samantha Meltzer-Brody, Ellen Chirwa, and Robert C. Stewart. 2019. "'Passing Through Difficult Times': Perceptions of Perinatal Depression and Treatment Needs in Malawi – A Qualitative Study to Inform the Development of a Culturally Sensitive Intervention." *PLoS ONE* 14 (6): e0217102. https://doi.org/10.1371/journal.pone.0217102.

Perrin, Andrew, and Monica Anderson. 2019. "Share of U.S. Adults Using Social Media, Including Facebook, Is Mostly Unchanged Since 2018." In *Pew Fact Tank*. Pew Research Center. Accessed July 2, 2023. https://pewrsr.ch/2VxJuJ3.

Scheuren, Fritz. 2004. *What Is a Survey*. 2nd ed. Alexandria: American Statistical Association. Accessed July 2, 2023. http://people.umass.edu/biostat690c/pdf/ASA%20What%20is%20a%20survey%2068%20pp.pdf.

Thomson, Catherine S., Sarah Woolnough, Matthew Wickenden, Sara Hiom, and Chris J. Twelves. 2010. "Sunbed Use in Children Aged 11-17 in England: Face to Face Quota Sampling Surveys in the National Prevalence Study and Six Cities Study." *BMJ* 340: C877. https://doi.org/10.1136/bmj.c877.

Wikipedia Contributors. 2023a. "Kelvin." In *Wikipedia, The Free Encyclopedia*. Last modified June 29, 2023. Accessed June 30, 2023. https://en.wikipedia.org/wiki/Kelvin.

Wikipedia Contributors. 2023b. "List of Counties by U.S. State and Territory." In *Wikipedia, The Free Encyclopedia*. Last modified March 25, 2023. Accessed June 30, 2023. https://en.wikipedia.org/wiki/List_of_counties_by_U.S._state_and_territory.

White, Howard D. 1994. "Chapter 4. Scientific Communication and Literature Retrieval." In *The Handbook of Research Synthesis*, edited by Harris Cooper and Larry V. Hedge, 41–55. New York: Russell Sage Foundation.

Yin, Robert K. 2016. *Qualitative Research from Start to Finish*. 2nd ed. New York: The Guilford Press.

# Part II

# Stand-Alone Methods

# 3 Questionnaire

**Chapter Overview Questions**

- What is questionnaire, and what major components are there in a survey?
- What types of questions are asked in questionnaires?
- What approaches can be taken in administering a questionnaire?
- What is response rate and how to ensure a reasonable one?
- What are the dos and don'ts in questionnaire development and using questionnaire in research?

## 3.1 What is questionnaire?

A questionnaire is a technique for data collection using a predefined list of questions. It enables the investigator to gather what is needed for addressing a particular research problem. The name of questionnaire well reflects the nature of this data collection technique. On the other hand, any study using a questionnaire as data collection instrument is customarily called survey research. Because of this tradition, questionnaires and surveys are often treated as synonyms in the research domain (Frankfort-Nachmias, Nachmias, and DeWaard 2015, 187–208). Some publications may distinguish questionnaires from surveys by treating questionnaires only as data collection instruments in survey research. Such a distinction, while meaningful, is not always made in the scholarly world because it appears more direct if we call "survey research" using a questionnaire for data collection "questionnaire research". This practice also resonates with the reasoning provided in Chapter 1: the data collection technique used in a study is often part of the research name as, for example, experiment in experimental research, interview in interview research, and questionnaire in questionnaire research. In other words, the research name is generally indicative of what data collection technique it adopts. Therefore, questionnaire and survey will be used interchangeably in this book.

Questionnaire is one of the most regularly used data collection tools in many disciplines. It can be used either alone, together with other techniques (e.g., interview, observation) or as one data collection technique in the so-called meta-methods (e.g., experiment). As a data collection technique, questionnaire has the following distinctive features:

- Low cost because a questionnaire, once created, can be distributed to a large number of participants regardless of their geographical locations.

DOI: 10.4324/9781003021711-5

- Easy implementation thanks to the assorted approaches that can be taken for administering a questionnaire. The advent of survey websites (e.g., Qualtrics, SurveyMonkey) further facilitates its distribution.
- Less pressure on respondents (i.e., participants in survey research) as they can complete the questionnaire at least at a pace of their own choosing if it is administered in person. If the questionnaire is distributed asynchronously (e.g., online or by mail), respondents can complete it at their leisure.
- Inability to gather real-time, in-depth data from respondents as well as no opportunity for asking clarification and follow-up questions.

Those features determine how the questionnaire should be developed and utilized in a study for data collection.

## 3.2 Questionnaire development

Questionnaire is composed of questions. The importance of questions in a questionnaire cannot be overemphasized. Creating quality individual questions to form a questionnaire thus requires a lot of efforts from the researcher. Whether a survey study will be successful depends primarily on the questionnaire development. The fundamentals of questionnaire development are depicted below.

### *3.2.1 Questionnaire components and question types*

A questionnaire encompasses two components: (1) a cover letter or an introductory note, and (2) a list of questions. Of the two, the cover letter or introductory note needs less time to be accomplished although the researcher ought to make sure that it is properly written. Generally speaking, a cover letter should be prepared if the questionnaire is designed for a large research project. Otherwise, an introductory note might be sufficient to serve the purpose. In addition, a closure that normally constitutes a sentence or two to thank the participants (e.g., "Thank you for your time!") will conclude the questionnaire.

In the cover letter or introductory note, the survey objective should be described using a language that is appropriate to the prospective people the researcher intends to invite for participation in the study. If the prospective participants are the general public, use plain English without any jargon and rarely used terms. Other components that should also be covered in the cover letter include confidentiality protection (i.e., ensuring that the participant's identity and other personal data collected would be kept confidential throughout the study), time needed (e.g., 15 minutes for completing the survey) and any incentive for research participants (e.g., a $5 Amazon coupon for college students, a research report summary for academic participants). The only purpose of the cover letter is to encourage and persuade prospective participants to complete the questionnaire distributed. The importance of the cover letter should be quite self-explanatory. Regardless of how interesting and attractive the questionnaire questions might be, those questions will not get answered if the cover letter or introductory note is not inviting and convincing.

The major part of a questionnaire is naturally the list of questions. How to create those questions proves to be both an art and a science. First of all, the researcher should consider what sources can be used to assist question preparation. The following three types of sources are commonly used in creating questionnaire questions: (1) related publications that may contain questions relevant to the intended study, (2) questionnaires developed in previous related research, and (3) conducting interviews or focus groups to solicit questions from a small set of prospective research participants. Among the three question source options, Option (2) seems to be the most straightforward as researchers only need to adapt or tailor it according to their own research problem. However, not many researchers can find an existing, relevant questionnaire. Instead they would have to develop one by adopting Option (1), Option (3), or a combination of both.

If a decision is made to develop a questionnaire from scratch, the researcher should consider the following two points carefully:

- What questions should be put in the questionnaire? Always first look closely at the research questions formulated for the study to see what data are needed to answer the questions. If a study plans to explore, for example, whether gender is a factor affecting one's salary, then data about gender and salary must be collected. It would be a good practice if all the questions created for the questionnaire are mapped against the research questions. When a question is found not related to any of the research questions formulated for a study through the mapping, delete it because it would only make your questionnaire unnecessarily longer than it should be.
- How would the collected data be analyzed? Even though it might seem too early to think about data analysis at the time of data collection, it is recommended in research that some thoughts be given to data analysis in order to avoid any issue that might arise or even prevent the researcher from using a specific data analysis technique later. For instance, Likert scale questions are commonly utilized in questionnaires for gathering data about participants' opinions and attitudes. The Likert scale, as described in Chapter 2, by default gathers data at the ordinal level which cannot be used for computing average or the mean (in statistical term). If the researcher wishes to use the Likert scale to collect interval data for computing means and other more advanced statistics, a statement indicating that "all the intervals of two adjacent points (e.g., the intervals between 1 and 2, 2 and 3, ... on a 5-point Likert scale) are treated as equal" must be included in the questionnaire itself. This statement would ensure that the data collected via the Likert scale are measured at the interval level rather than the default ordinal level.

Before actually writing individual questions for the questionnaire, it is helpful to get familiar with the types of questions a questionnaire typically has (see Table 3.1). All examples in the table are taken from the questionnaire the National Research Council (NRC) administered to doctoral students for assessing research doctorate programs (National Research Council 2006, 141–58). The full questionnaire along with its introductory note is selected as one of the two case studies for this chapter.

Among the four question types listed in Table 3.1, the first three can be grouped as closed-ended questions, whereas the fourth one is the open-ended question. Closed-ended questions are designed to solicit definite answers from respondents. By contrast, open-ended questions have no limits on how respondents would answer them. As briefly mentioned in Chapter 1, questionnaires contain mainly closed-ended questions although they do not exclude the possibility of having open-ended questions to let respondents express their own thoughts and views beyond what is already asked in the questionnaire. Below is a closer look at each of the four types of questions.

*Table 3.1* Common Question Types

| | *Type* | *Usage* | *Example* |
|---|---|---|---|
| Closed-ended Question | Factual question | Gathering data about the respondent self (e.g., age, gender, education) | A1.[1] When did you first enroll in this doctoral program? |
| | | Gathering data about other information (e.g., services offered) | A13. If you had a fellowship, scholarship, traineeship, or assistantship, with what degree of support did it provide you? |
| | Opinion/attitude question | Gathering data about the respondent's opinion/attitude regarding others or issues | B3. Do you feel supported by your advisor in your current career goals? |
| | | Gathering data about the respondent's self-perception | B4. When you entered your doctoral program, for what type of employer did you believe you would work when you graduated? |
| | Filter question | Determining if the respondent should answer subsequent questions | C11. Do you have access to career advice? |
| Open-ended question | | Gathering data beyond what the closed-ended questions already cover | D2. In the space below, please provide any additional comments you would like to make about program or university resources available to you: ____________ |

#### 3.2.1.1 *Factual questions*

This type of questions is used for collecting data about respondents' demographic, education, affiliation, and other factual data. In general, factual questions are easy for respondents to answer because they do not need to think how to answer each. Even so, we strongly recommend the following when preparing factual questions:

- Provide answer choices to facilitate respondents to complete the questionnaire efficiently. For instance, list "male", "female", and "other" as answer options for a question on gender. Give "married", "living in a marriage-like relationship", "widowed", "divorced", "separated", "never married" as answer choices for a question on marital status.
- Use predefined, grouped categories as answer options for collecting sensitive or personal information from respondents. For example, ask for age group instead of specific age for age-related questions: 17 and younger, 18–22, 23–27, 28–32, 33–37, ..., 73–77, 78, and older. How many age groups should be created? That all depends on the need of the study. The above example sets five year as category interval. Nevertheless, any age interval would be legitimate as long as all possible ages are included as answer options (i.e., collectively exhaustive) and no category overlaps with another (i.e., mutually exclusive). In addition, it makes more sense in data collection instrument development if each category carries equal intervals. We can always change initial groups (e.g., combining existing groups) to ones that would work better for a chosen research problem after data collection is over. Factual questions on income or other similar data should also use predefined, grouped categories as possible answer choices.

Collectively exhaustive and mutually exclusive are two fundamental principles we should always observe when developing answer categories for factual questions. This statement may sound simple and straightforward. However, it can be easily or naively overlooked. It is not uncommon for a novice researcher to create age groups such as "18 and younger, 18–22, 22–26, ... 96–100". These age groups fail to observe the collectively exhaustive and mutually exclusive principles in that, for instance, both 18 and 22 are repeated. Yet there is no age group for respondents who live over 100 years old. To address the latter issue, one more age category (i.e., 101 and older) must be formed to ensure that the age groups are collectively exhaustive. Sometimes we may not be able to list all the possible answer categories for a given question (e.g., Question A10 from the NRC 2006 Doctoral Student Questionnaire: "From which of the following sources have you received travel funds for research presentations?"). We shall then provide a category of "Other (please specify): ..." to make sure that all possible categories are covered.

Factual questions usually account for a good percentage of all the ones in a questionnaire. It is thus important to develop them in a way that respondents would feel comfortable and effortless to answer. This would subsequently help increase the response rate of a questionnaire, a topic to be discussed shortly.

#### *3.2.1.2 Opinion/attitude questions*

Opinion/attitude questions often represent a large group of questions in a questionnaire. They are designed to solicit answers from respondents regarding their opinions and attitudes about others or issues and topics relevant to the study. Various scales, indexes, inventories, or tests, all resembling the questionnaire but serving different purposes, are commonly adopted for gathering such data. We will have a brief look at those questionnaire-like instruments at the end of Section 3.2.2.1. Among them, scales are used most frequently in creating opinion or attitude questions. Although there are existing scales or other similar instruments available for use in some cases, researchers may need to develop scale questions for their own projects in other times.

A scale contains predetermined units of measurement for a statement or question. Likert scales, named after American psychologist Rensis Likert and taking various formats, constitute the most pronounced measure for creating opinion/attitude questions. Below is a summary of Likert scale formats with examples taken from the NRC 2006 Doctoral Student Questionnaire wherever possible.

I. A statement/question followed by $n$ ($n = 3, 4, 5, \ldots, n$) units of measurement

Example: C17. On a scale of 1 to 5, how supportive are students in your program of one another?

5 Very supportive
4
3 Somewhat supportive
2
1 Not supportive

The same answer options can also be listed in other arrangements. The first variation below provides more category names, while the second gives category labels without any numbering. The third one, by comparison, only labels the two ends of the scale. In the latter two cases, the researcher would need to convert the labeling into a numeral when coding the data for analysis purpose.

(1) 5-Very supportive 4-Supportive 3-Somewhat supportive
2-Somewhat unsupportive 1-Not supportive

(2) Very supportive Supportive Somewhat supportive
Somewhat unsupportive Not supportive

(3) Very supportive ←-------------------------------------------------------→ Not supportive

II. A statement/question followed by $n$ ($n$ = 3, 4, 5, ..., $n$) units of measurement with equal intervals between two adjacent points on the scale

Example: Green energy is environment friendly.

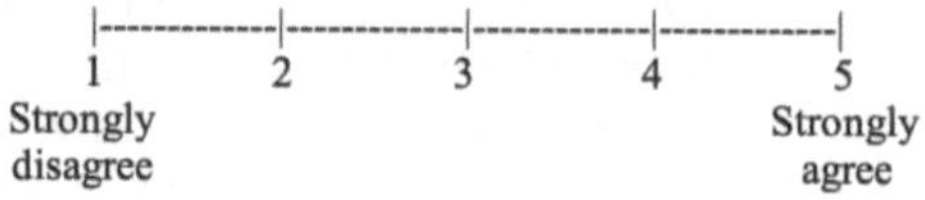


III. A statement/question is listed and the respondent is asked to use a numeral to answer the question

Example: C17. On a scale of 1 to 5, how supportive are students in your program of one another? ______

Those are the three major formats for Likert scale questions although other minor variations do exist. The researcher would ultimately choose a Likert scale format according to the research problem. Besides, several additional considerations should be made in devising Likert scale questions.

- Number of points on the scale: Theoretically, 3 or more points can be placed on the Likert scale. The most often chosen number is 5, followed by 7. A number lower than 5 is not used regularly because it does not allow respondents to adequately express their opinions or attitudes. On the other hand, a number higher than 7 offers too many options and may discourage respondents to answer the question efficiently. In addition, an odd number rather than an even number is usually preferred for the scale question so that the middle point would divide respondents' positive and negative viewpoints into two equal parts.
- Labeling of points on the scale: From the researcher's perspective, it is optional to label each point or category on the Likert scale because such data would eventually be converted into numbers before data analysis. But the category labeling would help respondents better understand the question simply because the labeling provides more information than a numeral alone.
- A neutral point on the scale: Of all the examples given above, there is no neutral point on the Likert scale. In the following example of "1) Strongly agree; 2) Agree; 3) No opinion; 4) Disagree; 5) Strongly disagree", "No opinion" is offered as a neutral point on the Likert scale. This is desirable if researchers would permit respondents to have such an option when they indeed have no opinion about what is asked. A similar practice is to list "N/A" (not applicable) or "Don't know" as an option next to the scale so that respondents would not be forced to answer a question if they cannot.
- Interval or ordinal data: As discussed in Chapter 2, Likert scale in nature collects ordinal data unless researchers purposely choose the interval level of measurement by explicitly asking respondents to treat the intervals between any two adjacent points on the scale as equal and by visually presenting the scale as such at the time of data collection. Format II of Likert scale depicted above is a visual example of interval Likert scale. Generally speaking, researchers should gather data using the interval Likert

scale whenever possible because they will have more techniques available for data analysis later.

- Which numeral represents the highest value on the scale: Regardless of how many points there are on the Likert scale, researchers need to consider whether 1 takes the highest value on a scale from "Very supportive" to "Not supportive" or $n$ ($n$ = 3, 4, 5, ..., $n$) should be the choice. The latter appears to make more sense because this selection is in accordance with the convention and common sense: Larger numerals have higher values. Otherwise, the Likert scale data collected would have to be converted according to the arithmetic rule at the time of data coding and analysis.

Once being clear about the afore-described scale options, we can phrase opinion/attitude questions in a questionnaire using Likert scale or another similar measure. Furthermore, opinion/attitude questions can also be asked of respondents about their perceptions of themselves. The following is an example of self-perception question taking from the NRC 2006 Doctoral Student Questionnaire:

> B4. When you entered your doctoral program, for what type of employer did you believe you would work when you graduated?

Self-perception questions apparently belong to the opinion/attitude category. They enable the researcher to find out from respondents how they perceive themselves with regard to topics and issues germane to a study.

#### 3.2.1.3 *Filter questions*

Filter questions serve as filters so that respondents are directed to right questions in a questionnaire. Question A9 from the NRC 2006 Doctoral Student Questionnaire is a filter question example.

> A9. Have you received travel funds for research presentations at regional, national, or international meetings?
>
> Yes
> No *(skip to A11)*
>
> *Ask A10 if A9 = Yes*

Filter questions usually take the form of Yes/No plus directions for what to do for each of the answer options. In the above example, skip Question A10 if the answer is No. Otherwise do Question A10. A quick comment on the wording of the direction for the Yes option: It would be clearer and more concise if simply putting "Do A10" right after Yes as done for the No option.

Always make sure that filter question directions do not unintentionally skip any questions in the questionnaire. In addition, use filter questions only if it is absolutely necessary. Too many filter questions in one questionnaire might cause confusion to the respondent.

#### 3.2.1.4 *Open-ended questions*

Open-ended questions are designed for the researcher to gather data beyond what the closed-ended questions in the questionnaire already cover. Respondents can answer them without any limit and hence open-ended. Question C23 in the NRC 2006 Doctoral Student Questionnaire represents an open-ended question example:

> C23. In the space below, please provide any additional comments you would like to make about your doctoral program, its characteristics or quality:

Open-ended questions are usually offered at the end of a questionnaire if there is only one such question. Sometimes multiple open-ended questions are asked when the questionnaire is long and divided into several sections like the NRC 2006 Doctoral Student Questionnaire. Even so, the open-ended question would still be placed at the end of a section instead of the beginning or middle of it so that all the questions in the section are already asked and the open-ended question truly solicit answers to other aspects of the study that are not covered via the closed-ended questions.

Furthermore, open-ended questions are mainly designed to collect qualitative data. For this reason, questionnaires may not carry any or many open-ended questions that are designed mainly for gathering quantitative data. In contrast, interviews comprise mainly open-ended questions and Chapter 4 is devoted to this data collection technique.

Although the fundamentals of question development are presented in this chapter on questionnaire, they are also applicable and relevant to other data collection techniques that involve question formulation (e.g., interview and focus groups). Such essentials in question development would not be repeated in other chapters of this book. Interested readers however can always refer to relevant sections in this chapter when needed.

### 3.2.2 Dos and don'ts in questionnaire development

A quality questionnaire can only be developed when the researcher makes sufficient efforts throughout the questionnaire development process as well as pays attention to the following dos and don'ts.

#### *3.2.2.1 Dos in questionnaire design*

- Build coherence in the order of questions asked. One common practice is to start with simple, demographic questions and proceed to more complex ones (e.g., Niu et al. 2010). However, some researchers prefer to ask demographic questions the last (e.g., National Research Council 2006) so that respondents would not feel that the survey questions get more complicated and thus are more likely to complete the questionnaire.
- Provide question context to help respondents better understand what each question asks about. For example, group all questions about demographic information and list them under a heading accordingly. Respondents would know that questions in this section are all about demographic information (e.g., age, gender, and race).
- Be clear, concise, and consistent. Always keep the questionnaire as short as possible. Only gather what is needed for the study and nothing more.
- Keep survey questions simple. "There must be no guesswork for the respondent when it comes to understanding exactly what information is being requested" (Schueren 2004, 43).
- Provide specific and unambiguous instructions for answering questions that are not self-explanatory. Never assume that respondents know how to complete the survey. For

multiple-choice questions, instruct the respondent to "Choose all that apply" if that is the intention. Otherwise, ask the respondent to only "Check one".
- Format the questionnaire properly and professionally. Poorly formatted questionnaire, including leaving too little or too much white space in the instrument, may negatively affect the response rate of the survey.
- Conduct at least one pretest of the questionnaire, as a form of validation, with several people from the target participants. Be certain that those who participate in the pretest will not take part in the full survey because they are already familiar with the instrument they pretest. Checking with colleagues and experts could be another option although it is preferable to do the pretest with a small subset of prospective respondents. The objective of the pretest is to discover errors and other similar issues in the questionnaire draft. Usually changes and corrections are made based on feedback from the pretesting before the questionnaire is finalized and administered. It should be pointed out that questionnaire pretesting is a must rather than an option unless the questionnaire is adopted from another published study in that such usage, including pretesting, by another study constitutes the questionnaire validation or quality control practice. But it is truly optional to check the reliability of questions in the questionnaire using measures such as Cronbach's Alpha because measures of this kind are for checking internal consistency of scale questions (What Does Cronbach's Alpha Mean 2020). If the researcher plans to develop a scale, an index, or other similar data collection instruments that will be repeatedly used beyond a single study, it becomes mandatory to apply such measures of reliability checking (Wildemuth 2017, 285–86).

Indexes (e.g., Behavior Attitudes Index), inventories (e.g., Cognitive Style Inventory), scales (e.g., Emotional Intelligence Scale), and tests (e.g., Personality Test), to a large extent, can be regarded as some forms of questionnaire. The major difference between questionnaires and those questionnaire-like data collection instruments (e.g., scales and tests) is that the latter normally are already developed, validated, and available for usage in a variety of research projects, while the former are usually created specifically for individual studies. Because of this difference, validations using techniques like Cronbach's Alpha, expert panels, and factor analysis are a quality control mechanism that scales and other questionnaire-like data collection instruments must take after they are created so that they can be used again and again in many studies of different disciplines. For example, a personality test is likely to be employed in a wide variety of research involving the variable of personality. The research could be about leadership personality in management, individuals' personality in psychology, or users' personality in information studies. The potentially wide applicability of a personality test determines the necessity of validating it after its creation.

By contrast, questionnaires are usually designed for specific individual studies. We would aim to create a quality questionnaire in the first place when we know our research problem rather than trying to perform quality control via validations afterward although pretesting the questionnaire remains indispensable. In addition, it is impractical and unrealistic to validate every questionnaire when we strive to uphold the highest standard in its development and when most questionnaires will be used basically once in a study.

Other than the repeated usage feature and validation requirement of scales and other questionnaire-like data collection instruments, they and questionnaires are both developed and utilized by following a similar set of guidelines. It is also common for a single study to employ both a self-devised questionnaire and one or more tests or similar data collection instruments developed by others. Therefore, tests and other questionnaire-like data collection instruments will not be discussed separately and any further in this book.

#### *3.2.2.2 Don't's in questionnaire design*

- Don't jump from question to question without observing the logical relations among the questions. It is usually not recommended to mix, for example, demographic questions with other types of questions in the presentation of one questionnaire.
- Don't create a response set (Frankfort-Nachmias, Nachmias, and DeWaard 2015, 225) by asking questions that may make the respondent to answer them in a specific direction regardless of their contents. This occurs most often in questions using a scale (e.g., Likert scale) or a similar set of answer options (e.g., Yes/No). Measures for preventing the response set issue from occurring include varying response categories for scale questions and changing question formats.
- Don't ask leading questions or questions that suggest answers the researcher anticipates or wishes to receive. An example of leading question is: "What do you think of the popular Facebook?" The word "popular" exhibits the researcher's positive view of Facebook – a social network platform originally developed in the United States. Quality questions should always be free of the questionnaire developer's value and judgment.
- Don't ask threatening questions or questions that might make respondents feel threatened. If a researcher would like to find out from respondents how to deal with plagiarism, for instance, don't phrase the question as: "Do you think if people committing plagiarism should be severely penalized?" Rather, ask this question as: "What measures do you suggest to handle those who commit plagiarism?"
- Don't ask double-barreled questions (i.e., each question touches upon more than one issue but allows for one answer only). Such questions normally combine two individual questions into one. The following example is a double-barreled question: "Do you think that students should take an information literacy course or the curriculum should not include information literacy as a required course?" The initial intention of the researcher is to gather data about these two different questions. However, combining the two in one question makes it a double-barreled question because the respondent would not be able to answer it appropriately with one "Yes" or one "No". The correct practice is to create one individual question for each case. Namely, "Do you think that students should take an information literacy course?" or "Do you think that the curriculum should not include information literacy as a required course?" Double-barreled questions basically force the respondent to choose one "Yes" or "No" in answering both questions contained within even though the true answer might be one "Yes" and one "No".
- Don't ask complex questions and questions with clauses or involved lexical structures. Simple and straightforward questions always work better in survey research because complex questions are more likely to cause confusion and misinterpretation.
- Don't ask questions that would tax the respondent's memory. According to Miller (1956), people's short-term memory has limited capacity. In addition, only a small portion of the information people receive through their five senses would get transferred from short-term memory to long-term memory and even a smaller percentage of that information would be successfully retrieved later when needed. For this reason, respondents may not be able to recall an answer to questions such as: "How many times did you shop online in the past year?"

- Don't use all uppercases if the questionnaire is in English because the public in general is not accustomed to reading all uppercases.

Some researchers put voluntary participation in the cover letter to allow target respondents to decide if they are willing to take part in a survey study (e.g., National Research Council 2006). While this practice is ethically correct, it is counterproductive for novice investigators to state so when the very purpose of having a cover letter is to encourage participation. The same also holds true for all other types of research that involve human participants. Researchers indeed should observe all ethical guidelines to keep confidentiality of collected data, maintain anonymity of participants, and take other similar measures. However, they need to be cautious to make a statement of voluntary participation when not having a viable mechanism to ensure adequate participation from potential candidates.

All researchers, when inviting humans to take part in their projects, should obtain informed consent (i.e., permission granted in the knowledge of the possible consequences in research participation) from them before data collection, which is one of the practices in accordance with research ethical guidelines but its implementations are varied. In survey research, attaining informed consent is usually done more implicitly than in interview and other types of research. Researchers ordinarily include a sentence (e.g., completing this survey constitutes your informed consent to participate in this study) in the cover letter or add a check box at the end of the cover letter as the National Research Council (2006) did in its survey. We will revisit the latter example in Section 3.4.1.1 when doing the first case study for this chapter. Rarely will researchers develop a separate form of informed consent for survey studies as done in interview and other similar research.

## 3.3 Using questionnaire in research

After the questionnaire is developed, it is time to consider how to administer it, obtain an acceptable response rate, and address other related topics.

### *3.3.1 Approaches to questionnaire administration*

There are various approaches to administering a questionnaire. Table 3.2 summarizes major approaches of questionnaire distribution along with their characteristics.

As shown in Table 3.2, the pros and cons of each questionnaire administration approach are obvious. The choice should always be made based on one's research project and available resources. For example, the on-site approach can be taken if the researcher intends to survey merely a small number of respondents who are also easily reachable in person. Otherwise, the survey site method would be preferred. Common survey sites in the United States include SurveyMonkey (https://www.surveymonkey.com) – a subsidiary survey platform under Momentive as of 2021, QuestionPro (http://www.questionpro.com), Qualtrics (https://www.qualtrics.com), and Google Forms (https://www.google.com/forms/about/). More researchers, as indicated earlier, adopt the survey site approach for questionnaire administration thanks to the wide application of web technology. An email invitation, functioning as a survey's cover letter, is usually sent to prospective respondents. The invitation also should provide a link to the site where the questionnaire can be accessed and completed.

*Table 3.2* Major Approaches to Questionnaire Administration

| *Approach* | *Feature* |
|---|---|
| Mail | • Reaching a large number of respondents who are geographically dispersed.<br>• Inexpensive to distribute the questionnaire compared to the telephone approach.<br>• No opportunity to obtain real-time responses and clarify any question the respondent may have.<br>• Used to be the primary means for questionnaire administration. |
| Telephone | • Allowing real-time responses and clarification questions.<br>• Expensive to implement due to the time required for appointment scheduling and questionnaire administration.<br>• Able to reach respondents who are illiterate or have vision problems as questions can be read to them.<br>• Only feasible if the respondent has a phone. Also becoming less adopted as more respondents replace the landline phone with a mobile phone. |
| Email | • No extra cost to administer the questionnaire compared with the mail approach.<br>• Fast and efficient in questionnaire administration.<br>• No opportunity to obtain real-time responses and clarify any question the respondent may have.<br>• Only feasible if the respondent has a valid email address.<br>• Quickly replaced by the survey site approach. |
| Survey site | • Becoming the dominant approach to questionnaire administration thanks to the wide application of web technology.<br>• Providing preliminary analysis of the data collected via the survey.<br>• Nominal cost compared to the mail approach.<br>• Limiting question and questionnaire presentation with preset formats.<br>• Only feasible if the respondent has access to the web and an email address which is used for sending out a participation invitation. |
| On-site/Self-administered | • Always a favored approach to questionnaire administration.<br>• Allowing clarification questions the respondent may have.<br>• Ensuring higher response rate compared to other methods of questionnaire administration.<br>• Unable to reach a large number of respondents or those respondents who are not in close proximity geographically. |

Letter polling is a personalized approach to questionnaire administration in which the cover letter is addressed to respondents individually with their names and certain information related to the survey, replacing the general addressing (e.g., dear participant) and a formed cover letter. The certain information related to the survey refers to, for instance, citation counts of scholarly publications in a questionnaire about research quality evaluation. This approach is feasible only if prospective participants' contact and certain information related to the survey are available to the researcher, and the number of participants is not too large (e.g., around 150). Although tailoring the cover letter for individual participants takes extra time and efforts, the response rate of letter polling would be much higher than that of other approaches depicted in Table 3.2.

### *3.3.2 Sample size and response rate*

Response rate is a perennial issue in survey research. Yet there is no so-called acceptable response rate because many factors (e.g., survey quality, distribution approach) play a role in determining that figure. Babbie (1991, 267) however explicitly offers his view on this topic: "I feel that a response rate of at least 50 percent is *adequate*[2] for analysis and reporting. A response [rate] of at least 60 percent is *good*. And a response rate of 70 percent is *very good*." Past practices on the other hand show that research in soft sciences (i.e., the humanities and social sciences) typically receives a response rate of 20% or higher, while studies in hard sciences (i.e., natural sciences) can generally expect a response rate of 50% and up. The difference between these two response rates by and large results from the different characteristics of these two survey applications. Surveys in soft sciences normally involve opinion/attitude questions that are harder to answer. In comparison, questionnaires in hard sciences contain mainly factual, straightforward questions that take less time and efforts to complete. Although exceptions do exist, researchers in soft sciences would be satisfied with a response rate of 20% or higher. Similarly, investigators in hard sciences are usually content with a response rate of 50% and up. Then, what shall the researcher do if the response rate is below an acceptable threshold?

In Chapter 2, we discussed how to determine sample size when a research project chooses to study only a subset of the population. A question often arises when the researcher gets a low response rate for a survey: Should additional sampling units be drawn from the population in order to increase the response rate? The answer is no because additional sampling would make the expanded sample not representative of the population for the study if a random sampling method is initially chosen. If a nonrandom sample is selected in a study, there is actually no representativeness issue to be talked about in the first place. In addition, nonrandom sampling is always not recommended because such practice would consequently affect the generalizability or external validity of the study.

What the researcher can do instead when facing low a response rate is to use follow-ups although, as indicated earlier, developing a quality questionnaire is always the most important measure for ensuring a reasonable response rate. Nothing would work if the questionnaire does not adhere to the general guidelines for survey development. A follow-up in survey research involves a revised cover letter and the original questionnaire. It is generally done after a certain period of time passes (e.g., one week for survey site distribution or two weeks for mail survey) after the first administration of the questionnaire. In the revised cover letter, add a paragraph or two to remind those who have not returned the questionnaire to complete it at their earliest convenience. If survey respondents are not identified in the administration process which is often the case, a sentence or two should be added in the revised cover letter to thank those who already complete the questionnaire and ask them to disregard the follow-up request. A second or even third follow-up might be necessary depending on the response rate of the previous follow-up. Each follow-up should be treated as seriously as the first distribution by, for example, providing a self-address and stamped envelope in the mail survey because respondents would not keep the questionnaire if they did not complete the survey at the previous request. When a survey site is used for administering the questionnaire, the link to the survey site should again be provided in the follow-up email cover letter.

Providing a self-address and stamped envelope in the mail survey is an established practice for safeguarding a reasonable response rate. Offering incentives to prospective respondents also would help improve the response rate. Nevertheless, a quality questionnaire is always the best measure the researcher can take to obtain a good response rate. It outperforms any other viable options (e.g., follow-ups, incentive offerings).

### 3.3.3 Dos and don'ts in questionnaire research

Although many scholars conduct survey research, it does not imply that it becomes an easy implementation and little effort is needed in this kind of undertaking. To the contrary, researchers still ought to pay attention to the dos and don'ts in survey research as outlined below.

First, select a representative sample of prospective respondents. Because survey research frequently targets a large number of participants (e.g., baby boomers' needs for health information), some random sampling technique should be adopted to ensure that the selected sample would mirror the population of potential participants regarding all the attributes the researcher plans to examine. If using a random sampling technique is not an option, still try to choose a nonrandom sampling technique based on its features to help maximize the representativeness of the sample in light of the target population. Section 2.5.2 details major sampling techniques, both random and nonrandom ones, along with their respective characteristics. Select a sampling technique that best suits the research project at hand.

Second, choose a questionnaire delivery method that is appropriate for the undertaken study according to the pros and cons summarized in Table 3.2 about major approaches to survey distribution. Administering a questionnaire via a dedicated survey site appears to be rapidly gaining popularity. However, the best approach for administering a survey should still be selected according to the research objective of a study as well as the features of a particular delivery approach. For instance, the on-site approach would be a favorable choice if all prospective respondents can be easily accessible at one setting. On the other hand, the survey site approach becomes a justified choice if potential respondents are geographically dispersed.

Third, apply all possible means to encourage respondents to complete the survey. Although developing a quality survey serves as the best insurance for a high response rate, other measures (e.g., follow-ups, incentives) should not be overlooked. High response rates can only be achieved if researchers try their best in every step of the survey research process, from questionnaire development to questionnaire distribution, to facilitate respondents to complete the survey.

Fourth, keep all the promises made during the survey research process. For instance, offer an incentive (e.g., Amazon coupon) to non-academic respondents once they complete the survey or send a summary of the research report to academic participants. No matter how long the time may have passed since the data collection, the promise made during data collection must be kept to win trust from respondents. This trust will be particularly helpful to the researcher if a follow-up or longitudinal study is to be carried out.

Regardless of how many dos and don'ts are presented in this section, in real research the investigator may still encounter problems or issues that are not described. A practical approach one can always take in questionnaire research is to plan carefully and implement rigorously. Always review and reflect at each step during the survey research process so that errors can be detected and corrected in time, while positive experiences can be learned to guide future research using the survey method.

## 3.4 Case studies of questionnaire research

Two case studies are made in this section with one focusing on the questionnaire only and the other on the questionnaire development as well as its implementation in a study. Both case study documents and other case studies selected for this book can be found via Google Scholar. A list of case study publications selected for this book is provided in Appendix.

### *3.4.1 Case Study 1*

National Research Council. 2006. "Assessment of Research Doctorate Programs: Admitted-to-candidacy Doctoral Student Questionnaire." In *A Guide to the Methodology of the National Research Council Assessment of Doctorate Programs*, edited by Jeremiah P. Ostriker, Paul W. Holland, Charlotte V. Kuh, and James A. Voytuk, 141–58. Washington: The National Academies Press. Accessed June 19, 2023. https://gradschool.unc.edu/facultystaff/nrc/NRC-Methodology-Guide.pdf

This case study focuses only on the questionnaire without discussion on other aspects of the survey research. In addition, this questionnaire is one of the three the NRC created for assessing doctoral research programs in the United States. The other two are the faculty survey and program survey respectively but neither is to be examined here in consideration of space. The student questionnaire is chosen because of its potential relevance to readers of this book.

This survey includes a question rationale at the very end, which is seldom seen yet very valuable to novice and seasoned researchers alike because it explains the how and why of question creation and questionnaire development. Research reports usually do not provide rationale of this kind. But it is highly recommended that researchers ask themselves why each question is created in the process of questionnaire development. As suggested in Section 3.2.1, try to map each survey question against all research questions formulated in a study. In doing so, redundant questions or gaps in questions can be detected and corresponding measures can be taken to ensure that every question has its purpose in the instrument and what is needed in the research would be collected through one or more survey questions. Some publications on survey research may suggest to purposely include redundant questions in order to cross check if a respondent answers the same questions consistently. This practice however should only be utilized by seasoned researchers when they have the expertise and resources or if they plan to create an instrument for repeated usage (e.g., Personality Test). Novice or student researchers definitely cannot afford to do so because of the time, resource, and other factors.

#### *3.4.1.1 Cover letter*

The cover letter starts with a brief introduction of the NRC, the organization that creates the questionnaire and conducts the survey research, followed by a description of the survey and its purpose. The importance of the survey is effectively emphasized by putting the first sentence in the fifth paragraph of the cover letter into boldface. Prospective student respondents are subsequently assured that their confidentiality of participation in the study would be protected. The last paragraph is about voluntary participation which, as explained earlier, only institutions like the NRC can afford to do because research individuals, especially novice ones, are not in a position to make such a statement due to their academic status. They have to do all they can to encourage prospective participants to take part in the study. At the very end of the cover letter, respondents are asked to check the box to show their consent to participate in the research. For novice or even seasoned researcher individuals, this again is not a common practice for the same reason as that with the voluntary participation. Research individuals might instead replace the box-checking with a sentence like "Completing this survey constitutes your informed consent to participate in the study" in the cover letter. Except the last two comments on voluntary participation and informed consent, this cover letter is appropriately prepared.

#### *3.4.1.2 Questionnaire*

Although the questionnaire is longer than average, it is divided into five parts which not only groups questions into segments but also provides some context to question answering. This organization enhances the structure and readability of the survey as well. For simplicity reason, individual questions in the survey are referred to in this section only by their labels (e.g., A1, B2).

Date is a common type of data researchers collect in a survey. Birthday and graduation day are two examples for such data. This survey clearly specifies how the respondent should list month and year (see A1). It also presents a format template to guide the respondent in filling in the month and year: two digits for month and four digits for year. Providing a format template for dates is necessary in all surveys because a wide variety of options exist for recording this type of data. The following samples some of the data format options in use:

- Word-number format (e.g., July 4, 1998)
- Number only format (e.g., 07/04/1998)
- Various other number-only formats for listing a date (e.g., 07/04/1998, 07-04-1998, 07.04.1998, 1998.07.04)
- Month before date or the other way around (e.g., 07/04/1998, 04/07/1998)
- Two digits or four digits for year (e.g., 07/04/1998 or 07/04/98)

The NRC survey sets a decent example in specifying a date format template even though other simpler formats can also serve the same purpose (e.g., mm/yyyy). Collecting data on dates without offering a format may not lead to anything devastating. However, the researcher would have to spend more time in coding and preparing the data for analysis when such time could have been utilized for other purposes.

Use an underline to highlight a word or phrase (see A2a) in order to put an emphasis on it. But do not overuse different styles (e.g., boldface, italic, underline) to a point where they overshadow the content of a questionnaire.

A drop-down list of specialization areas (see A2a and A2b), like a list of predefined items described in Section 3.2.1 on factual question development, can facilitate respondents to quickly answer the question with ease. It also helps data coding and analysis at a later stage of the research process because respondents may use different terms, for example, to name their areas of specialization if there is no drop-down list. The same is true whenever researchers intend to gather data that have a predefined set of answers. Regardless of what constitutes the list of predefined answers, the two principles of categorization (i.e., collectively exhaustive and mutually exclusive) must be observed in creating the list. Another point to be kept in mind is to provide the "Other" option (e.g., A10, B1, B2) to allow the respondent to put down a category that is not already included in the list.

Instructions for filter questions (e.g., A5, C5, C7) in this survey, as pointed out earlier in the discussion on filter questions, are phrased as if the survey were administered verbally in person or via telephone. That however is not the case. Therefore, it seems simpler and more straightforward if the instruction (e.g., *Ask C6 if C5 = Yes*) is changed to, for example, "Do C6" and put it right after the Yes option. There are several such instructions in the survey. They should be revised to further enhance the quality of the questionnaire.

In Part D of the survey, there are only two questions about education and research resources. However, D1 contains nine sub-questions, each for one different type of resources. This does save presentation space. As a result, Part D appears significantly shorter than other parts of the questionnaire particularly when D2 is an open-ended question that the respondent may choose not to answer. This questionnaire places all demographic questions at its end, while the other practice is to put all such questions in the beginning of a survey (e.g., Niu et al. 2010).

The NRC questionnaire overall proves to be well thought out, carefully constructed, and without major issues. It can serve as an example for novice researchers to model after in their initial attempts in questionnaire development.

### *3.4.2 Case Study 2*

Niu, Xi, Bradley M. Hemminger, Cory Lown, Stephanie Adams, Cecelia Brown, Allison Level, Merinda McClure, Audrey Powers, Michele R. Tennant, Tara Cataldo. 2010. "National Study of Information Seeking Behavior of Academic Researchers in the United States." *Journal of the American Society for Information Science and Technology* 61 (5): 869–90. https://doi.org/10.1002/asi.21307

This research adapted the survey by Hemminger et al. (2007) and distributed it to five academic institutions in the United States. The study provides the survey instrument as Appendix A. However, the cover letter for the questionnaire is not included in the report. The following case study will examine the questionnaire and its use for data collection in the research project.

#### *3.4.2.1 Questionnaire*

The questionnaire, as the authors explain, is designed "to quantify academic scientists' transition to electronic communications and how this affects different aspects of information seeking" (Niu et al. 2010, 871). The word "quantify" in this quote nevertheless raises a small issue because the last three questions in the survey (i.e., Q27–Q29) are open-ended questions. Answers to open-ended questions are qualitative in nature and can hardly be quantified. In addition, some questions are not suitable for quantification because of the variables involved. Take Q10 as an example, is it a common and established practice to ask the respondent to calculate research collaboration in percentage? The answer should be obvious in that research collaboration takes many forms (e.g., discussing a research idea, developing a grant proposal, conducting a research project, co-authoring a research report). Apparently, any attempts to quantify research collaboration would be counterproductive and unable to truthfully reflect the reality.

The questionnaire is divided into six parts, effectively making the otherwise extensive survey more structured and manageable. Demographic and background questions are asked in the first part of the instrument, completely the opposite of the NRC survey in which questions on demographic data are presented the last. As discussed earlier in this chapter, each presentation order has its advantages as long as it is an educated decision. However, Q4 in this part of the survey asks for the respondent's specific age, which should be avoided in order to protect the respondent's privacy. The best way to correct this problem, as explained in Section 3.2.1, is to list age groups (e.g., 24 and below, 25–34) for respondents to check.

As recommended in the discussion on factual question development, the "Other" option should be provided in case the predefined answers for a question do not exhaust all possibilities. This survey meets the requirement of "collectively exhaustive" in some questions (e.g., Q18.1) but deviates from the requirement in others (e.g., Q11, Q26). In Q11, for example, what if the respondent's usage frequency of web-based scientific databases is twice a week and thus does not fit into any of the categories listed? Always provide the "Other (please specify): ..." option when not all the possibilities are exhausted in the enumeration. In addition, attention should be given to category development so that all the categories are collectively exhaustive and mutually exhaustive. In Q26, for instance, one of the most important criteria for authors to select a journal for manuscript publication is perhaps the subject scope of the journal. Yet this category is not listed and nor is an option of "Other (please specify): ..." provided. In contrast, some of the other options (e.g., ability to include links, color, graphics, multimedia) listed in the survey are clearly secondary in the journal selection process. Furthermore, the two options regarding fees (i.e., author having to pay page charges for long articles or color figures; page charges for long articles or color figures) are redundant. Both could have been combined into one option for open access to avoid the violation of the "mutually exclusive" principle in categorization.

Multiple memory-taxing questions (i.e., questions that tax on the respondent's memory) are found in the survey (e.g., Q16, Q19.3, Q24) where participants are asked to recall the number of times certain activities (e.g., reading a publication, visiting a library) taking place in last month (e.g., Q24) or even last year (e.g., Q23). Discussion is made earlier in this chapter on why memory-taxing questions should be avoided. As known, human beings are good at recognition but notoriously unreliable with recall from memory. Even if respondents are able to answer those memory-taxing questions, the accuracy of their answers remains to be a question due to the limitations of human memory and people's inadequate ability for retrieving data from long-term memory. In addition, such questions might ask for "details that may never have been committed to memory" (Scheuren 2004, 42).

There is also redundancy in some of the questions in the survey. For instance, Q19 repeats all the sub-questions in Q16 relating to one's personal collection (e.g., personally subscribed journal, personal digital library). Moreover, Q16 assumes a Yes answer to Q19 that is presented later in the questionnaire. It could be that the authors wish to cross check the consistency of the participants' responses. Nevertheless, such questions could cause confusion to the respondents or even make them stop completing the survey. The double-listing of each question number (e.g., 2: 2.) might be an issue originating from the self-built web platform, which however appears both improper and unnecessary.

On the other hand, the questionnaire overall serves well the purpose of collecting data for the study in that it comprehensively covers various aspects of academic researchers' information-seeking behavior in the electronic environment. The instructions given in smaller rectangular boxes are specific and should be helpful to the respondents, especially those with examples (e.g., Q13, Q15). As researchers, we always want our survey instructions as clear as possible so that they can be understood correctly and with ease.

#### 3.4.2.2 *Data collection using questionnaire*

The questionnaire was administered via a self-built web platform without utilizing any commercial survey sites such as SurveyMonkey. The research project is a major scholarly undertaking by any measure when the target population is faculty and graduate students in five research universities. Participants for this study were recruited from departments of the five selected institutions by email, letters, and flyers. Incentives utilized include pizza parties or iPods/cash awards for top-performing departments. Since the study reports that about 40% of the respondents were between 20 and 30 years of age and mostly graduate students, those material incentives are appropriate. However, non-material incentives such as a research summary could have been used for the faculty respondents because those scholars would be more interested in learning about the research results than having a slice of pizza.

The average response rate of this survey across the five participating universities is 14.4% with a range from 5% to 26% (see Table 1 in Niu et al. 2010). It is good that the authors recruited participants at the department level as opposed to the college or university level. Nevertheless, it is unclear from the report if the authors had thought about using department directories to send out personalized invitations to faculty members and graduate students to complete the survey. Blanket invitations are significantly different from personalized ones as far as the response rate is considered. Some random sampling technique (see Section 2.5.2) could have been applied in recruiting participants from individual departments to ease the load on data collection. In addition, the authors did not seem to do any follow-up of the survey after learning the low response rate. A total of 2,063 respondents are truly a large number. But this number is derived from a population of more than eight times larger. Those memory-taxing questions (e.g., Q16, Q24) and questions requiring a quantitative answer (e.g., Q10, Q19.3) in the survey, to a certain degree, contribute to the low response rate the study received as well.

As a survey study, this research represents many features of the survey method and could be beneficial to the readers in terms of the dos and don'ts in questionnaire development and for implementing this data collection instrument in scholarly undertakings.

**Chapter Summary**

- A questionnaire is a technique for data collection via a list of questions specifically created for a study. It also includes a cover letter or introductory note to invite and encourage the respondent to complete the survey.
- The majority of survey questions are closed-ended that can be further divided into factual, opinion/attitude and filter questions. A small number of open-ended questions may also be asked in a survey for collecting qualitative data.
- Major approaches to administering a survey encompass mail, email, telephone, survey site, and in person. The survey site approach is becoming the dominant method for distributing a questionnaire.
- Survey response rate is affected by many factors among which the quality of a questionnaire is the most crucial. Using incentives and follow-ups could help increase the response rate.
- See Sections 3.2.2 and 3.3.3 for dos and don'ts in questionnaire development and research. Keep them in mind when conducting survey research.

## Notes

1 The question labeling (e.g., A1, B3) is directly taken from the questionnaire the National Research Council (2006) administered to doctoral students for assessing research doctoral programs. The full questionnaire along with its introductory note is available at https://gradschool.unc.edu/facultystaff/nrc/NRC-Methodology-Guide.pdf.

2 The italics are in the original text. The word "rate" in square brackets is added by the present author according to the context.

## References

Babbie, Ear R. 1991. *The Practice of Social Research*. 6th ed. Belmont: Wadsworth.

Frankfort-Nachmias, Chava, David Nachmias, and Jack DeWaard. 2015. *Research Methods in the Social Sciences*. 8th ed. New York: Worth Publishers.

Hemminger, Bradley M., Dihui Lu, K.T.L. Vaughan, and Stephanie J. Adams. 2007. "Information Seeking Behavior of Academic Scientists." *Journal of the American Society for Information Science and Technology* 58 (14): 2205–25. https://doi.org/10.1002/asi.20686

Miller, George A. 1956. "The Magical Number Seven Plus or Minus Two: Some Limits on Our Capacity for Processing Information." *Psychological Review* 63 (2): 81–97. https://doi.org/10.1037/h0043158

National Research Council. 2006. "Assessment of Research Doctorate Programs: Admitted-to-candidacy Doctoral Student Questionnaire." In *A Guide to the Methodology of the National Research Council Assessment of Doctorate Programs*, edited by Jeremiah P. Ostriker, Paul W. Holland, Charlotte V. Kuh, and James A. Voytuk, 141–58. Washington: The National Academies Press. Accessed June 19, 2023. https://gradschool.unc.edu/facultystaff/nrc/NRC-Methodology-Guide.pdf

Niu, Xi, Bradley M. Hemminger, Cory Lown, Stephanie Adams, Cecelia Brown, Allison Level, Merinda McClure, Audrey Powers, Michele R. Tennant, Tara Cataldo. 2010. "National Study of Information Seeking Behavior of Academic Researchers in the United States." *Journal of the American Society for Information Science and Technology* 61 (5): 869–90. https://doi.org/10.1002/asi.21307

Scheuren, Fritz. 2004. *What is a Survey*. 2nd ed. Alexandria: American Statistical Association. Accessed July 2, 2023. http://people.umass.edu/biostat690c/pdf/ASA%20What%20is%20a%20survey%2068%20pp.pdf.

"What Does Cronbach's Alpha Mean?" 2020. UCLA: Statistical Consulting Group. Accessed July 2, 2023. https://stats.idre.ucla.edu/spss/faq/what-does-cronbachs-alpha-mean/.

Wildemuth, Barbara M., ed. 2017. *Applications of Social Research Methods to Questions in Information and Library Science*. 2nd ed. Denver: Libraries Unlimited.

# 4 Interview

**Chapter Overview Questions**

- What is interview?
- What types of questions are commonly asked in interview?
- What are the three kinds of interviews?
- What approaches can be taken to conduct interviews?
- What steps are involved in a typical interview?
- What are the dos and don'ts in developing an interview guide and performing an interview?

## 4.1 What is interview?

An interview is a technique for data collection by which the researcher asks each participant questions relating to a research problem. The participant would in return answer the questions asked, making the interview process resemble a conversation. It is in this sense that Berg (2009, 101) defines interview "simply as a conversation with a purpose".

Like questionnaire, interview is also frequently used in research for data collection. Compared with questionnaire, interview has the following features:

- Getting real-time responses from participants, which is impossible when a written questionnaire is used for data collection.
- Having opportunities to ask probe questions for gathering in-depth data and seeking clarification upon learning participants' responses.
- Asking follow-up questions for soliciting data on any new idea or topic the interviewee brought up or when the participant gives imbalanced or oversimplified answers in the interview process.
- Gathering nonverbal or body language data (e.g., facial expressions, body posture, gestures, eye movements) that cannot be captured in questionnaire.
- Collecting data from people who cannot write or read (e.g., young kids, seniors, the illiterate).
- Soliciting more data from participants because people can talk faster than they are able to write.

DOI: 10.4324/9781003021711-6

- Incurring more cost in data collection and processing (e.g., transcribing interview recordings) although voice recognition or voice-to-text technology has been improving. Interview is thus usually done with a limited number of people.
- Losing anonymity in face-to-face interviews which might affect participants' willingness to answer sensitive or embarrassing questions honestly.

When choosing interview as a data collection technique, we should keep in mind all the features of interviewing. In addition, job interviews and other similar types of interviews are not considered because they are beyond the scope of this book. As interviews involve questions, it seems necessary to look at what kinds of questions we typically ask in an interview.

## 4.2 Interview questions and types

Question types for questionnaire are depicted and discussed in Section 3.1 that include closed-ended questions (including factual, opinion/attitude, and filter questions) and open-ended questions. While interview mainly uses open-ended questions, interview questions are further categorized as main questions, probe questions, and follow-up questions (Rubin and Rubin 2005, 152–200). All interview questions, regardless of their types, are formulated to address the study's research questions.

### *4.2.1 Main questions*

According to Rubin and Rubin (2005, 152), main questions are formulated to elicit the overall experiences and understanding of the participant but might not provide the requisite depth to address the research problem. The latter is typically done via probe and follow-up questions. The interviewer would ask main questions that are broad in nature to start the conversation. An example of main question is "What does happiness mean to you?". This question allows the researcher to gain an overview of how participants perceive happiness in the beginning of the interview.

Main questions can also be asked as tour questions, dubbed from Spradley's (1980, 80–81) "grand tour" in observation research. In responding to such questions, interviewees are expected to describe their activities step-by-step. For instance, "I am interested in what students do at this university. Could you tell me about a typical day?". Multiple main questions can be created separately in one interview to gather various data needed for answering one or more research questions in a study.

### *4.2.2 Probe questions*

Besides main questions, interview researchers also need to ask probe questions, or probes in short, to obtain further explanation about an interview question already asked. For example, "What do you think has caused this problem?" after querying first about a school budget cut. The interviewer can also seek clarification based on an answer already provided. For instance, "Did you say that you learned the news from a friend or someone else?" after the participant refers to the news in interviewing.

Probes can be verbal or nonverbal. Verbal probes might be short and simple, such as "Could you tell me a little more about that?", "Then what?", or "And ...?" until the participant resumes talking. Nonverbal probes include nodding, waiting a bit for the interviewee to continue talking, or leaning forward to express interest. Asking suitable probe questions can help one conduct an interview more effectively. Rubin and Rubin (2005, 164) suggest that researchers create a warehouse of possible probes and then pick whichever ones they need at the moment. Rubin and Rubin (2005, 164–69) also categorize and describe different kinds of probes.

- A continuation probe encourages the interviewees to keep talking on the present subject. A continuation probe might involve little more than saying, "Mmm hmm. So ..." or it might simply echo part of what has just been said but with a questioning intonation (e.g., "You went to that meeting anyway?").
- An elaboration probe asks for more detail or explanation of a particular idea or concept the interviewee has said. Apply the sample wordings below when doing elaboration probes: "Such as ...?", "Could you give me an example?", and "Anything additional can you tell me about ...?".
- An attention probe is meant to let the interviewee know that you are listening carefully while encouraging the participant to elaborate. "That is interesting." and "Can I quote you on that?" are some common wordings used in asking attention probes.
- A clarification probe asks the interviewee to explain something that you do not follow. This kind of probes can be as simple as, for example, "What?" or use a longer wording such as "I beg your pardon?" and "Could you run that by me again?".

These different types of probes, when properly used, can greatly facilitate the interview process. Occasionally, an interview goes off track and a steering probe is needed to bring the interview back to its intended path. For example, "Sorry that I digress to the staffing issue by asking you about the teacher-student ratio. Let's explore what has caused the school budget cut". Unlike main questions for interview, most probes are often phrased during an interview when the researcher sees a need for further inquiry after getting the participant's response. For novice researchers, however, it is recommended to prepare some possible probes in advance to minimize the challenges of creating probes on site. Even so, the researcher should always be prepared to instantaneously ask probes because real-time interaction between the researcher and interviewee is one of the salient features of using interview as a data collection technique.

### *4.2.3 Follow-up questions*

With main questions and probes, the researcher should be able to get answers that together address the research problem and questions. Getting a better, balanced understanding of the participant's answers is what the researcher can accomplish through follow-up questions (Rubin and Rubin 2005, 173). Follow-up questions are built on what the participant has said and often asked in the following occasions.

- The participant suggests an idea, an event, or an issue that relates to the research project at hand. For example, the interviewee said that the company would not hire graduates from that college because of some concerns. This is new to the researcher. A follow-up question could be "What concerns?".
- The participant provides a partial or imbalanced answer to an interview question. In answering a question about what factors have an impact on electronic health record (EHR) system utilization by nurses, for example, the participant only talks about the

challenges in using an EHR system and totally ignores the benefits of such usage. A follow-up question for obtaining a balanced and thorough answer could be "What kinds of benefits do you think the nurses could have in using the EHR system?".
- When the participant's response to an interview is too general or simplistic, a follow-up question should be asked to obtain a better and more detailed answer. For instance, "Could you tell me more about the challenges you face in using the EHR system?".
- Occasionally, participants may have inconsistency in their responses. For example, "You mentioned that John Doe withdrew his application for the position, but I heard that he was not offered the position after the search company found out that he faked his resume. Do you think that he withdrew his application in anticipation of not being offered the position?". Asking follow-up questions about inconsistent responses particularly requires a gentle, nonthreatening approach.

Follow-up questions in most cases are worked out and asked during the same interview in which the topic of interest is mentioned. This is the scenario for which this book is written. On the other hand, follow-up questions might be asked in subsequent or follow-up interviews in ethnographic and longitudinal research that involves more than one interview. The researcher in those cases would have time to carefully look at participants' responses before formulating follow-up questions accordingly for the next interview.

### *4.2.4 Summary of interview questions types*

Probes and follow-up questions could sound similar but they are different in that probes are for ensuring the interview to stay on track and for soliciting in-depth responses main questions cannot generate. In comparison, follow-up questions enable the researcher to get a better and balanced understanding of the participant's answers. Any new ideas introduced in participants' responses signal a need for follow-up questions as well. Apparently, good interviews usually maintain a balance among main questions, probes, and follow-up questions. How to strike this balance requires practice and experience. In concluding their discussion on three types of interview questions, Rubin and Rubin (2005, 200) write:

> Taken together, these three kinds of questions bring about the thoroughness that is one of the hallmarks of responsive interviewing. Main questions are worded to cover the whole research problem; probes ensure that unclear answers are explained and questions are fully answered; and follow-up questions ensure that missing or implied information is tracked down, that contradictions are addressed if not resolved, that alternative explanations are examined, and that you learn about all sides of an argument and different perspectives on an event.

Sometimes, extra questions and throw-away questions are also created for interviews (Berg 2009, 114–15). Extra questions are those questions that are similar to main questions but worded differently. They are formulated to check the reliability of responses or measure the possible influence a change of wording might have. This practice, however, is not recommended for novice researchers because extra questions will take participants' longer time to answer which might negatively affect the interview quality.

Throw-away questions refer to those questions asked to serve as ice-breakers in the beginning of an interview, make the interview process smoother, or build rapport with participants throughout the interview. Throw-away questions, as the naming suggests, are not used for gathering data about the research problem and shall be discarded after the interview. Therefore, throw-away questions could be casual (e.g., "Is it easy for you to find this room [for interview]?") or incidental as long as they help the researcher to ask main questions, probes, and follow-up questions as planned.

### *4.2.5 Dos and don'ts in formulating interview questions*

The quality of interview questions to a great extent determines the successfulness of an interview. In addition to what is already discussed above (e.g., the balance among different types of questions) and all the points covered in Section 3.2.2 regarding the creation of questions for questionnaire (e.g., asking no complex, leading, threatening, or double-barreled questions), keep in mind the following dos and don'ts when formulating interview questions.

- Prepare questions according to your research problem and questions. But don't pose your research questions directly to interviewees. Instead, develop interview questions that would enable you as the researcher to gather data for addressing the research question. Interview questions should in general be more specific than the research question. It usually needs several or more interview questions to gather data for answering one research question. Take Ng'oma et al. (2019), one of two case studies for this chapter, as an example, one of the research questions they attempt to answer in their study is to identify the challenges that perinatal women with depression encounter although the authors imbed that research question in their problem statement without formulating it explicitly and separately (Ng'oma et al. 2019, 2–3). In order to gather data for answering this research question, Ng'oma and colleagues prepare one main question (i.e., Question 12 "Tell me more about your experiences during this pregnancy/delivery and postpartum.") and three subsequent probes (i.e., "Challenges during pregnancy/delivery and postpartum"; "Family experiences/family life"; "Relationship with husband/spouse and significant others ...") in the interview guide for the perinatal women participants. Question 12 seeks an overall answer from the participants about their negative perinatal experiences (i.e., challenges) while the three probes further explore challenges coming from pregnancy/delivery/postpartum, from family life and experiences, and from spouse. Attached to the third probe is one further exploration regarding how much, if any, material or emotional support the participant receives from her spouse. All together five interview questions are asked to gather data for answering one research question: What challenges do the perinatal women with depression meet in that period of their life? This example illustrates how to prepare interview questions for a research question and typically how many of them should be created for one specific research question.
- Formulate questions that encourage interviewees to provide a detailed answer (e.g., "What makes you to choose and buy an iWatch?"). Don't ask questions that lead to a yes-or-no answer (e.g., Did you buy an iWatch?).
- Create questions that are suitable to be asked verbally because that is how interview questions are raised. Don't use complex sentence structures in your questions.

- Ask "what influenced ...", "what caused ...", "what contributed to ...", or a similar phrase when you intend to find out why something happened. Don't ask the why question directly because people are comfortable talking about their experiences but often do not know how to respond to the more abstract question of why. Indirect questions may work better. For example, you could ask: "What contributed to the increase of the deficit size in 2019?". In contrast, interviewees may find it hard to answer a question of the same nature but worded differently: "Why was there an increase of the deficit size in 2019?".
- Ensure that each interview question is clear, unambiguous, and free of jargon for the general public.

## 4.3 Interview types and interview guide

As briefly mentioned in Section 1.4.8, there are three kinds of interviews (i.e., structured, semi-structured, and unstructured) depending on how interview questions are prepared and if probes and follow-up questions are allowed. A detailed discussion of each interview types is provided below along with their features and applications.

### *4.3.1 Structured interview*

Structured interviews, also known as standardized or formal interviews (Briggs 2000), get the name from the way in which the interview is conducted. All the questions to be asked of interviewees are formulated beforehand in the form of an interview guide which will be discussed in Section 4.3.4. The questions are asked in the same order in all interviews. Besides, the interviewer can only ask questions from the interview guide and is unable to pose any probes or follow-up questions at the time of interviewing (Gorman and Clayton 2005, 127).

Because of those characteristics, many publications call structured interviews as interview surveys (e.g., Briggs 2000; Frankfort-Nachmias, Nachmias, and DeWaard 2015, 206; Scheuren 2004, 22). Indeed, structured interviews are analogous to questionnaire surveys but they are not synonyms because of the following reasons: (1) Structured interviews still have the opportunity to gather nonverbal data even if they are done by phone. Intonation, pause, and other similar nonverbal signals all can only be collected in structured interviews, but not in questionnaire surveys. (2) Interviews are normally conducted with a small number of participants because they are costlier than questionnaires to implement. (3) When opinion polls are made via phone, they are essentially questionnaires completed by phone. Reading questions to participants via phone does not turn a questionnaire survey into an interview. The many polls reported by the Pew Research Center and other organizations never include nonverbal data in their analyses and reports.

Although structured interviews possess less distinctive features if compared with semi-structured and unstructured interviews, they are adopted when more than one researcher conducts the interview, people other than the researcher do the interview, or participants (e.g., small kids, the illiterate) cannot complete a written questionnaire.

### *4.3.2 Semi-structured interview*

Semi-structured interviews mean that the researcher asks some pre-formulated questions from an interview guide and is also able to pose some probes and follow-up questions at the time of interviewing for clarification and further exploration according to the interviewee's responses. Semi-structured interviews are sometimes known as semi-standardized interviews (Berg 2009, 104). In contrast to structured interviews, semi-structured interviews have more flexibility in the following aspects: (1) The researcher can rephrase interview questions in a way that is the most appropriate to each interviewee. If the interviewee, for example, is from the academia, more formal language might be used in the interview. Otherwise, less formal language would be more suitable. (2) The researcher can change the order of questions already listed in the interview guide and create additional ones in the form of probes and follow-up questions depending on the interviewee's verbal and nonverbal feedback in the interview process. Nonverbal feedback includes any reaction or response (e.g., a nod, a pause, a shrug) that is not verbalized. Researchers are encouraged to take advantage of the nonverbal data right there and then to facilitate the ongoing interview. Some of the nonverbal data can also help analyze and interpret the interviewee's responses. Nonverbal data are usually recorded in conjunction with verbal data during the interview process. Researchers can make a quick note to record any nonverbal data. If the interview is recorded using a video camera or via a video conferencing platform, nonverbal data will automatically be captured to aid data analysis later.

The flexibility in semi-structured interviews clearly poses more challenges to the interviewer because all new question creation and changes in question wording or ordering must be done in real time, then and there, without the chance of taking a break. The researcher must stay highly observant and respond rationally yet without any hesitation in semi-structured interviews. Despite the challenges, semi-structured interviews are perhaps the most adopted one among the three different kinds of interview in the scholarly world (e.g., Chu and Ke 2017) thanks to its flexibility in application.

### *4.3.3 Unstructured interview*

Unstructured interviews, also known as unstandardized or informal interviews (Berg 2009, 104), maintain a close tie with ethnography or field study. For this reason, unstructured interviews are called ethnographic interviews if used in ethnographic research as a data collection technique. Some authors (e.g., Bell 2005; Rubin and Rubin 2005) tend to label unstructured interviews as qualitative interviews in consideration that it is mainly applied in ethnography. This labeling, however, does not seem appropriate because of two points. First, we do not have a category of quantitative interview as opposed to qualitative interview. Second, interviews indeed are chiefly for gathering qualitative data. But in no circumstance are interviews, including unstructured interviews, excluded from soliciting quantitative data when necessary.

The most distinctive feature of unstructured interviews is the absence of a typical interview guide (see Section 4.3.4). This is determined by the fact that participants in unstructured interviews are in general more heterogeneous than homogeneous. They possibly have different vocabularies and assume different roles in communities (Berg 2009, 106). In addition, the context in which an unstructured interview is conducted could also vary from participant to participant. Therefore, the interviewees are less likely to answer a same set of questions from an interview guide. The researcher instead should create and adapt main, probe, and follow-up questions appropriate to each participant within a given research framework. On the other hand, the researcher may use an aide-mémoire to help prepare and facilitate the unstructured interview. An aide-mémoire includes a short list of loosely defined questions (MaCann and Clark 2005) or a small set of keywords written on a compact piece of paper (Yin 2016, 147). In either case, the chosen questions or keywords are directly relevant to a specific interview. Main questions can be asked based on the initial short list of loosely defined questions or a small set of keywords. Probes and follow-up queries would be formulated depending on the interviewee's responses. An aide-mémoire thus can both serve as preparation for an unstructured interview and function as a set of reminders. The researcher has the flexibility to decide which element in the aide-mémoire would be covered, changed, or ignored during the interview process while achieving some consistency in interviews of different participants (Wildemuth 2017, 240). Having an aide-mémoire might also produce a positive impact on the unstructured interview in that when the researcher refers to the aide-mémoire from time to time the participants would view this as a manifestation of preparedness and seriousness. They consequently might be more willing to actively participate in the interview (Yin 2016, 147).

The interactive, flexible, and adaptive nature of unstructured interviews sets higher standards for the interviewer who must be particularly well prepared before the interview and decide when to ask probes and follow-up questions by staying responsive to participants' responses during the interview process. Among the three kinds of interviews, unstructured ones are the most demanding for the interviewer to carry out. Researchers would adopt unstructured interviews for gathering comprehensive, balanced, and in-depth data on a chosen problem without the possibility of having an interview guide.

Table 4.1 summarizes and compares several elements the researcher must consider when determining which interview type to adopt: structured, semi-structured, or unstructured. They include time needed to conduct an interview, degree of control in the interview process, requirements for the interviewer, and necessity of having an interview guide.

The four elements in Table 4.1 must be considered along with the various features that are related to each interview type already discussed when we make a selection decision of interview types. Choose a type that is the most appropriate as one assurance to conduct a quality interview.

### *4.3.4 Interview guide*

In our discussion of different interview types, we know that an interview guide plays a crucial role in both structured and semi-structured interviews. An interview guide includes a list of questions for the interviewer to ask along with proper techniques (e.g., when to ask a probe) for ensuring interview quality. It functions as the recipe for making the "interview" dish. An interview guide is also known as an interview schedule (Berg 2009, 109) or interview protocol (Boyce and Neale 2006).

Similar to questionnaires, an interview guide includes questions. But interview questions are mostly open-ended ones that are further categorized as main questions, probes,

*Table 4.1* Considerations for Interview Type Selection

| *Consideration* | *Structured* | *Semi-structured* | *Unstructured* |
|---|---|---|---|
| Time needed | Less | Moderate | More |
| Degree of control | Little | Some | A lot |
| Requirements for interviewer | Few | Some | Many |
| Interview guide | Must have | Yes | Use aide-mémoire |

and follow-up questions. The balance among the three types of questions is decided by the research problem and questions as well as interviewees' responses. The central task in developing an interview guide involves three elements:

- Create a list of main questions based on the research problem and questions. These questions form the backbone of all the interview questions and shall help the researcher gather essential data for the study. Pay close attention to the dos and don'ts in formulating main interview questions outlined in Section 4.2.
- Determine what are asked in probes and follow-up questions, and when and how frequently they should be asked. We describe and contrast probes and follow-up questions in Section 4.2. Although both types of questions are generated largely based on interviewees' responses, we shall ask probes when we need to get a deeper understanding (e.g., "Can you elaborate on that idea?") or seek clarifications (e.g., "I am not sure I understand what you are saying. Could you explain that further?") of participants' answers. Follow-up questions in comparison are formulated when we intend to follow any new idea or issue the interviewee brings up (e.g., "That topic is new in our conversation. Could you say more about it?") or attempt to obtain a complete answer or a balanced view about a question already being asked (e.g., "Now I already know one side of this argument. Could you tell me your side as well?"). While it is possible to include some probes in the interview guide, follow-up questions are typically formulated during the interview process. On the other hand, use probes and follow-up questions only if they help data collection. Never overuse them for the sake of having probes and follow-up questions or to the extent that they become burdensome for the interviewee to answer.
- Arrange all the questions in the guide logically and coherently. Avoid jumping from topic to topic in an interview. Some authors (e.g., Boyce and Neale 2006; Rubin and Rubin 2005) suggest that questions about the interviewee's opinion should be asked after questions for factual information. The general order for the researcher to take in an interview is to ask one main question, followed by probes and/or follow-up questions if necessary, before moving to the next main question.

Using the cooking recipe analogy again, we can see that the list of questions in the interview guide resembles a list of ingredients. The dish we plan to make will determine what ingredients we need. The different types of interview questions are similar to the different kinds of ingredients while the order of interview questions is identical to the order in which the ingredients are used in cooking.

It is recommended that all the questions (i.e., main and/or probe questions) created for the interview guide be mapped against the research questions of a study. When an interview question is found not related to any of the research questions, delete it because it would only make the interview guide unnecessarily longer than it should be. Likewise, add one or more interview questions if there are not adequate questions for covering a particular research question.

What else should be included in the interview guide besides main and probe questions? They are part of the so-called interview tactics or techniques which include what to say in the beginning of an interview, what to say if the interviewee is too talkative or too quiet, and what to say at the end of the interview.

Other components that an interview guide may have include pleasantries in the interview process. Some of the pleasantries can be in the form of throw-away questions to function as ice-breakers in the beginning of an interview (e.g., "How was the traffic on your way here?") or as gentle remarks if the interviewee digresses or says too little. Those components may not be written in the interview guide. However, they are indispensable for the researcher to have in order to conduct a smooth, quality interview.

Appendix 1 in Boyce and Neale (2006), one of the two case studies for this chapter, provides an interview guide example. It also covers informed consent which the researcher must obtain before conducting an interview. More about informed consent in interviews will be presented in Section 4.4.2.

After the interview guide is developed, it should be pretested with a couple of people from the prospective participants as we do with the questionnaire. Make sure that those who participate in the pretest should not take part in the actual interview because they are already familiar with the instrument they pretest. Checking with colleagues and experts could be another option although it is preferable to do the pretest with a small subset of prospective interviewees. The objective of the pretest is to discover errors and other similar issues in the draft interview guide. Usually changes and corrections are made based on feedback from the pretesting before the interview guide is finalized and used in an interview. As indicated in Section 3.2.2.1 about questionnaire validation, we always place our focus on developing a quality interview guide and pretest the guide for improvement purpose because validating an interview guide with techniques such as Cronbach's Alpha, expert panels, and/or factor analysis is impractical and unrealistic in research.

## 4.4 Using interview in research

What we have discussed till this point in this chapter is preparation for conducting interviews (e.g., interview question creation, interview guide development). Now we will focus on how to do an interview.

### *4.4.1 Interview approaches*

Before conducting an interview, we need to consider which approach we will choose to do it. Face-to-face, by telephone, via social media app (e.g., WhatsApp, Facebook Messenger, WeChat), and video conferencing (e.g., Zoom) represent the four major interview approaches. Table 4.2 summarizes and contrasts these approaches in several aspects.

Face-to-face has been the most common and preferred interview approach because it takes full advantage of using interview as a data collection technique and can gather both verbal and nonverbal data (e.g., facial expressions, body postures, gestures, eye movements) from the interviewee. However, the face-to-face approach is feasible only if both the researcher and the interviewee can meet in person which might be an issue in some circumstances. In comparison, the physical distance is not an issue for all other three

*Table 4.2* Major Approaches to Interview

| *Consideration* | *Face-to-Face* | *Telephone* | *Social Media App* | *Video Conferencing* |
|---|---|---|---|---|
| Distance issue | Yes | No | No | No |
| Tech-dependent | No | Some | Yes | Yes |
| Data collected | Verbal + Nonverbal | Mainly Verbal | Verbal + Some nonverbal | Verbal + Some nonverbal |
| Transcription needed | Yes | Yes | Possibly | No |

approaches (i.e., telephone, social media app, video conferencing). Interviews using any of the three approaches can be conducted regardless of where the researcher and the interviewee are. They could be a few blocks apart or in any corner of the world as long as there is the internet.

On the other hand, interviews by telephone, social media app, or video conferencing are all technology dependent, which means they are not viable interview options if participants do not have access to those technologies. Another drawback these approaches share is their inability to gather all the nonverbal data in an interview. Most telephone interviews, without the video call function (e.g., iPhone FaceTime), can only capture limited amount of nonverbal data such as intonation and pauses. The social media app and video conferencing approach can do better than the telephone approach in gathering facial expression and some other body language. But neither approach is optimal for the researcher to fully observe the interviewee's body language. In addition, the video conferencing approach may incur additional cost for its advanced features such as extended usage time length, video recording, and voice-to-text transcription. Among all the four interview approaches, the video conferencing approach is the one that supports voice-to-text or interview data transcription – a time-consuming activity in any interview research. The social media app approach might be able to transcribe interview data with the help of a third-party app. All the tech-dependent interview approaches, especially the social media app and video conferencing ones, are for sure expected to evolve further in the future.

Text messaging can also be used for interview and a complete script of the interview data is readily available although this perhaps should be the last option for the researcher to adopt due to its multiple limits (e.g., asynchronous and impossible to gather nonverbal data). All interview approaches are feasible for structured and semi-structured interviews, while the face-to-face approach remains to be the most appropriate for unstructured interviews that need the in-person contact to help establish rapport between the researcher and the interviewee.

### *4.4.2 Interview procedures and techniques*

Besides developing an interview guide or aide-mémoire and choosing an interview approach, the researcher also needs to select and recruit participants before starting an interview. Unlike in questionnaire research, interview studies mostly select participants via a nonrandom sampling method (e.g., convenience sampling, purposive sampling) because interview research usually is conducted with a small number of participants and often cannot use a random sampling method to choose participants.

Interview researchers will send out a letter (less likely) or a formal message (more likely), given the communication means we have nowadays, to prospective participants and invite them to take part in an interview. What needs to be covered in the invitation include a brief description about yourself, your research purpose, expected interview length, confidentiality of the study, and any incentive for participants. Some researchers may repeat such information in the introduction step during the interview process, to be discussed below, or include it in the interview guide (e.g., Boyce and Neale 2006). The wording in the invitation should be as inviting as possible. When a prospective participant accepts the invitation, an appointment needs to be made with the person for an interview.

As depicted in Section 4.3, there are structured, semi-structured, and unstructured interviews. Although each interview type requires different techniques and needs different degree of attention, there are some common steps in conducting an interview (e.g., Wildemuth 2017, 252) as presented below.

1) Introduction

   This step is standard in every kind of interviews. The introduction includes expressing gratitude to the interviewee for accepting your invitation, reiterating certain required components from the invitation (e.g., your research purpose, expected interview length, confidentiality of the study). As briefly mentioned at the end of Section 3.2.2.2, informed consent for interviews normally takes the shape of a prepared form through which participants grant permission to the researcher for collecting and using data in research. The informed consent can be part of an introduction (e.g., Appendix 1 in Boyce and Neale 2006) or a separate document whose template is usually available from an organization's IRB (institutional review board). If an interview is not conducted face-to-face, the participant's verbal agreement to be interviewed would constitute an implied consent (Berg 2009, 89). Of the three options for obtaining informed consent from participants, using an interview consent form is the most formal and also regularly done. When the researcher applies for IRB approval, the IRB application package will include an interview consent form in which the researcher can fill in her or his name and a brief research project title. Before an interview starts, the researcher will present the interview consent form to the participant who will read, sign, and date it. The researcher will sign and date it as well to complete the process of informed consent.

   The introduction for structured interviews is uniform and with little variation from participant to participant. The researcher may simply read a written introduction to the participant although that is not recommended. Always try to talk to the interviewee whenever possible. For unstructured interviews, the introduction should be tailored to individual participants as one way of building some initial rapport with each of them. The researcher in this case should avoid reading an introduction to the participant. Instead the introduction should be made as a conversation that involves some warm-up skills to be covered in the next stage of interview. Semi-structured interviews in the introduction step would strike a balance between the structured and unstructured interviews. The introduction should always be given in plain English regardless of which type of interviews the researcher carries out.

2) Warm-up

   Due to the nature of structured interviews, warm-up is typically integrated with the introduction part. But for unstructured interviews, multiple techniques should be applied to do the interview warm-up. The researcher can start a conversation about the environment in which an interview is conducted, about some topics (e.g., weather, sports) the interviewee might be interested, or about matters that would gradually lead to interview questions. For example, if an interview is about finding out how strategic planning is done in a college, possible questions for warm-up can be the following: "When was the last time when the college made strategic planning?" "Have you ever involved in any strategic planning in a college?". All the warm-up should be made with adequate preparation so that only proper conversations are carried out in this period. As its name suggests, warm-up is a step in the interview process that aims to establish rapport between the researcher and the interviewee. In some cases, the rapport building takes time and takes place even before the actual interview starts (Fontana and Frey 2000, 655–56).

3) Main body of the interview

   This step is dedicated to asking questions and gathering data for a given research project. The researcher would use an interview guide in structured and semi-structured interviews. In the case of unstructured interviews, the researcher would employ an aide-mémoire in the interview process. The types and sequence of questions are either defined in the interview guide or noted in the aide-mémoire. The typical order in which the researcher asks questions in an interview is to pose one main question, followed by probes and/or follow-up questions if necessary before moving to the next main question. For structured interviews, the researcher only needs to go by the interview guide which is relatively straightforward. For semi-structured interviews, the researcher may need to rephrase some questions, change the order of some questions, as well as create follow-up questions when needed. For unstructured interviews, the researcher is expected to word most questions on site and should stay the most responsive throughout the interview process among all three kinds of interviews in order to fully benefit from using interview as a data collection technique. Consult and follow all the guidelines discussed in Section 4.2 for creating and asking different kinds of interview questions at this stage of interviewing.

4) Cool-off and closure

   After all the questions from the interview guide or aide-mémoire are covered, the researcher can cordially ask if the interviewee has any anything else to add or any question to ask. The researcher could also comment casually by saying, for example, "It has been nice talking with you". This step ends the entire interview process. Commonly the researcher would take the opportunity to thank the interviewee for their time and participation before bidding good-bye to the participant.

Although those four steps in the interview process are described individually, it is by no means that the researcher has to show these steps discretely in an interview. Rather, the interview should be conducted as an integral and seamless process that covers all the four steps.

### *4.4.3 Common implementations in research*

When shall we use interviews for data collection in research? We do not need to conduct an interview to find out, for example, how frequently people take shower, use social media, or purchase a particular electronic gadget. Such data can be more easily gathered using a questionnaire. But if we want to know what people think about their personal hygiene, why they are so attached to social media, or what motivates them to buy a particular electronic gadget, interview is the appropriate method to use (Rubin and Rubin 2005, 2). If what you need to find out cannot be answered simply or briefly, if you anticipate that you may need to ask people to explain their answers, give examples, describe their experiences, or if you want to get real-time responses and gather nonverbal data from the participant, then you should choose interviews for data collection.

Interviews, especially unstructured interviews, are often used in conjunction with other data collection techniques (e.g., participant observation) in ethnography or field study. Another common implementation of interview in research is its usage together with questionnaire. There are two ways of adopting both interview and questionnaire in a same study. One is first using a questionnaire to collect demographic (e.g., age, gender, race) and other factual (e.g., usage frequency, electronic gadget ownership) data, and then conducting an interview to gather data on reasons, motivations, clarifications, explanations, and other similar kinds of data. The other implementation of choosing these two data collection techniques together is usually seen in research projects where interviews are conducted to create questions for a questionnaire. The questionnaire is subsequently distributed for data collection. In the interview-first-questionnaire-second implementation, it usually implies that the researcher needs to develop a questionnaire from scratch and no similar questionnaire has been created in the past. Of the two implementations of using interview and questionnaire in combination, the questionnaire-first-interview-second approach is utilized more often. More discussion on research designs involving interview will be provided in Chapter 14, Research Design via Method Selection.

### *4.4.4 Dos and don'ts in interview research*

Although interview is a conversion with purpose (Berg 2009, 101), doing it properly still requires a lot of preparation and practice because interview is both an art and a science. Up to this point in this chapter, many dos and don'ts in interview research are already presented. Below is a recapitulation of conducting interview research along with elaboration on some major points.

First, do planning. Planning before the start of an interview includes sending out invitations, choosing an interview site, and scheduling interviews. Always choose an interview site (e.g., home, office) that is natural to the interviewee and would make the participant feel at ease to be interviewed.

Second, do act professionally. Arrive at the interview site ahead of the scheduled time. Dress properly for the interview, and avoid overdressing or underdressing. Stay attentively throughout the interview process. Avoid looking at the clock or your watch. Avoid looking sideways. Be positive, encouraging, and positive. React with interest, and never express any disapproval or shock. Never argue. Always remember to thank the interviewee profusely. Use common sense and observe etiquette in the entire interview process.

Third, do conduct the interview proficiently. Make full preparation for the interview, and avoid turning the interview into a chat or random talk. Ask only one question at a time. Ensure each question is clear and unambiguous. Give your interviewees the opportunity to answer as they see fit. Do not put words in the interviewee's mouth. Listen to what is said, and what is not said. Take notes of important points and nonverbal data while using a recording device for capturing the entire interview with interviewees' consent. Take advantage of real-time feedback in order to decide when to ask probes and follow-up questions as well as when to move to next question. Do not interject stories or comments during the interview.

Fourth, do keep all the promises made for the interview research. Complete an interview within the estimated time frame. Offer your incentive as promised. Do the same if you agree to reimburse, for example, interviewees' public transportation cost. If the incentive for academic participants is a summary of your research report, do prepare and send one to them no matter how many months or years may have passed since the data collection. All the promises made during data collection must be honored to win trust from participants. This trust will be particularly helpful to the researcher if a follow-up or longitudinal study is to be carried out.

Fifth, do complete the after-interview work promptly. Transcribe your interview data as soon as possible if, for example, the automatic transcription feature of video conferencing is not available to you. Even with the automatic transcription support, review and edit the transcribed data while the interview is fresh in your mind. For the same reason, data analysis should also be done soon afterward. This book, however, is not designed to specifically cover interview data analysis although how to choose appropriate techniques for analyzing qualitative data, of which interview data are part, will be discussed in Chapter 13, Techniques for Data Analysis and Their Selection.

It is never possible to exhaust a list of dos and don'ts for interview research. Instead, we should always plan carefully, prepare adequately, and implement rigorously when doing interviews. In addition, review and reflect at each step during the interview research process so that errors can be detected and corrected in time while positive experiences can be learned to inform future research using the interview method.

## 4.5 Case studies of interview research

Two case studies are made in this section with one focusing on the interview guide only and the other on both the interview guide as well as interview implementation in a study.

### *4.5.1 Case Study 1*

Boyce, Carolyn, and Palena Neale. 2006. *Conducting In-Depth Interviews: A Guide for Designing and Conducting In-Depth Interviews for Evaluation Input*. Pathfinder International Tool Series, Monitoring and Evaluation – 2. Watertown, MA: Pathfinder International.

The Sample Key Stakeholder Interview Guide is listed as Appendix 1 in Boyce and Neale (2006). Boyce and Neale (2006) is not a research article that actually utilizes the interview guide as its data collection instrument. Rather, it is provided as an interview guide example. The interview guide can be adopted, according to Boyce and Neale (2006), to evaluate the African Youth Alliance (AYA) program. The AYA program was launched in 2000 by Pathfinder International, the Program for Appropriate Technology in Health (PATH), and the United Nations Population Fund (UNFPA) with funding from the Bill & Melinda Gates Foundation (African Youth Alliance 2005).

The interview guide comprises three parts: introduction, questions, and closing components. It is formatted especially for illustration purpose with the left column outlining the key components in each part whereas the right column listing the corresponding contents for each part. Because this interview guide is designed for key stakeholders of the AYA program, Boyce and Neale (2006) assume that interviewees are familiar with all the acronyms and abbreviations appeared in the interview questions and thus do not spell them out. In actual research, we should always spell out every acronym or abbreviation the first time one is used in an interview guide unless they are well-known acronyms or abbreviations such as NASA and GPS.

The introduction of the interview guide covers all the basic components in an interview introduction. It includes a thank-you, self-introduction of the interviewer, what the interview is about and what it is for, estimated length of the interview, permission for taking notes and recording the interview, confidentiality, opportunity for questions, and informed consent. Like what is discussed in Section 3.4.1.1 for the questionnaire cover letter, this interview guide also places an emphasis on voluntary, informed consent by specifying towards the end of the introduction that "Remember, you don't have to talk about anything you don't want to and you may end the interview at any time". This statement is politically correct but realistically impractical, particularly for novice researchers who have no "research capital". Although we should not force interviewees to complete an interview, we must do our best to encourage them to participate in the entire interview.

The informed consent for interview always needs to be signed and dated. As this interview guide is for the AYA program stakeholders, some interviewees are likely less than 18 years old. The option of legal guardian is hence provided as part of the informed consent. The only element that is absent from the otherwise comprehensive and concise interview guide is a description about incentives for participants. Perhaps the nature of this interview does not warrant any incentive.

There are in total 11 questions in this interview guide along with three comments in the left column. While the second and third comments are recommended practices for interviewing, the first point (i.e., no more than 15 open-ended questions) may not be applicable in every interview because the number of questions an interview should have is determined by the research questions of a project. Some research may need well over 15 open-ended questions! In addition, the two abbreviations, not spelled out in the interview questions, YFS and NTCD, respectively, stand for Youth-Friendly Services and Nontraditional Condom Distributor. All the questions in this part will be referenced by their original numbers in the guide in the following discussion.

Question 1 is a main question by type. However, the question tries to cover all three aspects of Youth-Friendly Services along with examples, which is too long for the interviewee to listen to, let alone answer it properly. If this question is split into three separate ones, each dealing with one aspect of Youth-Friendly Services (e.g., strategies), it will be more manageable for the interviewee. Moreover, Question 1, as it stands, seems more suitable for questionnaire rather than for interview. Try to think why by referencing to Section 3.2.1.

Similar to Question 1, Question 2, Question 6, and Question 7 should each cover YFS strategies, inventions, and tools separately. Lumping three aspects of the topic (i.e., YFS strategies, inventions, and tools) in one question would overwhelm the interviewee who in turn is probably unable to answer all three aspects of the question completely.

Question 7 is a probe question but it asks why, which should be avoided in interview question formulation because people often do not know how to respond to the abstract question of why (see Section 4.2). A different version of the question should instead be asked. For example, "What makes you think a particular YFS strategy should be discontinued?".

Question 8 is another main question. Although it lists three barriers that interviewees might encounter in the AYA program, this practice actually turns the question into a closed-ended one no matter how many of the three given options the interviewee might choose. The interviewee is offered no opportunity to answer Question 8 openly. Question 8, like Question 1, appears more appropriate for questionnaire instead of interview. Even if it were a questionnaire question, it should provide one additional answer option "Other (please specify): ..." so that the interviewee can indicate any barrier not already specified in the question. Question 10 has the same issue as Question 8 by enumerating a finite number of possible answers.

Question 9 is supposedly a probe question for Question 8. But it is too broad and general for the interviewee to respond when the answer to Question 8 includes multiple barriers. It is unrealistic to expect the interviewee to remember all the barriers listed in Question 8 and then respond accordingly. A more feasible approach would be to rephrase Question 9 as: "You mentioned in answering the previous question that X was a barrier. How did you overcome it?".

Question 11 sounds like a probe question for "... further efforts such as these". However, it is unclear what "these" refer to in this question. Are they the options (e.g., increased use of services by youth) listed in Question 10? Additional context should be provided.

This interview guide is for structured interview and contains only main and probe questions. The last part of the interview guide, closing components, is properly worded. Always remember to thank the participant adequately before completing an interview.

### 4.5.2 Case Study 2

Ng'oma, Mwawi, Samantha Meltzer-Brody, Ellen Chirwa, and Robert C. Stewart. 2019. "'Passing Through Difficult Times': Perceptions of Perinatal Depression and Treatment Needs in Malawi – A Qualitative Study to Inform the Development of a Culturally Sensitive Intervention." *PLoS ONE* 14 (6): e0217102. https://doi.org/10.1371/journal.pone.0217102.

This interview study is the first phase of a larger research project on developing a psychosocial intervention for perinatal depression. It is designed to explore experiences of perinatal women with depression, the challenges they meet, their coping strategies, and their service needs. The study finds that financial difficulties, relationship problems (polygamy, lack of support, neglect, and infidelity), traumatic events (intimate partner violence and loss), and fear of birth outcomes are causes of depression. A family-focused psychosocial intervention (e.g., support group discussion with individual perinatal women and their family members in their homes) should be implemented to address the challenges perinatal women face in this period of their life.

In order to achieve their research objective, the researchers of this study specifically create two interview guides: one for perinatal women with depression and the other for maternal health care workers who are referred to as "key informants" in the report. Informants are frequently used as a synonym for participants in ethnographic research even though the current study is not ethnography. Both interview guides are available as supporting information and can be downloaded from the article site free of charge. We will first examine the two interview guides, followed by an exploration of how they are used in the research for data collection.

#### *4.5.2.1 Interview guides*

Both interview guides were developed based on the study objective and informed by the existing literature (Ng'oma et al. 2019, 4–5). This is a textbook implementation regarding how an interview guide should be created. There is no introduction provided in the two interview guides perhaps because what needs to be conveyed in the introduction (e.g., who the interviewer is, interview purpose, confidentiality) is already explained in the recruitment process of study participants. We will look at the recruitment processes for interviews of both perinatal women and health care providers in the next section.

The overall structure of both interview guides is the same. Each consists of two parts: Section A for demographics and Section B for questions related to perinatal depression. One distinctive feature of both interview guides is that each contains many probes, and all probes are formulated before the interviews because three different people served as interviewers in the study. On average three to four probes are asked after every non-demographic, main question. There are also probes for most demographic questions in the interview guide for perinatal women with depression. Judging from both interview guides, the interviews are unquestionably in depth.

The interview guide for perinatal women with depression is labeled as open-ended although it is unclear whether it means that the interview questions are open-ended or the interview guide is semi-structured. If the former, Question 1 about age is apparently a closed-ended question. If the latter, no mention was made in the research report that the interviewer could ask probe or follow-up questions beyond what is listed in the interview guide although the interviews appear more semi-structured than otherwise. The interview collects data on religion (Question 5) and HIV status (Question 6), but Ng'oma and her colleagues make no reference to either in their results and discussion of the study.

The interview guide for maternal health care providers in comparison has fewer questions, especially fewer probes overall. However, the researchers assume that "a health care issue" is the answer to Question 8 "How are your views regarding depression in pregnancy or postpartum period?". What would the interviewer probe if an interviewee's answer is not "a health care issue"?

Apart from the afore-mentioned points, either main questions or probes in both interview guides are formulated closely related to the research topic. They are worded according to the participants' background and vocabulary. For example, the interviewers use terms such as "counseling", "talk therapy", and "medication" with perinatal women about how health care workers can offer help to treat their depression. In contrast, terms such as "pharmacological" and "nonpharmacological" are included in the interview guide for health care providers for the same question.

Furthermore, well over 15 questions (including probes) are developed for both interview guides, far exceeding the limit of "no more than 15 open-ended questions" Boyce and Neale (2006, 12) suggest in their Appendix 1 already examined in Case Study 1 of this chapter. This shows that how many interview questions should be asked are determined by the research problem and questions of a study, rather than by a preset limit. Pretests of both interview guides were performed at a primary health care clinic that is not the two chosen study sites. Changes were made to both subsequently "until the study team was confident in the performance" (Ng'oma et al. 2019, 5). The pretest procedure sets a fine example for ensuring the quality of data collection instruments in interview research.

#### 4.5.2.2 *Data collection using interview guides*

The 22 perinatal women participants in the study were recruited from the waiting rooms of two maternal health care facilities, one urban and one rural, in Malawi. Including both an urban and a rural site in the study can enhance the representativeness of prospective participants. The nurses working in the two facilities' waiting rooms first talked with perinatal women during their routine health check-up sessions and directed those women who showed interest to the research team. The research team thoroughly explained to the prospective participants the study and other components (e.g., benefits and risks of the study) that are commonly specified in the interview introduction. A total of 189 perinatal women were willing to take part in the research and also met the study inclusion criteria. They all signed a consent form for the study. As this study is about perinatal women with depression, the 189 perinatal women were screened using the Edinburgh Postnatal Depression Scale (EPDS), translated from English to Chichewa, a Bantu language spoken in parts of Malawi, where it is an official national language along with English. The language translation is necessary as that is the language the perinatal women speak. Of those screened, 37 were found positive for depression. Fifteen of the

37 women did not participate in the study because of different reasons (e.g., in need of immediate treatment for depression). At last, 22 perinatal women with depression were invited for the interview.

Purposive sampling plus self-selection was used in recruiting the 22 perinatal women participants for the study. We discussed purposive sampling, a nonrandom sampling method, in Section 2.5.2. Self-selection is also known as volunteer sampling where participants volunteer to take part in a study when asked. In the first step of participant recruitment, only those perinatal women who showed interest in the study were invited to proceed to the next step. This practice actually satisfies the voluntary consent requirement for research involving human participants, thus becoming an ideal scenario for obtaining voluntary consent from research participants. Purposive sampling was again used for recruiting the ten maternal health care workers for the study by following a consenting process. Most, if not all, interview studies adopt nonrandom sampling methods (e.g., purposive) rather than random sampling in selecting participants because of the nature of interview research and a limited number of participants involved.

All the 32 interviews were conducted by the first author of the report and two research assistants who are not listed as coauthors but do have nursing experience in mental health from November 2017 to April 2018, a time span that is atypically long for most non-ethnographic research. The 22 interviews with the perinatal women were carried out using Chichewa, the local language, in quiet, private rooms within the waiting areas at the two selected sites, while the 10 interviews with the maternal health care workers were performed in English in their offices. The interviewers took notes of nonverbal data about the participants. All interviews were audio-taped and later transcribed. In addition, the interviews with the 22 perinatal women were translated from Chichewa into English. This is absolutely required when more than one language is used by participants. All the 32 participants were provided with refreshments. The researchers also gave an equivalent of 4.5 USD as transportation and lunch reimbursement to the 22 perinatal women participants, which is a generous amount when the annual average take-home earning in Malawi is an equivalent of around 7608 USD[1] in 2019.

In the research report, the authors stated that "[w]e conducted 22 in-depth interviews of perinatally depressed women ... until we reached data saturation ..." (Ng'oma et al. 2019, 4). Data saturation is reached when the interviewer begins to hear the same comments again and again. That is a point where the interviewer can stop data collection and start data analysis. In the present case, however, the number of interview participants was determined before the interviews instead of when the interviews were still in process. Data saturation is therefore mentioned in the report more for decoration than for application. Nevertheless, data saturation is often utilized in, for example, ethnographic interviews to decide when the interview data collection process can end.

This interview study is exemplary in almost every aspect of the implementation process: starting from the development of the two interview guides and pretesting each, to the recruitment of research participants and performing all the interviews. How the collected interview data were analyzed is beyond the scope of this book. But that topic of analyzing qualitative data obtained from the interview and other techniques will be discussed in Chapter 13, Techniques for Data Analysis and Their Selection.

**Chapter Summary**

- An interview is a technique for data collection through which the researcher asks each participant questions relating to a research problem and the participant will in turn answer the questions asked.
- Interview questions are mostly open-ended, which can be further categorized as main, probe, and follow-up questions. Main questions are worded to cover the entire research problem. Probes ensure that unclear answers are explained and questions are fully answered, while follow-up questions are for tracking down missing information and for addressing new points the participant brings up.
- There are three kinds of interviews: structured, semi-structured, and unstructured. They all depend on how interview questions are prepared and if probe and/or follow-up questions are allowed.
- Interviews can be conducted face-to-face, by phone, via social media app, or video conferencing platform. Interviewing face-to-face remains to be the preferred approach.
- An interview commonly includes an introduction, warm-up, question-and-answer between the researcher and participant, cool-off, and closure.
- Plan carefully, prepare adequately, and implement rigorously, and make sure to collect nonverbal data during interviews while keeping in mind the other dos and don'ts for developing an interview guide and performing an interview.

## Note

1 Malawi Salary. 2020. "Net Salary." In *The Malawi Average Salary Survey*. Accessed March 22, 2020. https://www.averagesalarysurvey.com/malawi.

## References

African Youth Alliance. 2005. *Youth-Friendly Services: Botswana End of Program Evaluation Report*. Watertown: Pathfinder International.

Bell, Andrew J. 2005. "Oh Yes, I Remember It Well!" Reflections on Using the Life-Grid in Qualitative Interviews with Couples." *Qualitative Sociology Review* 1 (1): 51–67. https://doi.org/10.1038/35098141.

Berg, Bruce L. 2009. *Qualitative Research Methods for the Social Sciences*. 7th ed. Boston: Allyn and Bacon.

Boyce, Carolyn, and Palena Neale. 2006. *Conducting In-Depth Interviews: A Guide for Designing and Conducting In-Depth Interviews for Evaluation Input*. Pathfinder International Tool Series, Monitoring and Evaluation – 2. Watertown: Pathfinder International. Accessed July 3, 2023. https://tinyurl.com/yc7ef52b.

Briggs, Charles L. 2000. "Interview." *Journal of Linguistic Anthropology* 9 (1–2): 137–40.

Chu, Heting, and Qing Ke. 2017. "Research Methods: What's in the Name?" *Library & Information Science Research* 39 (4): 284–99. https://doi.org/10.1016/j.lisr.2017.11.001.

Fontana, Andrea, and James H. Frey. 2000. "The Interview: From Structured Questions to Negotiated Text." In *Handbook of Qualitative Research*, edited by Norman K. Denzin and Yvonna S. Lincoln, 645–72. 2nd ed. Thousand Oaks: Sage.

Frankfort-Nachmias, Chava, David Nachmias, and Jack DeWaard. 2015. *Research Methods in the Social Sciences*. 8th ed. New York: Worth Publishers.

Gorman, Gary E., and Peter Clayton. 2005. *Qualitative Research for the Information Professional: A Practical Handbook*. 2nd ed. New York: Neal-Schuman.

MaCann, Terence V., and Eileen Clark. 2005. "Using Unstructured Interviews with Participants Who Have Schizophrenia." *Nurse Researcher* 13 (1): 7–18. https://doi.org/10.7748/nr2005.07.13.1.7.c5996.

Ng'oma, Mwawi, Samantha Meltzer-Brody, Ellen Chirwa, and Robert C. Stewart. 2019. "'Passing Through Difficult Times': Perceptions of Perinatal Depression and Treatment Needs in Malawi – A Qualitative Study to Inform the Development of a Culturally Sensitive Intervention." *PLoS ONE* 14 (6): e0217102. https://doi.org/10.1371/journal.pone.0217102.

Rubin, Herbert J., and Irene S. Rubin. 2005. *Qualitative Interviewing: The Art of Hearing Data*. 2nd ed. Thousand Oaks: Sage.

Scheuren, Fritz. 2004. *What Is a Survey*. 2nd ed. Alexandria: American Statistical Association. Accessed July 2, 2023. http://people.umass.edu/biostat690c/pdf/ASA%20What%20is%20a%20survey%2068%20pp.pdf.

Spradley, James P. 1980. *Participant Observation*. New York: Holt, Rinehart and Winston.

Wildemuth, Barbara M., ed. 2017. *Applications of Social Research Methods to Questions in Information and Library Science*. 2nd ed. Denver: Libraries Unlimited.

Yin, Robert K. 2016. *Qualitative Research from Start to Finish*. 2nd ed. New York: The Guilford Press.

# 5 Observation

**Chapter Overview Questions**

- What is observation?
- What are the major observation types?
- What can be observed in research?
- What are the three forms of observations done in research?
- What are the major dos and don'ts in conducting observation research?

## 5.1 What is observation?

We all spend a large amount of time observing the world around us. In that sense, observation is familiar to us even if we have never considered it a way of doing formal research (Fox 1998, 2). When observation is utilized in research, it becomes a method for gathering data via carefully and attentively watching and systematically making notes on what is being studied. "Observers neither manipulate nor stimulate their subjects. They do not ask them research questions, pose tasks for them or deliberately create new provocations" (Adler and Adler 1994, 378). Observers are likely not only to employ all their five senses in data collection but also to draw on their broad common sense or cultural knowledge. These features of observation effectively separate it from other research methods such as questionnaire (see Chapter 3) and interview (see Chapter 4). It appears apparent that observation is an activity that we do on a regular basis, and yet it has special requirements when we adopt it as a research method.

First, observation in research is beyond simply using our five senses, the faculties of sight, smell, hearing, taste, and touch. We should also process what is received by our senses and make sense of it based on our experience, common sense, related knowledge, and the research problem under investigation.

Second, observation is more than just recording what is observable at the site. Attention should likewise be given to any tacit aspects of what is being observed (e.g., what is not said or done, hidden relationship among participants) when taking observation notes.

Third, observation should be done in context rather than in isolation. The environment in which the observation is made and the context in which participants' activities and behavior occur all need to be considered when we use this method for data collection.

In addition, observation allows researchers to observe from their own perspectives. Researchers can observe and record what occurs in real time without the need for

DOI: 10.4324/9781003021711-7

participants to recall from memory as in doing questionnaire and interview research. Researchers can also observe what is happening in the natural setting and in context, which enriches the data collected. Observation can be used for verifying what people say with what they actually do, which is exactly Northcote and Livingston (2011), one of the case studies selected for this chapter, explore in their study. It permits the study of people who are unable (e.g., very young children, senior citizens) or unwilling (e.g., those who do not want to take part in questionnaire or interview research) to provide reports of their own activities. Furthermore, observation can lead to serendipity, enabling researchers to observe what is unexpected.

Observation, on the other hand, raises concerns in application as a research method (Gorman and Clayton 2005, 104–5; Morgan et al. 2017). Many behaviors are intermittent (e.g., information desk at an unpopular tourist site) or hard to be observed (e.g., participants' decision-making processes). Observable events could occur during the researcher's absence. The context of behaviors (e.g., participants' background and experience), though relevant, might be hard to be captured via observation. Being observed could be bothersome and prone to the Hawthorne effect. Among all the concerns in doing observation research, ethical concerns always stand out when researchers have to keep a delicate balance between conducting a study and respecting participants' privacy or personal space at the same time. Observation is time-consuming and might involve subjectivity of the observer as well. It may cause inconvenience to those who happen to be at the observation site.

Despite those concerns, observation has been regarded as the fundamental base of all research methods in the social and behavioral sciences (Adler and Adler 1994, 389). Observation can be applied alone, as part of another research method (e.g., observing nonverbal language in interview), or in conjunction with other research methods (e.g., used together with such methods as interview and research diaries in ethnography or field study). This chapter is devoted to examining observation as a stand-alone method, while its imbedded and integrated usage is covered in related chapters of this book.

## 5.2 Observation types

When observation is chosen as a research method, several questions must be addressed before the study begins. First, to what degree shall researchers get involved with people and in activities they observe? Different degrees of involvement define different types of observation, from nonparticipant observation to participant observation (Spradley 1980, 58–62). Second, shall the observation be made structurally using an instrument of predetermined categories to guide the recording process (Glazier 1985)? If the answer is yes, it is called structured observation. Otherwise, it is unstructured observation. Third, shall the observation be done obtrusively or unobtrusively? This question introduces another criterion for categorizing observations, namely, obtrusive observation or unobtrusive observation. Table 5.1 lists a taxonomy of observations according to the three categorization criteria.

As denoted in Table 5.1, nonparticipant observation means that researchers take no part in what they observe. A typical example of nonparticipant observation is for researchers to use a one-way mirror in a specifically arranged observation room where the observer can see what goes on there but not vice versa. Nonparticipant observation is also possible without a one-way mirror as long as researchers stay unnoticed and unobtrusive

*Table 5.1* Taxonomy of Observations

| *Categorization Criterion* | *Category Value* | *Observation Type* |
|---|---|---|
| Degree of Involvement | High | Complete Participant |
| | ↕ | Active |
| | | Moderate |
| | Low | Passive |
| | No | Nonparticipant |
| Structured? | Yes | Structured |
| | No | Unstructured |
| Obtrusive? | Yes | Obtrusive |
| | No | Unobtrusive |

in the observation process. Nonparticipant observation is often done in experimental psychology, market research, usability studies, and other pertinent areas. Participant observation, by comparison, consists of more subcategories depending on researchers' degree of involvement in the observation. We will elaborate on this topic in Section 5.2.1.

According to Gorman and Clayton (2005, 103–4), "Structured observation samples a predetermined event or activity, using a prearranged instrument or form into whose categories the observer records whether specific activities take place, when and how often". If an instrument of pre-determined categories is employed for guiding the data recording process, the observation is structured. All other observations without such an instrument belong to the unstructured type. Further discussion on structured observation is provided in Section 5.2.2.

Obtrusive observations mean that observations are done overtly, and participants know they are being observed. Therefore, researchers need to be extra vigilant about not being intrusive in this kind of observation. Although unobtrusive observations are less intrusive and made covertly, researchers doing so should still adhere to the principle of respecting participants' privacy and other rights. Furthermore, unobtrusive observations for the most part remain in the realm of the surface level (Berg 2009, 171–72) because unobtrusiveness in observation limits researchers' ability of gathering in-depth data. In addition to observing human beings and human activities unobtrusively, we can also adopt the same method to observe physical traces for data collection (Berg 2009, 287–89). Such physical traces include both erosion measures (e.g., wear and tear of furniture, remainder of a hand sanitizer bottle) and accretion measures (e.g., fliers on the wall, trash community residents throw away). All of them can enable the observer to gather certain data needed in a study. For example, observing the wear and tear on furniture can determine if the furniture is frequently used. Attentive observers should exploit those unobtrusive measures in their research.

After getting familiar with the major types of observation depicted in Table 5.1, it is time for us to consider which type is suitable for our research. Generally speaking, our research problem is the main factor that determines the observation type to be selected. Then the selection chiefly focuses on the first two groups of observation type (i.e., participant versus nonparticipant, structured versus unstructured) in Table 5.1. Besides the individual type (e.g., participant, structured) in the groups, all combinations of any two cross-group types listed in Table 5.1 would constitute an additional observation kind (e.g., participant unstructured observation) as shown in Figure 5.1.

Figure 5.1 shows that there are in total eight different kinds of observation, four individual (e.g., participant, structured) and four amalgamated (e.g., participant-unstructured,

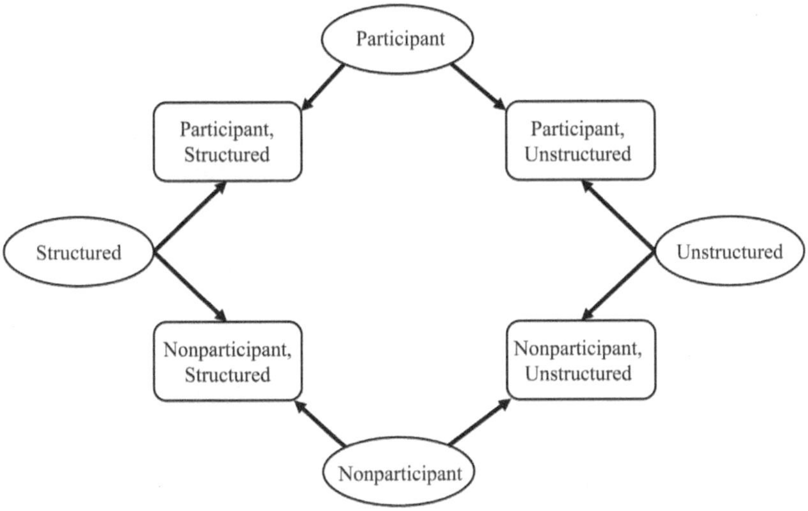


*Figure 5.1* Major Observation Types

nonparticipant-unstructured), although we tend to emphasize either the "participant" or "structured" option among all the eight available choices because of their prominent features. That is the very reason why those two kinds of observation will receive additional coverage in Sections 5.2.1 and 5.2.2, respectively. As for the obtrusive versus unobtrusive group, we are certain that participant observation is obtrusive, while nonparticipation observation is mostly unobtrusive. However, an observation can be either obtrusive or unobtrusive, regardless of whether it is structured or unstructured.

### *5.2.1 Participant observation*

Participant observation is a collective name that can be further categorized by the extent of researchers' involvement in what is observed, forming a continuum from passive participation to complete participation (Gorman and Clayton 2005, 106; Spradley 1980, 58–62).

- Passive participation – The researcher is present at the site of observation and takes notes but stays unobtrusive and avoids interacting with participants. Passive participation usually takes place in public places that are easily accessible to everyone (e.g., bus stops, libraries, parks, shopping malls). Nash (1981) reported an observational study in which the research team members explored the effect of a prolonged winter on selected aspects of urban life in the Twin Cities of Saint Paul and Minneapolis, Minnesota. They made observations in four kinds of outdoor public places: parks, streets including bus stops, ski resorts, and recreation centers with skate rings. What the research team members did at those public places was exactly passive participant observation. They did go to parks, wait for and take buses, and so on. However, they did not proactively engage or interact with other people at those sites. If researchers are approached in passive participant observation (e.g., being asked for directions), they would respond. Passive participant observation can also take place when the researcher does not hold the qualification to take part in what is observed (e.g., accountants, surgeons). Unlike the passive participant observation made in public places, observation done in non-public spaces (e.g., an office) is typically obtrusive.

- Moderate participation – In moderate participant observation, researchers try to balance their roles as an observer and a participant at the same time. The researcher occasionally joins the activities and events while performing observation and taking observation notes. For example, the researcher watches a football game with friends, cheering for their favorite players and commenting on the game. Meanwhile, the researcher observes how friends interact with each other during the game-watching and makes notes accordingly.
- Active participation – The observer in this case possesses the right qualifications (e.g., knowledge and skills) to fully take part in what is observed but is not a natural member or participant in the setting. A relevant example is electronic sports (aka esports), in which people compete by playing video games individually or as a team. In active participant observation, the researcher knows how to play esports and joins one team as a guest member in order to carry out observation. The researcher participates in esports competition as a team member while making observations. The note-taking, however, cannot be done at the site and has to be completed right after the competition is over.
- Complete participation – Complete participation is similar to active participation except that the researcher is already a natural member or participant in the setting. Taking the same example of esports, the researcher in complete participant observation is already a team member before making observation and consequently does not need, as in active participant observation, to get accepted and acquainted with other team members or go through any procedures to get settled before playing esports. This latter part comprises the first several steps in conducting ethnographic or field research, a topic for Section 11.3.1. A real-life example of complete participant observation can be found in Nash (1975), who rode several bus lines of the Metropolitan Tulsa Transit Authority (MTTA) for two years for the purpose of commuting. The author observed, among other things, operation of the transit system, bus schedules, bus stops, bus scenes, bus riders, the bus driver, where to sit, interactions among riders as well as between the bus driver and riders, and encounters between regular riders and new riders. Nash even observed bus riders' activities, feelings, and thoughts ordinarily associated with communities or stable groups.

While your research problem would help you decide how involved you should be in your participant observation, the characteristics of each participant observation subtype need to be taken into account as well in the selection process. Participant observation, as one kind of observation, is unique in that observation is done differently depending on the extent of researchers' involvement in the process.

In addition to the four subtypes of participant observation described above, there is one more participant observation called the Tavistock-Bick model, which was initially conceived and developed at the Tavistock Clinic in London, UK, by child psychoanalyst Esther Bick (Datler et al. 2014). The Tavistock-Bick observation is designed as one component of training courses for psychoanalytic clinicians. Students learning the observation method observe infants from their birth to two-year old for one hour every week when the infants are at home with at least one parent. Students also attend a seminar, facilitated by a seminar leader, over two academic years to present and discuss their weekly observation reports with peers (Brown 2006). The use of this observation model is occasionally extended to nursing homes and other settings.

No matter which type of participant observation the researcher may choose, how to play the dual-role of observer and participant at the observation site is a challenging yet imperative issue. We will discuss this topic in depth in Chapter 11, Meta-methods: Ethnography/Field Study, as participant observation is widely used as one major method in ethnographic or field research.

### *5.2.2 Structured observation*

Structured observation has three distinctive features (Martinko and Gardner 1985, 676): "(a) the method relies on observation by a person other than the subject; (b) the method must rely on the use of category systems; and (c) the method does not use randomized activity sampling procedures". The first feature of structured observation shows that the researcher cannot be a participant by wearing two hats in data collection, while the second characteristic clearly highlights the observation structure via using an instrument of pre-determined categories to guide the recording process. Such pre-determined category systems serve as observation guides or observation protocols (Stausberg 2011, 384). The third point Martinko and Gardner (1985) encapsulate above nonetheless is not merely true to structured observation. Rather, most observation of human participants and activities, structured or otherwise, cannot be done by adopting a random sampling procedure due to the very nature of such observation. A quick note is necessary about the aforementioned first point of Martinko and Gardner (1985), namely, the observer cannot be a research subject or participant in structured observation. However, the researcher should still be able to participate in what is observed in structured observation as long as s/he remains a passive participant because passive participant observation is one of the four subtypes of participant observation (see Section 5.2.1).

While having a guide for data collection is a must in structured observation, how to acquire one becomes an essential part in research. Compared to questionnaire or interview guide development, the creation of an observation guide is easier and more straightforward because such instruments commonly take the form of columns and rows, listing all the categories that should be observed and recorded in the data collection process. These categories in the observation guide are devised and selected according to the research problem under consideration. The observation guide can be either developed from scratch or selected and adapted from one used in a prior related study. If the former, more efforts are needed by conducting a range of activities from identifying mutually exclusive and collectively exhaustive categories to testing the validity of the complete observation guide. If the latter, the focus is placed on selecting a guide that serves the purpose of the research project at hand and making some adaptations if necessary. A structured observation study may need more than one observation guide, as in research done by, for example, Mueller, Volery, and von Siemens (2012) – one of the two case studies for this chapter.

Table 3 in Mueller, Volery, and von Siemens (2012) presents a structured observation guide. It mimics the observation template from an earlier study. The authors observed entrepreneurs' workday behavior in their start-up and growth stages. The observation guide includes categories such as description of action, participants involved, and location. The other observation guide in the same study (see Table 4) is created by Mueller, Volery, and von Siemens (2012) themselves for recording categories and subcategories of observed entrepreneurs' work content. We will examine both guides closely in Section 5.5.

Structured observation, on the other hand, does not prohibit the researcher from making detailed or extensive notes beyond the categorized items already listed in the observation guide. Some pitfalls that researchers should be cautious about in structured observation include observer fatigue and observer drift (Stausberg 2011, 390). As structured observations normally mean continuous monitoring and recording for long hours (e.g., around 11 hours on average per day in Mueller, Volery, and von Siemens 2012), observer fatigue appears to be a side effect of the long observation process consequently. Observer drift refers to gradual, systematic changes over a period of time made by observers in their application of criteria for recording observation. As a result, they may use certain categories more than others or choose a different category after learning more about what is observed in their structured observation. These pitfalls can be mitigated by scheduling observation hours realistically, having clearly defined observation categories, and active collaboration between the researcher and participants. Moreover, data collected from pure structured observation can look fragmented and decontextualized, which demonstrates the need for using structured observation together with other kinds of observation (e.g., participant observation) and data collection techniques (e.g., interview).

## 5.3 Using observation in research

Observation is a complex research method because it often requires researchers to play a number of roles and to use a number of techniques, including their five senses, to collect data (Baker 2006, 172). We are going to discuss in this section how to use observation in research generally without linking observation solely to ethnographic or field research as numerous publications do. The rationale for this arrangement is twofold: (1) Observation is not exclusively utilized in ethnographic or field research only, and (2) Chapter 11 is devoted to ethnography and its synonymous field study.

### *5.3.1 Observation preparation*

Because of the complexity in observation, preparation is required rather than optional. Most observation research involves people and their activities. Observation preparation thus starts with selecting participants and sites for the study. Various techniques (e.g., using personal and professional connections) should be applied to get permission for accessing chosen sites and obtain consent from research participants. Unlike interview, observation is done at the participant's location for a prolonged period of time and the location is a target of observation as well. Therefore, all the ethical concerns must be considered when preparing for observation.

Observation preparation should also include a sampling scheme for date and time. Usually, a purposive sampling technique is chosen for determining the appropriate dates and time periods that would enable researchers to observe and gather what is needed for their research. For instance, Mueller, Volery, and von Siemens (2012) observed each of the 12 entrepreneurs in their research for four full work days, averaging around 11 hours per day. The four days were suggested by the entrepreneurs, not necessarily consecutive but representative of their workday activities. Observation date and time can be selected using a systematic sampling method sometimes (e.g., in a laboratory setting). Such observation is called systematic observation, in which strict time intervals trigger the relevant sample of observation periods (Yin 2016, 139).

In Section 5.2.2, we describe how to develop or adopt observation guides for structured observation which is part of research preparation. Although other types of observation (e.g., participant observation) may not need an observation guide, researchers should still make a plan for what and how to do their observation. Such practices are highly recommended or even required for researchers doing observations, especially for those who are new to this method.

### *5.3.2 What to observe*

Broadly speaking, anything, anyone, or any activity at the observation site can be a target for observation. Yin (2016, 152–53) groups observable items as follows:

- Characteristics of individual people, including their dress, gestures, and nonverbal behavior;
- Interactions between and among people;
- Actions taking place, whether human or mechanical;
- Physical surroundings, including visual and audio cues as well as props like wall hangings, posters, plaques, books on bookshelves, furniture, and other items associated with a specific participant or that participant's organization.

Spradley (1980, 78) identifies nine dimensions for researchers to observe in participant observation:

- Space: the physical place or places
- Actor: the people involved
- Activity: a set of related acts people do
- Object: the physical things that are present
- Act: single actions that people do
- Event: a set of related activities that people carry out
- Time: the sequencing that takes place over time
- Goal: the things people are trying to accomplish
- Feeling: the emotions felt and expressed

Of the nine dimensions, *act, activity,* and *event* are activities aggregated with different granularity. In contrast to other dimensions, *goal* and *feeling* are hard to observe. While *feeling* might generate behavior that becomes observable, data on *goal* need to be gathered via techniques such as interview. *Time*, together with *goal* and *feeling*, is not included in the observable items Yin (2016, 152–53) enumerates. Fox (1998, 12) suggests *reflection* as the 10th dimension for recording the observer's response to any of the nine on Spradley's list above. Based on the preceding elaboration, the following seven categories of entities can be observed for data collection if possible:

- Space – the physical place or places
- Actor – the people involved, their associated characteristics, and behavior
- Activity – the things people do, including interactions between and among people
- Object – the physical things that are present
- Time – the sequencing that takes place over time
- Feeling – the emotions felt and expressed
- Reflection – the responses to any of the above

A matrix can be created for the afore-listed seven categories of items to generate questions for data collection by following the template of the 9 × 9 matrix Spradley (1980, 82–83) develops with his nine dimensions of observation. Each cell in the matrix shows a sample question that can be asked during observation to facilitate data collection. Table 5.2 displays a matrix of sample observation questions with the cells in the lower half of the matrix filled. The upper half of the matrix simply mirrors the lower half and thus is not explicitly listed.

As indicated earlier, the questions in each cell of Table 5.2 are just suggestions. Researchers can create their own when considering two or more of the seven observation dimensions. Most of the questions in Table 5.2 are related to two dimensions, one from the column heading (e.g., *actor*) and the other from the row heading (e.g., *activity*). In the question of "*When* do *actors* show *feelings*?", for example, three dimensions will be observed: *time*, *actor*, and *feeling*. *Reflection* on any of the other six dimensions (i.e., *space, actor, activity, object, time,* and *feeling*) should only be noted using keywords, shorthand, and the like at the observation site so that the researcher's attention can be focused on what is happening. Detail notes of reflection should be made immediately after the observation is over.

It must be pointed out that the matrix is used to merely help suggest what can be observed in general. Researchers would refer to it when needed in their observation study. In no case does it imply that observers should cover all the questions in the matrix in one study. What should be observed in a study is always guided by its research problem.

### *5.3.3 How to observe*

Before embarking on an observation project, the researcher should have a plan that includes essentials for how to observe because "simply 'hanging around' in an unfocused way can be notoriously time-consuming, unproductive, exhausting and sometimes embarrassing or risky" (Mason 2002, 90). If an observation is structured, for instance, the researcher should have a guide ready. The guide contains a list of pre-selected categories defining what is to be observed.

Observation is always done at the site, regardless of whether the site is a laboratory or a field. It is therefore crucial for researchers to select an observation post or position themselves carefully and strategically in order to monitor and record what needs to be observed without being intrusive. Where to position oneself is an issue that the researcher must address in unobtrusive observation as well as obtrusive observation. Glazier (1985), for example, finds that the most ideal position for making structured observation in his study is behind the participant and slightly to the right. Naturally, it is likely that the observer may need to change locations during the data collection process, especially in participant observation. But the location should be chosen for facilitating observation while minimizing intrusiveness.

This is also the time for researchers to use all of their five senses for observation, which means that they will do all the activities, such as listening, watching, communicating, interacting, and reflecting, whenever appropriate. There are three forms of observation in the data collection process by observation purpose: descriptive observation, focused observation, and selective observation (Spradley 1980, 33).

Descriptive observation is made with the intention of getting an overview of the site and what goes on there in general. In reference to Table 5.2, researchers inspect the *space,*

*Table 5.2* Matrix of Sample Observation Questions

| | *Space* | *Actor* | *Activity* | *Object* | *Time* | *Feeling* | *Reflection* |
|---|---|---|---|---|---|---|---|
| **Space** | What is present at all the places? | | | | | | |
| **Actor** | Where do actors place themselves? | Who are all the actors? | | | | | |
| **Activity** | What activities happen at places? | What activities do actors carry out? | What are all the activities? | | | | |
| **Object** | Where are objects located? | How are objects used by actors? | How are objects used in activities? | What are all the objects? | | | |
| **Time** | How do places change over time? | When do actors show feelings? | When do activities occur? | What effects does time have on objects? | What are all the time periods? | | |
| **Feeling** | What impact do places have on feelings? | What feelings do actors have? | How do feelings affect activities? | What feelings are caused by objects? | What impact does time have on feelings? | What are all the feelings? | |
| **Reflection** | What reflection can be made about places? | What reflection can be made about actors? | What reflection can be made about activities? | What reflection can be made about objects? | What reflection can be made about time? | What reflection can be made about feelings? | What are all the reflection? |

*actor, activity,* and *object* dimensions in their descriptive observation. Sample questions to be asked in this form of observation include the following:

- Who are at the setting?
- What activities are taking place there?
- What objects are there at the setting?

Spradley (1980, 76) further breaks down descriptive observation into grand tour and mini-tour observation while making the following analogy of these two observation sub-forms: "Every grand tour observation is like a large room with numerous doors into smaller rooms, each door to be opened by a mini-tour question-observation" (Spradley 1980, 79). Grand tour observation is for gathering data about the target site on the whole with questions such as "What is going on in the cafeteria?". In contrast, mini-tour observation deals with particular aspects of something already observed by asking questions like "What kinds of food are offered there?".

Focused observation is performed to gather in-depth data on any of the seven dimensions chosen to be examined in a study. It could be a deeper look at the various aspects (e.g., one kind, one part) of a dimension. For example, "Is the young engineer who just entered the room a new member of the team"? Focused observation can also be used to gather data on reasons, consequences, and the like at the site. Sample questions could be: What causes the long waiting line at the cafeteria? What happens when the waiting line gets longer and longer? The researcher may consider specific aspects of a dimension listed in Table 5.2 (e.g., space, actor, activity, object, and time) in focused observation. As indicated when discussing what to observe in Section 5.3.2, we can each time go beyond observing one or two dimensions listed in the matrix (see Table 5.2). We can instead observe three or even four dimensions simultaneously to gather in-depth data about interrelationships among them (e.g., "What effects does Activity A have on the actors at this site"?). Observation can become fairly complex even if only two or three of the seven dimensions (i.e., space, actor, activity, object, time, feeling, and reflection) are involved. Focused observation, thus, can help researchers concentrate on what they should observe. The descriptive observation, performed before focused observation and guided by the research problem, should assist observers in selecting focuses in the data collection process.

Selective observation is made after most descriptive observation and some focused observations are done, aiming to gather additional data to clarify or supplement what is already collected. Compared with descriptive and focused observations, selective observation is at the end of a funnel for collecting observation data (see Figure 5.2).

Spradley (1980, 128) likens the data collection process using the three forms of observation to a funnel, which is plain but informative. It is through selective observation that researchers collect what is missed, overlooked, or hidden during descriptive and focused observations. Selective observation also makes the process of collecting observation data complete. For example, if we plan to observe a five-member team working on a renewable energy project but one of them is absent for personal reason on the day for observation, we go ahead with the scheduled observation while making additional observation after that particular team member returns to work. This is selective observation. Selective observation is needed when, for example, participants' attendance becomes impossible unexpectedly, data collected via descriptive and focused observations are found unclear and/or incomplete, as well as when other similar circumstances occur.

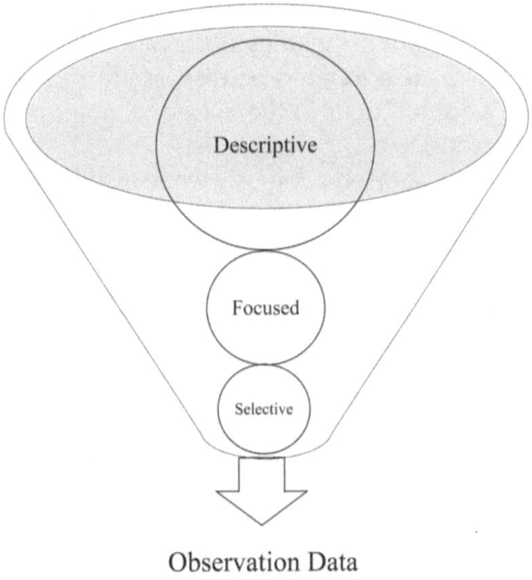


*Figure 5.2* Funnel of Descriptive, Focused, and Selective Observations

One question may arise at this point about how to allocate time to the three forms of observation. Spradley (1980, 108) provides an illustrative answer to this question. Although the context of Spradley's illustration is participant observation in ethnography, what is presented in that figure seems quite applicable to observation in general because participant observation, after all, is one category of observation. According to that illustration, we can estimate that the approximate percentage of time for doing descriptive, focused, and selective observations is, respectively, 66, 20, and 14. That is, about 66% of total observation time is spent on descriptive observation, and so on. As for the sequence of these three forms of observation, their starting time is staggered in the order of descriptive observation, focused observation, and selective observation. The percentage of time spent on descriptive observation decreases, that on focused observation remains roughly the same, and that on selective observation increases as the observation process progresses. The quantitative delineation of these three forms of observation, however, serves only as a sketch about their sequencing and time allocation in the data collection process. The exact implementation of them in a particular observation study should ultimately be determined by the research problem. Furthermore, few observation studies would specifically report on these three forms of observation, which is confirmed in both case studies selected for this chapter (Mueller, Volery, and von Siemens 2012; Northcote and Livingston 2011).

How to record what has been observed represents a vital part in observation research since only recorded observation data can be utilized for analysis later. Note-taking is the mainstay in recording observation data, while the observation guide with pre-defined categories functions as an effective data collection instrument in structured observation. Videos, audios, and photographs are also helpful in recording observation data to supplement observation notes taken, although we should be aware that none of those aids can capture the entire observation process like the notes.

Taking notes during observation needs attention and efforts. Yin (2016, 163–83) details recording data in field study, which includes observation as one data collection technique. As recording field notes is a topic to be discussed in Chapter 11, Meta-methods: Ethnography/Field Study, we shall focus in the current chapter mainly on the fundamentals of note-taking in observation.

First, observation notes must be taken copiously while avoiding writing down everything. Always capture and note down what is related to the study at the observation site if possible. Second, take observation notes with the language used by participants and appropriate to the observation. For example, recording data using children's language if the observation is made at elementary school classrooms. Avoid using jargons or language that would not be familiar to the participants. Third, follow the verbatim principle (Yin 2016, 167) by recording word-by-word of what is said at the time of observation. Avoid summarizing and paraphrasing essentials in dialogs and conversations. Fourth, be concrete and objective in taking observation notes to "record a vivid image rather than a visual stereotype" (Yin 2016, 166). Fifth, avoid interpreting and reflecting on what is observed at site due to the time constraint except for making a quick note or jotting down a few keywords that serve as a reminder for late contemplation. Sixth, use sketches, drawings, or diagrams to help record accurately and comprehensively what is observed. The English adage "a picture is worth a thousand words" holds true in cases where you want to depict an observation scene, seating of a group discussion, a series of activities, and the like. Seventh, proper procedures should be pursued (e.g., getting consent from participants, transcribing video and audio recordings as soon as possible) if videos, audios, or photographs are chosen to help record observation data. Photographs are especially a valuable means for capturing a snap-shot of an observation scene, which can help researchers to recall what is observed and add details to their notes. When smart phones become an indispensable gadget of most or all researchers, taking photos during observation appears to be a preferred technique for recording data and may replace the sketches or drawings briefly mentioned above.

Ideally observation notes are taken while the observation is ongoing, although such practices may not be feasible sometimes because of multiple reasons (e.g., concerns about intrusiveness, the observer being a participant). In those occasions, researchers should do the note-taking right after the observation is over. Furthermore, the notes taken early should be reviewed and refined afterward to improve their quality. This is also a suitable time to reflect on all that is observed. Although what is discussed above on note-taking is primarily for unstructured observation, those note-taking techniques are relevant to studies using structured observation which often needs to record data beyond pre-developed categories in the observation guide or from the so-called open-ended observation (Mueller, Volery, and von Siemens 2012, 999–1000).

## 5.4 Dos and don'ts in observation research

As with any other research method, investigators who wish to use observation for data collection need to learn and practice it in order to apply it properly and effectively in research. In this section, we attempt to outline some dos and don'ts in observation.

- Planning – Plan ahead in choosing observation sites and participants, as well as in selecting observation types (see Figure 5.1) based on their respective features and the research problem. Avoid doing observations without planning because such practices will result in any outcome but achieving the intended research objective.

- Preparation – Make preparation for what and how to observe. If structured observation is chosen, develop and/or adopt one or more observation guides for data collection. Train observers with no prior experience if people other than the researcher are recruited to help with data collection because observation skills in research can only be learned through apprenticeship or training at site (Baker 2006). Northcote and Livingston (2011), one of the case studies for this chapter, described how they trained their observers before the research began. Never perform an observation without any preparation, as that will affect both the validity and reliability of the observation study.
- Observation – Ease oneself into the context, and place oneself carefully at the observation site. If participant observation is chosen, maintain a suitable level of involvement in observation as planned. Don't forget about one's own role as observer. Use descriptive, focused, and selective observations intelligently, seamlessly, and flexibly. Perform observation with all five senses. Don't overlook the context and other specifics at the observation site. Don't lose one's focus throughout the observation process.
- Note-taking – Quite a few dos and don'ts for note-taking are already presented above. Record what is observed as copiously, appropriately, objectively, and concretely as possible while exploiting videos, audios, photos, sketches, and/or other supplemental recording aids when needed. Don't interpret what is observed on the site in order to stay focused on observation.
- Other related points – Dress and behave appropriately. Be approachable and friendly. Uphold research ethics in observation research by obtaining permission and consent for observation while staying nonintrusive. Try to become invisible without jeopardizing ethical concerns. Glazier (1985) suggests several techniques to minimize intrusiveness in observation that include avoiding eye contact, keeping one's head down to focus on filling out the data collection instrument in structured observation or concentrate on taking notes during meetings, and positioning oneself less visible to the participant if possible.

What has been discussed so far, of course, does not exhaust all the dos and don'ts in observation. However, the observation study will be fruitful and run smoothly when those major dos and don'ts are followed whenever applicable.

## 5.5 Case studies of observation research

Two studies are to be examined closely in this section, one using participant observation and the other employing structured observation for data collection.

### *5.5.1 Case Study 1*

Northcote, Jeremy, and Michael Livingston. 2011. "Accuracy of Self-Reported Drinking: Observational Verification of 'Last Occasion' Drink Estimates of Young Adults." *Alcohol and Alcoholism* 46 (6): 709–13. https://doi.org/10.1093/alcalc/agr138.

The research by Northcote and Livingston (2011) was conducted to verify if self-reported alcohol consumption by young adults (aged 18–25 years) is accurate when compared with observation data. As Northcote and Livingston needed assistance in data collection, they purposively sampled and recruited 11 students as observers, being referred to as fieldworkers in the report, for this study from three universities in Perth, Australia. A questionnaire and follow-up interview were also used in recruiting observers by asking about the drinking levels of their friends. Pay attention to the two points in the selection of observers. First, like most observation studies, the purposive sampling technique is chosen for recruiting observers who subsequently "recruited" their friends as participants for the research project. Second, a questionnaire with a follow-up interview were utilized to ensure that the chosen observers had friends who drink. All of the 11 observers, but 1, were students in a degree program in anthropology at the commence of the data collection. The education background of those observers should be helpful in gathering data for the project, as observation is a key research method in anthropology.

The observers received training in a series of workshops, including live observation exercises in pubs and similar settings, before data collection, which is a highly recommended practice because doing observation in research is a skill that can only be learned through apprentice and training on site (Baker 2006). Northcote and Livingston trained their observers to work in pairs, one as a natural member of the studied social group, while the other as a "guest" of the network. Some observers were permitted to work solo based on their own experience and trust from the project leaders. It is desirable to have a pair of observers at site, especially one of the two being a natural member of the studied social group, to ensure the comprehensiveness of data collection and easy access to the observed social group. However, 69% of observation reports were prepared by solo observers (Northcote and Livingston 2011, 710). The authors did not explain explicitly why less than one third of the reports were written by observation pairs. They, however, hinted that, based on post-study interviews, participants' feelings of self-consciousness were greatly diminished when there were only solo observers. This finding indicates the subtle role obtrusiveness plays in observation.

The observers were asked to note down the time, type, and size of each alcohol drink as well as movements in and between venues, social interactions such as conversations and conflicts. The data about the latter points, otherwise unrelated to the present research, were collected for a broader study (Northcote and Livingston 2011, 709). Additional data about the size of alcohol drinks (e.g., leftover drink in cans and glasses, partial refills of unfinished glasses, and sharing of drinks) were also recorded. This level of details in observation proves necessary, giving nuances to the observation data. The observers "tended to focus on a few key participants who were likely to stay together as a group throughout the duration of the session and hence remain in their field of view" (Northcote and Livingston 2011, 710). That is a clever observation strategy to take after they realized that tracking as many participants as possible was not practical because group sizes fluctuated during the course of sessions.

The observers were not permitted to consume alcohol during research, although light drinking did occasionally occur. Due to this requirement, the observation study by Northcote and Livingston more or less fits into the category of moderate participant observation. It is not that important to categorize the degree of involvement in observation as long as observers are able to accomplish their data collection successfully. Sometimes, it is hard to make an accurate categorization of participant observation by level of involvement because what goes on at the observation site perhaps cannot be placed neatly into one of the four established categories: passive, moderate, active, or complete participation. Although the authors presented in the abstract and the text larger numbers of participants and social life occasions, data finally used in analysis were obtained from 48 study participants, observed in 129 sessions that took place in houses (37%), hotels (34%), nightclubs (23%), outdoor events (4%), and other venues (2%, including restaurants, a function center, and an overseas flight). The venues chosen for conducting observations are representative of the places where young adults usually have drinks. Selecting proper sites for observation is one of the major components in observation research design.

The selected participants gave verbal permission to be observed prior to the start of an observation session, a step that should never be overlooked in participant observation, while nonparticipant observation researchers would seek permission indirectly (e.g., from supervisors or group leaders). However, no information was reported in the publication regarding whether the observation data were recorded during or after the sessions. This might be an oversight in the final report. But absence of such data may affect the reliability of the study. The participants were "surveyed" one or two days following the observation session about the number of alcohol drinks consumed. The self-reported data include number of drinks consumed pre- and post-observation as well as during observation. Two brief comments are due on the "survey" technique used in the research. First, this study confirms that observation is rarely used alone for data collection in a research project due to the types of data it can gather. Second, Northcote and Livingston did not specify whether the "survey" was in the form of interview or questionnaire, even though interview was likely the choice based on the description presented elsewhere. But interview, strictly speaking, is not survey, as explained in Chapter 4.

The observations were designed and made methodically to find out the differences between self-reported drink numbers and observed ones. The gathering of additional drink size data (e.g., leftover drink in cans and glasses) and checking with participants for data accuracy are all examples of careful planning and rigorous execution of the study. The study would be more solid if it reports on, for instance, how observation data were recorded.

### *5.5.2 Case Study 2*

Mueller, Susan, Thierry Volery, and Björn von Siemens. 2012. "What Do Entrepreneurs Actually Do? An Observational Study of Entrepreneurs' Everyday Behavior in the Start-Up and Growth Stages." *Entrepreneurship Theory and Practice* 36 (5): 995–1017. https://doi.org/10.1111/j.1540-6520.2012.00538.x.

Mueller, Volery, and von Siemens (2012) observed six entrepreneurs in the start-up stage (i.e., firms incorporated no more than 12 months prior to observation) and six entrepreneurs in the growth stage (i.e., companies of 5–10 years old at the time of observation) to capture the nature of their behavior. Entrepreneurs' behavior is defined as "the concrete enactment of an individual task or activity required to initiate, grow, or transform a business venture" in this study (Mueller, Volery, and von Siemens 2012, 996). There is no mentioning of any observation training as the three authors appear to be seasoned researchers using the observation method. In addition, structured observation is commonly less challenging to be done than participant observation by contrast.

Like the participant observation study by Northcote and Livingston (2011) examined in the previous case study, purposive sampling was again used for choosing the 12 entrepreneurs in the research. This shows the predominant use of purposive sampling in observation research. All the 12 sampled entrepreneurs and their firms operate in the German-speaking region of Switzerland, Austria, and Germany. When choosing participants and sites, we must consider accessibility, language, and other factors to ensure the feasibility of carrying out an observation study. The purposive sampling Mueller, Volery, and von Siemens (2012) did included additional criteria for selecting individual entrepreneurs from each group. For the six start-up entrepreneurs, they were all founders of their companies, whereas the six growth entrepreneurs must be both the founder and manager of an established, expanding business. All these requirements, along with other selection criteria, denote that purposive sampling is made after all related factors are taken into account so that the chosen entrepreneurs are representative of the target population the researchers have in mind even when random sampling is not feasible.

All the 12 entrepreneurs were each observed for four days of work for a total of 542 hours of their activities. The average work hours per day were 11.8 for entrepreneurs in the start-up stage and 10.6 for those in the growth stage, which implies that the same number of hours of observation was done by one of the researchers. The four days, suggested by each of the 12 entrepreneurs, were representative of their day-to-day activities but not necessarily consecutive. The 48 days of observation by three researchers consequently scattered over a period of more than two years, from January 2009 to March 2011, showing that the researchers placed the importance of having representative days for observation over the desire of completing data collection quickly. Prolonged time commitment is one of the outstanding features of observation in that not only researchers have long hours of observation sessions but also the entire observation process may last months or even years. We should never overlook this feature in conducting observation research.

One of the authors, Mueller, Volery, or von Siemens, followed each entrepreneur for a whole day in the study. But they did not mention if the same author observed the same entrepreneur for all four days. The first observation guide (i.e., Table 3) the researchers used is a typical data collection instrument for structured observation, while the second one (see Table 4) appears less obvious because it is for "open-ended observation" to seek certain types of structured data (Mueller, Volery, and von Siemens 2012, 999–1000).

Table 3 in the research report, modeled after an instrument in an earlier study (Mintzberg 1973), represents a common structured observation guide that includes date, starting time, description of action, participants involved, location, scheduled versus unscheduled action, and self-initiating of action. The researchers noted down every action the entrepreneurs took at work using the observation guide on site without specifying if any aid (e.g., audio recording) was used in recording observation data. Mueller, Volery, and von Siemens (2012, 1003) believed that "a new action began whenever the location, the medium of communication, the individuals present, or the focus of the action changed". A total of 4,479 units of actions were observed in the study.

The 4,479 actions, obtained through open-ended observation, form a data set for Mueller, Volery, and von Siemens to create Table 4. Open-ended observation, as opposed to structured observation for gathering data with an instrument of predetermined categories, can yield data that are categorized after observation. Table 4 Categories and Subcategories – Work Content in Mueller, Volery, and von Siemens, as noted, is a product of open-ended structured observation. It can serve as an observation guide for future research on similar topics, although it functions more like a coding schema for categorizing the observed actions collected for this study. The 4,479 observed actions were categorized by content (i.e., activities, function, exploration versus exploitation) and communication aspects (communication status, communication partner, and media) according to the report. However, only the categories of action content are listed in Table 4, while those for action communication aspects are presented in Table 6 together with results because the latter were again adapted from an earlier study. Compared with the questionnaire or interview guide, it is more straightforward to develop a guide for structured observation as there are fewer requirements for establishing the categories in the guide besides the collectively exhaustive and mutually exhaustive principles. As seen, the researchers adapted two out of the three guides from prior research and created only one (i.e., Table 4) by themselves.

As a structured observation study, the researchers gathered purely categorial data without any contextual or other similar data that the observation method can provide to address the research problem. This is a limitation we discussed in Section 5.2.2. It is also uncommon that observation is the only data collection technique this study utilized for data collection, given its limits in gathering various kinds of data. Nevertheless, Mueller, Volery, and von Siemens (2012) did an admirable job in choosing appropriate research participants for their observation study.

**Chapter Summary**

- Observation is a method for gathering data via carefully and attentively watching and systematically making notes on what is being studied. It requires researchers to use all their five senses and draw on their knowledge and common sense in addition to a number of other techniques.
- There are three major criteria, but not mutually exclusive, for categorizing observation types. (1) Participant versus nonparticipant observations based on if the researcher takes part in what is observed. (2) Structured versus unstructured observations based on if the researcher records observation data using a form of pre-determined categories. (3) Obtrusive versus unobtrusive observations based on if participants are aware that they are observed. Participant observation and structured observation are most often performed.
- There are seven dimensions (i.e., space, actor, activity, object, time, feeling, and reflection) for researchers to observe individually or in various combinations (e.g., actor-space, actor-activity-space-time) in a study.
- Three forms of observation are made in research with different purposes and scopes of data collection. Descriptive observation is made with the intention of getting an overview of the site and what goes on there in general. Focused observation is performed to gather in-depth data on any of the seven dimensions chosen to be examined in the study. Selective observation is conducted after most descriptive observation and some focused observation are done, aiming to gather additional data to clarify or supplement what is already collected.
- Make proper preparation before starting observation research. Avoid being intrusive during observation, plus some other major dos and don'ts in conducting observation research discussed in Section 5.4.

## References

Adler, Patricia A., and Peter Adler. 1994. "Observational Techniques." In *Handbook of Qualitative Research*, edited by Norman K. Denzin and Yvonna S. Lincoln, 377–92. Thousand Oaks: Sage.

Baker, Lynda M. 2006. "Observation: A Complex Research Method." *Library Trends* 55 (1): 171–89. https://doi.org/10.1353/lib.2006.0045.

Berg, Bruce L. 2009. *Qualitative Research Methods for the Social Sciences*. 7th ed. Boston: Allyn and Bacon.

Brown, Joanne. 2006. "Reflexivity in the Research Process: Psychoanalytic Observations." *International Journal of Social Research Methodology* 9 (3): 181–97. https://doi.org/10.1080/13645570600652776.

Datler, Wilfried, Margit Datler, Nina Hover-Reisner, and Kathrin Trunkenpolz. 2014. "Observation According to the Tavistock Model as a Research Tool: Remarks on Methodology, Education and the Training of Researchers." *Infant Observation* 17 (3): 195–214. 20p. https://doi.org/10.1080/13698036.2014.977558.

Fox, Nick. 1998. *How to Use Observations in a Research Project*. Sheffield: Trent Focus Group. Accessed July 5, 2023. http://www.webcitation.org/5ZndSmDde.

Glazier, Jack. 1985. "Structured Observation: How It Works." *College & Research Libraries News* 46 (3):105–8. https://doi.org/10.5860/crln.46.3.105.

Gorman, Gary E., and Peter Clayton. 2005. *Qualitative Research for the Information Professional: A Practical Handbook*. 2nd ed. London: Facet Publishing.

Martinko, Mark J., and William L. Gardner. 1985. "Beyond Structured Observation: Methodological Issues and New Directions." *Academy of Management Review* 10 (4):676–95. https://doi.org/10.2307/258038.

Mason, Jennifer. 2002. *Qualitative Researching*. 2nd ed. London: Sage.

Mintzberg, Henry. 1973. *The Nature of Managerial Work*. New York: Harper and Row.

Morgan, Sonya J., Susan R. H. Pullon, Lindsay M. Macdonald, Eileen M. McKinlay, and Ben V. Gray. 2017. "Case Study Observational Research: A Framework for Conducting Case Study Research Where Observation Data Are the Focus." *Qualitative Health Research* 27 (7): 1060–68. https://doi.org/10.1177/1049732316649160.

Mueller, Susan, Thierry Volery, and Björn von Siemens. 2012. "What Do Entrepreneurs Actually Do? An Observational Study of Entrepreneurs' Everyday Behavior in the Start-Up and Growth Stages." *Entrepreneurship Theory and Practice* 36 (5): 995–1017. https://doi.org/10.1111/j.1540-6520.2012.00538.x.

Nash, Jefferey E. 1975. "Bus Riding: Community on Wheels." *Urban Life and Culture* 4: 99–124. https://doi.org/10.1177/089124167500400106.

Nash, Jefferey E. 1981. "Relations in Frozen Places: Observations on Winter Public Order." *Qualitative Sociology* 4 (3): 229–43. https://doi.org/10.1007/BF00988379.

Northcote, Jeremy, and Michael Livingston. 2011. "Accuracy of Self-Reported Drinking: Observational Verification of 'Last Occasion' Drink Estimates of Young Adults." *Alcohol and Alcoholism* 46 (6): 709–13. https://doi.org/10.1093/alcalc/agr138.

Spradley, James P. 1980. *Participant Observation*. New York: Holt, Rinehart and Winston.

Stausberg, Michael. 2011. "Structured Observation." In *The Routledge Handbook of Research Methods in the Study of Religion*, edited by Michael Stausberg and Steven Engler, 382–94. New York: Routledge.

Yin, Robert K. 2016. *Qualitative Research from Start to Finish*. 2nd ed. New York: The Guilford Press.

# 6 Theoretical Approach

**Chapter Overview Questions**

- What is the theoretical approach?
- What are the three major traditions for theoretical approach research? What are their corresponding characteristics?
- What does constructionism mean in theoretical approach research?
- What is data saturation of the theoretical approach?
- What are the dos and don'ts of using the theoretical approach in research?

## 6.1 What is theoretical approach?

The theoretical approach is a technique for gathering data through conceptual analysis, theoretical examination, or other similar activities. Some theoretical approach studies aim to develop a model, framework, or theory in a respective domain, while others intend to address their research questions by performing theoretical or conceptual analysis of collected data. Although there are few studies that explicitly label their research method as theoretical approach, a lot of research projects in the humanities and social sciences actually employ this method which is often called "qualitative approach", "hermeneutic research", or other similar terms.

The theoretical approach has several distinctive features as a research method. First, this method places its focuses on "theoretical" in that the data collection performed in research using this technique is at the theoretical or conceptual level without getting into concrete entities (e.g., participants' age and race) as we do in many studies such as experiment and survey research.

Second, theoretical sampling, a phrase borrowed from grounded theory that Glaser and Strauss (1967) have developed but with a different connotation, is utilized for selecting data to be gathered in theoretical approach research. In this context, theoretical sampling means the selection of ideas, concepts, themes, theories, and the like to be collected according to a study's research objective. However, few researchers would overtly indicate this sampling technique in their reports unless their research has a focus on grounded theory, phenomenology, or another similar emphasis. As mentioned earlier, theoretical sampling in grounded theory, phenomenology, or other related research has a very different implication which we will expound in Section 6.2.1. Theoretical sampling, as a sampling technique, differs from purposive sampling already described in Section

DOI: 10.4324/9781003021711-8

2.5.2.3.2 because theoretical sampling only samples ideas, concepts, themes, theories, and other conceptual elements that will be relevant and meaningful in the research. In contrast, purposive sampling can sample anything, theoretical or otherwise, as long as what is sampled fits into the kinds of data the researcher plans to gather. That is, what can be sampled in purposive sampling is not limited to theoretical, intangible data only.

Third, data collection and data analysis in research using the theoretical approach are intertwined and inseparable. The pendulum of the theoretical approach is constantly in motion, oscillating between data collection and data analysis in the research process. Although theoretical sampling is the technique for choosing what is to be gathered in the theoretical approach, conceptual analytical methods (e.g., comparison, categorization, deduction, induction, reasoning) must be applied to complete the data collection process. It is through analyzing the sampled data that data collection is done.

Fourth, publications reporting theoretical approach research usually do not have some of the common components (e.g., literature review, methodology) as in research reports using other methods (e.g., questionnaire, interview). Instead each of such research reports takes a format of its own. The general format for reporting research depicted in Table 2.1 (i.e., introduction, literature review, methodology, results, discussion, and conclusion), prevalent in experiment, interview, observation, questionnaire, or other kinds of studies, is not adopted and used in presenting theoretical approach research.

The attentive reader may have a question at this point: Is the theoretical approach identical to content analysis, a topic for Chapter 9, as a data collection technique? A simple answer is no due to the following two features of the theoretical approach. (1) The theoretical approach usually treats a source item as a whole for data collection purpose. Articles, books, newspapers, reports, personal files, records, social media posts, and the like can all serve as data source items, from which theoretical sampling is performed to select what is needed for a study. In comparison, content analysis is done with words, phrases, sentences, or any segments in source items, including the whole entity, in data collection. In other words, the theoretical approach is applicable at the item level, whereas content analysis is utilized at various levels of granularity in data collection. (2) That being said, the difference between the theoretical approach and latent content analysis, one of the two types of content analysis (and the other being manifest content analysis) seems indeed subtle. Latent content analysis examines what is hidden in the content, while the theoretical approach focuses on theoretical or conceptual analysis of the target content which might be latent or manifest. Latent content analysis and the theoretical approach bear some similarity because they both perform conceptual analysis on the target content. But the theoretical approach has two implementations: one is for performing conceptual/theoretical analysis using existing theories and the like, while the other is for creating a model, theory, or framework. In addition, latent content analysis is devoted to working on what is hidden or underlying, unobservable, and uncountable. By contrast, the theoretical approach does not have the same restriction in its application as long as the target content is theoretical or conceptual.

Although studies using the theoretical approach typically start their reports with a review of related literature or are often regarded as "grounded" in previous research, the theoretical approach is not just a review because the latter only does evaluation without any further conceptual analysis, modeling or theory building. Take Saarijärvi, Kannan, and Kuusela (2013), one of the case studies for this chapter, as an example, it would be merely a review if the authors stop after their review of different approaches to value co-creation.

All these characteristics of the theoretical approach show that this method is unique yet widely adopted in research of the humanities and social sciences. Acquiring this technique will enable us to carry out research at the theoretical or conceptual level.

## 6.2 Origins and traditions of the theoretical approach

Adopting the theoretical approach for data collection which has to be done hand-in-hand with data analysis, researchers can follow various theoretical origins and traditions in their data analysis and other steps of the research process. Grounded theory, phenomenology, and hermeneutics, even though not mutually exclusive, are some examples of such theoretical origins and traditions that contribute to theoretical approach research. As a result, a single theoretical approach study may involve one or more origins and traditions explicitly or implicitly. An overview of the three widely applied paradigms, grounded theory, phenomenology, and hermeneutics, is provided below to demonstrate their different characteristics in the context of research using the theoretical approach.

### *6.2.1 Grounded theory*

Grounded theory was developed and introduced by two sociologists, Barney G. Glaser and Anselm L. Strauss, based on their collaborative research on dying hospital patients (Glaser and Strauss 1967). Grounded theory has been extensively adopted by researchers in the field of sociology and beyond since Glaser and Strauss' work. It is called grounded theory because a "theory" or research outcome is "grounded" in the data collected and analyzed or the development of a theory occurs after the data are collected and examined. The opposite scenario of the grounded theory approach is that a study begins with a theory and data are collected and analyzed to confirm or reject it afterward. People regard the latter variously as "the quantitative research paradigm", positivism, and so on. Charmaz (2000) indicates that "Glaser and Strauss's (1976) work was revolutionary because it challenged (a) arbitrary divisions between theory and research, (b) views of qualitative research as primarily a precursor to more 'rigorous' quantitative methods, (c) claims that the quest for rigor made qualitative research illegitimate, (d) beliefs that qualitative methods are impressionistic and unsystematic, (e) separation of data collection and analysis, and (f) assumptions that qualitative research could produce only descriptive case studies rather than theory development" (511).

Grounded theory is apparently proposed to challenge the hegemony of the quantitative research paradigm in the social sciences. But as time passes, even the two creators have different views of grounded theory as reflected in the statement made by Charmaz (2000, 510): "Barney G. Glaser and the late Anselm Strauss, with his more recent coauthor, Juliet Corbin, have moved the method in somewhat conflicting directions". Croucher and Cronn-Mills (2015, 196–98) even coin the "Glaserian approach" and "Straussian approach" to illustrate the differences between the two founders of grounded theory. These differences, however, are not to be examined here as they are beyond the scope of this book. What will be elaborated below is that grounded theory, as interpreted by many researchers, clearly deviates from the path its first creators had in mind in several aspects.

First, is grounded theory a technique for data collection? The answer is no because, as briefly explained in Section 1.4.15, grounded theory researchers still need to use interview, focus groups, research diaries, and/or other techniques for data collection. Grounded theory itself cannot enable researchers to gather data for their projects. Glaser

(1978, 4–5) maintains that grounded theorists cannot shop their disciplinary stores for preconceived concepts and dress their data in them. Any existing concept must earn its way into the analysis. In other words, analysis must be performed to code the collected data into concepts in grounded theory research rather than having concepts first and collecting data about these concepts. Later, Charmaz (2000) echoes the same point by indicating that "[g]rounded theory methods specify analytic strategies, not data collection methods" (514). Corbin and Strauss (2008) also point out this fact by detailing the data analysis process in grounded theory research in most of their 14-chapter book.

Second, what does "theory" mean in grounded theory? Is it true that one or more theories will be developed in all grounded theory research? Before addressing these two questions, let us look at how Robert K. Merton, Barney G. Glaser's doctoral advisor, noted the problem that the term theory might refer to many different things: "including everything from minor working hypotheses, through comprehensive but vague and unordered speculations, to axiomatic systems of thought" (Merton 1968, 39). All these different types of "theories" are reported in grounded theory research, some of which are merely research results. For instance, Green and Schleien (1991) present the following three theories in their grounded theory study: "Firstly, adults with mental retardation living in community residential facilities often do not develop meaningful, reciprocated friendships with peers. Rather, they mistakenly perceive staff, family, and other externally motivated or obligated people to be their friends. Secondly, adults with mental retardation do not typically meet and make new friends during recreation participation. At best, they may use the recreation domain to further nurture a relationship that was established previously. Finally, friendship development between adults with mental retardation and their nondisabled peers may be inhibited by the absence of social skills necessary to participate cooperatively in recreation" (29). It is evident that the "three theories" Green and Schleien created are no more than research findings we typically see in many scholarly publications. Grounded theory research may lead to the creation of one or more theories. Nevertheless, as indicated in relevant publications, not all grounded theory research would establish one or more theories as its final outcome. For this very reason, grounded theory as a research method is somehow a misnomer.

Third, is grounded theory only used in qualitative research? The response to this question is negative although the grounded theory approach is applied in qualitative studies more often than in quantitative research projects. Charmaz (2000) aptly notes: "Researchers can use grounded theory methods with either quantitative or qualitative data, although these methods are typically associated with qualitative research" (510).

In short, grounded theory refers to research grounded in data collected and analyzed instead of relying on one or more existing theories, models, or frameworks. It places an emphasis on constant comparison of data that are usually gathered via interview, focus groups, research diaries, and other techniques (Glaser and Strauss 1967). Those salient features of the grounded theory approach are particularly exhibited through the so-called theoretical sampling (Corbin and Strauss 2008, 143–58).

Theoretical sampling in grounded theory research is far more than a sampling technique as described in Section 6.1 from the perspective of the theoretical approach. It refers to performing systematic selection of conceptual entities (e.g., ideas) to efficiently obtain the concept indicators needed to develop categories for the emerging theory (Breckenridge and Jones 2009; Corbin and Strauss 2008, 143–58; Draucker et al. 2007; Engler 2011). It involves sampling to test, elaborate, and refine a category. Further sampling is needed to develop the categories, their relationships, and interrelationships with the aim of theory development (Coyne 1997). The entire data collection and analysis of grounded theory research, including theoretical sampling, rely on multiple techniques Patton (2002) categorizes and groups for nonrandom or qualitative sampling as well as the three different types of coding (i.e., open, axial, and selective coding) that numerous grounded theory researchers have elucidated (e.g., Glaser and Strauss 1967; Strauss and Corbin 1990). In general, the data collection and analysis process, including theoretical sampling, in grounded theory research consists of the following steps (Draucker et al. 2007):

- Selective sampling for initial recruitment of participants using techniques such as criterion sampling, maximum variation sampling, and snowball or chain sampling.
- Open coding of what is collected from the participants via interview, observation, focus groups, and other data collection techniques.
- Theoretical sampling (i.e., selection of ideas and concepts that contribute to the creation of an emerging theory) through axial coding of the same dataset used previously for open coding. Techniques for sampling at this point comprise intensity sampling, typical case sampling, extreme/deviant case sampling, and stratified purposeful sampling.
- Theoretical sampling through selective coding of the same dataset used previously for open and axial coding. Techniques for sampling at this stage include confirming and disconfirming case sampling.

As shown, the different sampling techniques listed above encompass the four basic nonrandom sampling techniques illustrated in Section 2.5.2.3 and many more. They all belong to the various qualitative sampling techniques Patton (2002) depicted and examined in his book. As for the three types of coding (i.e., open, axial, and selective), a major qualitative data analysis technique, will be presented in Section 13.2.1.

In addition to the afore-described data collection and analysis procedure for grounded theory research, theoretical sampling can also be employed for selecting additional data sources, adding new interview questions, recruiting more participants, and the like when needed. It should be apparent by now that theoretical sampling is not simply a sampling technique in grounded theory research. Rather it is integrated into the entire process of grounded theory studies!

Then what is the relevance of grounded theory to theoretical approach research? The relevance lies in the orientation of theoretical approach research toward its data collection and analysis process. Namely, theoretical approach research must be "grounded" in data collected via the theoretical sampling technique depicted in Section 6.1 and analyzed using conceptual analytical methods such as comparison, categorization, deduction, induction, and reasoning. Moreover, as indicated in Section 6.1, data collection and data analysis in theoretical approach research are integrated and inseparable. All these features can be traced back to the grounded theory approach.

### *6.2.2 Phenomenology*

Phenomenology, established by a German philosopher Edmund Husserl, seeks to describe experience as it presents itself to subjective consciousness. Consciousness in phenomenology is basically a medium between a person and the world. Phenomenology is interested in the activities of consciousness (e.g., perception) and the objects (e.g., smartphones) that present themselves to consciousness. Phenomenology does not dictate phenomena but rather it wants to understand how phenomena present themselves to consciousness and the elucidation of this process is a descriptive task. Thus, phenomenology is descriptive, not interpretive (Giorgi 2012). As seen, there are three keywords in phenomenology: experience, consciousness, and description. That is, any phenomena to be examined in research must be first experienced and then described from the participant's consciousness without reference to scientific theories or prior learning. The last point bears a clear resemblance to grounded theory discussed in Section 6.2.1.

Husserl created the school of phenomenology in the early years of the 20th century. By the time of Husserl's death in 1938, phenomenology had achieved acceptance not only in philosophy but also in anthropology, sociology, and psychology. This acceptance has since grown to the point where phenomenology is a widely adopted approach in most of the social sciences (Cibangu and Hepworth 2016; Relph 2014). Phenomenology empowers researchers to investigate qualitatively the different ways in which participants experience something physically or spiritually.

In addition, Cibangu and Hepworth (2016) delineate the three most important concepts that have been considered to characterize phenomenology: (1) intentionality which entails directness, extendedness, and context as opposed to indirectness, fixity, reclusion, and isolation; (2) intersubjectivity which addresses issues such as multiplicity, pluralism, diversity, minority, unity, community, organization, negotiation, and dialog. Without intersubjectivity, research is set to result in skewed and shortsighted assumptions; and (3) reduction or bracketing, the process by which the participant or researcher question their taken-for-granted presuppositions, misconceptions, and biases that preclude the fuller acquisition and actualization of knowledge. Knowledge in the context of phenomenology means facts and information acquired through experience.

Same as grounded theory, phenomenology is also not a data collection technique because it depends largely on interviews to gather data that focus on participants' direct and pure experience, bracketing away the interpretations that people make of them (Spickard 2011). Interviews conducted for data collection in phenomenology can be called phenomenological interviews. Gentles et al. (2015) likewise affirm that phenomenology is a qualitative approach in which researchers aim to develop new understandings of human lived experience, relying on first-person accounts generally obtained through interviews of research participants.

Giorgi and Giorgi (2003) create a model of how to use the phenomenological approach in research. "For the scientific analysis, one first obtains descriptions of experiences from others, then one enters into a scientific phenomenological reduction while simultaneously adopting a psychological perspective of the experience, then one analyzes the raw data to come up with the essential structure of the experience, which is then carefully described at a level other than that of the original description" (Giorgi and Giorgi 2003, 247). In this model, Giorgi and Giorgi emphasize the psychological perspective of the experience because they both are from the field of psychology. Based on the model of Giorgi and Giorgi (2003), Spickard (2011) derives the following four concrete steps of the phenomenological approach:

- Locate and interview participants who have shared a particular experience.
- Help research participants focus on exactly how this experience presented itself to their consciousness, leaving aside what they think was "really" happening. For example, when people point a finger to their own faces for a couple of minutes, they would feel some sensation at the pointed spot of their faces. That sensation is what they

experience in phenomenology which does not consider why doing so makes them feel the sensation.

- Compare and analyze these accounts to identify the basic structures of the experience. The researcher needs to have good enough interview material to determine which features of the experience are idiosyncratic and which are central or common. The researcher must read beneath each participant's account to find the patterns that it represents.
- Re-describe and summarize the experience by boiling down to these basic structures.

The features of phenomenology stand out when it comes to data analysis and result interpretation. Theoretical sampling is also utilized when one works with phenomenological data as done in grounded theory research. What we may not know when adopting the phenomenological approach is whether participants reconstruct their experience in phenomenological interviews. This is an issue that phenomenology researchers should address as reconstructed experience is no longer pure and direct.

Theoretical approach research can benefit from phenomenology in that the latter signifies the importance of having pure, immediate experience of participants recorded and analyzed in research. Researchers adopting the theoretical approach ought to ascertain the data they gather and represent what participants truly experience. In the analysis of such data, the focus is placed on the basic structures or components underlying the collected data by using the common set of qualitative analytic techniques such as comparison and reasoning. The three concepts of intentionality, intersubjectivity, and reduction Cibangu and Hepworth (2016) have synopsized would further help researchers to ensure the credibility and other quality attributes in theoretical approach research.

### *6.2.3 Hermeneutics*

When interpretation is developed into a scientific approach, it is given a Greek name – "hermeneutics" – a concept derived from hermeneuein, which means to "express", "translate", and "interpret" (Gilhus 2011). Friedrich Schleiermacher, a German philosopher, theologian, and biblical scholar, believed that every problem of interpretation is a problem of understanding. He even defined hermeneutics as the art of avoiding misunderstanding. Schleiermacher explored the nature of understanding in relation not just to the problem of deciphering sacred texts but also to all human texts. He also stated that in order for an interpreter to understand the work of another author, they must familiarize themselves with the historical context in which the author published their thoughts. In short, interpretation is understanding and vice versa (Schwandt 2000).

Although Schleiermacher placed hermeneutics in the historical context because of his background, devotion, and profession, hermeneutics today considers all kinds of texts and contexts. The central principle of hermeneutics is that it is only possible to know the meaning of an act, event, or statement within the context of the discourse from which it originates. Context is critical to comprehension and interpretation since an act, event, or statement that carries a particular meaning to one person or culture may be viewed differently by another. For example, green and red are two colors used to indicate whether a stock market is up or down all over the world. In Western countries, the green color represents that the stock market is up, while red denotes the market is down. But an entirely opposite interpretation of these two colors is made in China due to its different cultural context.

Hermeneutics has been broadly applied in the humanities, especially in law, history, and theology as a historical and critical technique for analyzing texts. Disciplines in the social sciences (e.g., sociology) also adopt this technique when analyzing qualitative data. The interpretation of a text must proceed by framing its content in terms of the overall organization of the work in hermeneutic research. During Schleiermacher's time, a fundamental shift occurred from merely understanding the exact words and their objective meaning to an understanding of the writer's distinctive character and point of view (Wikipedia Contributors 2023). In hermeneutic research today, what can be analyzed and interpreted includes not only texts but also various other kinds of qualitative data (e.g., interview transcripts and observation notes). However, the principle of interpretation in context remains unchanged over time. For the sake of simplicity, the word "text" is used in this book to refer to all data that can be interpreted hermeneutically.

The hermeneutic circle, created by Martin Heidegger who was a German philosopher and a seminal thinker, describes the process of understanding a text hermeneutically. It refers to the idea, as illustrated in Figure 6.1, that one's understanding of the text as a whole is established by reference to the individual parts and one's understanding of each individual part by reference to the whole. Neither the whole text nor any individual part can be understood without reference to one another, and hence it is a circle. This circular character of interpretation shows the cyclical and recursive nature of interpreting a text. It also stresses that the meaning of a text must be found within its cultural, historical, literary, and social contexts (Wikipedia Contributors 2022).

Hermeneutics has different branches with unique emphases (e.g., objective hermeneutics, critical hermeneutics, philosophical hermeneutics) but they are beyond the scope of our discussion on hermeneutics in this book. Besides, hermeneutics is sometimes regarded as one school of phenomenology: hermeneutic phenomenology (Dowling 2004). As depicted in Section 6.2.2, the phenomenological approach looks to uncover meanings of phenomena experienced by study participants through the analysis of their descriptions. Phenomenology has its focus on the participant's lived experience and obtains commonalities and shared meanings among experienced descriptions, whereas hermeneutics assumes that participants experience the world through language and this language provides both understanding and knowledge (Byrne 2001). The relationship between phenomenology and hermeneutics displays the non-mutual exclusiveness between these two research approaches we cover in this chapter.

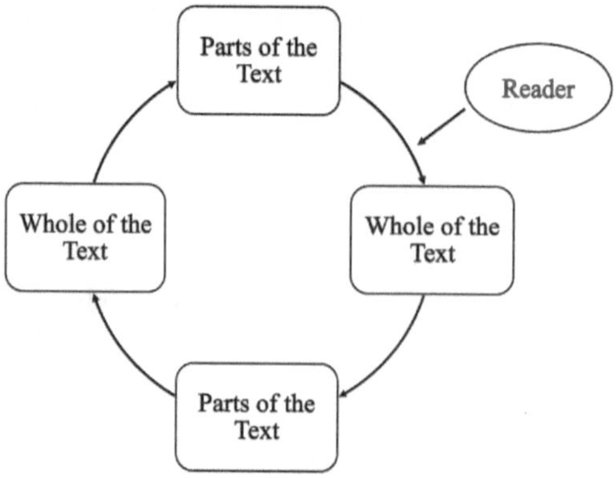


*Figure 6.1* The Hermeneutic Circle

As indicated early, the source material of hermeneutics is texts and other kinds of qualitative data collected in research. Hermeneutics consists of a reading that moves back and forth between the parts and the whole of a text, between its structure and meaning, between the reader's horizon and the horizon of the text, and between the text and its contexts. These processes are described as different varieties of a hermeneutic circle (Gilhus 2011). The hermeneutic approach is characterized metaphorically as dialogical because it presupposes a continuous exchange between the researcher and the source material. The act of interpretation should always include systematic reflection over the hermeneutical process and one's own starting points in this process.

Gadamer (1989) asserts that we must not attach blindly to our own fore-meaning when attempting to understand the meaning of something held by another. He argues that we remain open to and also embrace the meaning held by the other person or text with our own meanings (Gadamer 1989, 268). What is essential throughout this hermeneutical process is that we are aware of our biases in order for the text to portray its uniqueness against our own fore-meanings. This point echoes the presuppositions, misconceptions, and biases that preclude the fuller acquisition and actualization of knowledge in phenomenology (Cibangu and Hepworth 2016).

Furthermore, Gilhus (2011) suggests some hermeneutic guidelines that include reading the text slowly and thoroughly without any prejudice or bias; applying everything one knows about the language and context of the text; keeping an eye on the possibilities of cultural comparison; being aware that textual meanings are always in flux; and asking whose interests are promoted in a text. Interpreting texts involves reading and rereading, keeping a dialog between parts and whole, posing questions and bringing in different contexts to tease out some of the layers of meaning that even the tiniest text always presents to its readers. The range of acceptable readings of texts is limited. The researcher hence should not exceed the boundaries of interpretation that are dependent on the texts' cultural, historical, literary, and social contexts.

Like grounded theory and phenomenology, hermeneutics cannot enable researchers to collect data. It is instead an analytic approach focusing on interpretation in context. This particular characteristic should underlie all qualitative research, including theoretical approach research. This is the very purpose that hermeneutics is presented and elucidated in this chapter as one of the three major approaches associated with studies adopting the theoretical approach.

### 6.2.4 *Grounded theory, phenomenology, and hermeneutics: A brief comparison*

Grounded theory, phenomenology, and hermeneutics each signifies a focus that theoretical approach studies can take in the research process. Grounded theory emphasizes that research outcome can only be obtained by grounding its analysis on data instead of an existing theory, model, or framework. Phenomenology by contrast centers around analytical description of participants' pure and direct experience using interview as the chief data collection technique. Hermeneutics, on the other hand, concentrates on interpretation of source material and collected data in related contexts. All the three approaches to research converge at the point where analysis is a cyclical and recursive process rather than a one-time activity, which in fact typifies an outstanding feature of all research, especially qualitative research, regardless of its research tradition or paradigm. The theoretical approach, a theme for the present chapter, is qualitative in nature and may employ one or more of these three analytical techniques in investigations even if none of them (i.e., grounded theory, phenomenology, and hermeneutics) appears explicitly in the research report.

Grounded theory, phenomenology, and hermeneutics each is utilized on a different scale according to the universe of research literature. By comparison, grounded theory is applied the most widely and hermeneutics the least, while phenomenology falls somewhere between the two. Worth noting is that one study may adopt more than one of the three approaches because they are not mutually exclusive at all in the first place. That is, these three approaches can be used in combination (e.g., hermeneutic phenomenology) in some cases.

## 6.3 Using the theoretical approach in research

One salient feature of the theoretical approach, as discussed in Section 6.1, is that data must be collected via analysis of source material. Oftentimes it is the analytical techniques used for data collection that provide different flavors of the theoretical approach. Grounded theory, phenomenology, hermeneutics, and analytic techniques without any distinctive labeling (e.g., comparison and reasoning) can all become such "flavors" in theoretical approach research. Because of the intertwined data collection and analysis process as well as the many analytical flavors the theoretical approach may have, there is no general procedure for using the theoretical approach in research. Instead the actual procedure is chiefly determined by the analytical technique a study adopts. This leads to the two outstanding limitations of the theoretical approach: subjectivity and research biases (Croucher and Cronn-Mills 2015). Therefore, extra caution must be exercised throughout the research process to mitigate those two limits. Measures to be presented in Section 13.2.3 (e.g., prolonged engagement in data analysis, peer debriefing, member checking) ought to be taken to overcome subjectivity and research biases in research using the theoretical approach.

As also demonstrated in the two case studies chosen for this chapter in Section 6.4, each theoretical approach study follows a different research procedure. However, one component that all theoretical approach research projects share is constructivism which will be described and discussed in Section 6.3.1.

### *6.3.1 Constructionism*

Constructionism is a research paradigm in which knowledge is not regarded as an insight into some objective reality, but instead constructed by humans, partly through social interactions (Moss 2016). Corbin and Strauss (2008, 9) postulate in their presentation of grounded theory that knowledge in constructivism refers to what researchers construct in their investigations. Researchers attempt to understand the intangible constructions or perspectives that individuals form about local, specific places, events, and issues (Moss 2016).

Schwandt (1998) elaborates that "[i]n a fairly unremarkable sense, we are all constructivists if we believe that the mind is active in the construction of knowledge. ... In this sense, constructivism means that human beings do not find or discover knowledge so much as construct or make it. We invent concepts, models, and schemes to make sense of experience and, further, we continually test and modify these constructions in light of new experience" (237). In addition, we do not construct our interpretations of our findings in isolation but against a backdrop of shared understandings, practices, language, and so forth (Schwandt 2000). The latter point remains true in all kinds of research.

From the educational perspective, constructionism is about how people learn. People construct their own understanding and knowledge of the world through experiencing things and reflecting on those experiences. When we encounter something new, we have to reconcile it with our previous ideas and experience, maybe changing what we believe, or maybe discarding the new information as irrelevant. In any case, we are active creators of our own knowledge. To do this, we must ask questions, explore, and assess what we know (Educational Broadcasting Corporation 2004).

In research adopting the theoretical approach or other methods, a similar mechanism of constructionism is followed. Research is conducted to answer questions. We may get results that are beyond our expectations or even contradictory to existing knowledge in the research process. It is at this point that we start to construct a full answer to the research problem by piecing together what is collected and analyzed. Constructionism of course is not simply putting together various segments obtained through data collection and analysis. Rather, it entails meticulous consideration, exploration, and integration of research results from the perspective of the research problem as well as prior studies before concluding the study. In that sense, constructionism is applied not only in theoretical approach research but also in all other kinds of studies.

Charmaz (2000), a well-known proponent of constructionism as one school of grounded theory, states that the researcher "constructs a picture that draws from, reassembles, and renders subjects' lives. The product is more like a painting than a photograph" (522). Indeed, no research process is a one-time activity like picture-taking. Researchers instead should undertake all the necessary steps to achieve their research objectives even if that requires additional efforts and involves hardship.

Therefore, constructivism means to obtain findings and draw conclusions by piecing together what is collected and analyzed (Gubrium and Holstein 2000) in theoretical approach research that aims to perform conceptual analysis, modeling, and theory building.

### *6.3.2 Data saturation*

Another issue that researchers must address when adopting the theoretical approach is data saturation. Data saturation, simply put, refers to the point where no new data emerge in the process of data gathering (Corbin and Strauss 2008, 143; Tuckett 2004). This concept in fact is relevant to data collection in many qualitative research projects. Some scholars even advocate that data saturation be considered in interview research as well. In most interview research, however, the number of interviews to be conducted is determined before data collection rather than during the data gathering process. Data saturation thus becomes unrelated to interview data collection. One major exception in this situation is when interviews are done in ethnography and field study as one of the multiple techniques for data collection. Further elaboration on this topic will be provided in Chapter 11, Ethnography/Field Study.

As presented in Section 6.1, the theoretical approach employs theoretical sampling in choosing what to be collected. Unlike many other data gathering techniques (e.g., questionnaire, experiment) that have a clear indicator to signal the completion of the data collection process, there is no such obvious sign post for finishing data collection with the theoretical approach. Instead it is the researcher's responsibility to determine if adequate data are collected for the study or if the point of data saturation is reached. In contrast, for example, receiving completed questionnaires from the initial administration and possible follow-ups marks the end of survey research's data collection process.

How to determine data saturation hence becomes one additional task for researchers who adopt the theoretical approach for data collection. The discussion on theoretical sampling in grounded theory can shed some light on the issue of data saturation although what is involved there concerns with theoretical saturation. Theoretical saturation is defined by Corbin and Strauss (2008) as: "In doing theoretical sampling, the researcher takes one step at a time with data gathering, followed by analysis, followed by more data gathering until a category reaches the point of 'saturation'" (145). Corbin and Strauss (2008, 147–48) further explain how data saturation and theoretical saturation are reached in theoretical sampling. It is straightforward for a researcher to continue gathering data until reaching the level of "data saturation". By contrast, arriving at "theoretical saturation" is not that easily attained in that theoretical saturation is more than a matter of no new categories or themes emerging. It also concerns the development of categories in terms of their properties and dimensions, including variation, and possible relationships to other concepts. In the case of theoretical approach, the data collection process would end when data saturation is reached or data needed for answering research questions are all gathered. However, meticulous attention should be paid to determining data saturation because data collected via the theoretical approach are conceptual or intangible. This may make the determination of data saturation more elusive. Additionally, theoretical approach research could lead to theory development which may require the researcher to consider the theoretical saturation issue Corbin and Strauss (2008) raise above.

### *6.3.3 Dos and don'ts in theoretical approach research*

As indicated early, there is no general procedure for conducting theoretical approach research as with many other kinds of research (e.g., survey research, experiment research) due to its distinctive features. Likewise, little common ground can be found among individual studies using the theoretical approach with regard to the dos and don'ts. The dos and don'ts recapitulated below thus are mainly about several general threads in theoretical approach research.

First, do prepare a solid plan for doing theoretical approach research as there are many decision points, from data collection to conclusion drawing, in the research process. This does not imply that research done with other methods (e.g., interview, observation) requires no planning or preparation. Rather theoretical approach research is obligated to make additional efforts in planning because of its unique characteristics. For instance, few guidelines exist in theoretical approach research for developing a data collection procedure, performing theoretical sampling, determining data saturation, and conducting data analysis which actually is integrated with data collection. Although there are always many unknowns in all research, the complexity involved in theoretical approach research demands careful planning before a study is undertaken. Lack of planning or unpreparedness in theoretical approach research will only lead to failure or even fiascos. Each study adopting the theoretical approach is different when there is, for example, neither any validated data collection instrument (e.g., questionnaire) nor an established data analysis procedure (e.g., coding schema) to follow or adapt in the research process.

Second, do seek a full understanding of any research tradition and paradigm selected for your theoretical approach research project in order to benefit from using one. Simply labeling your study as grounded theory, phenomenology, and/or hermeneutics will not automatically enable you to achieve your research objective. Instead do practice what these approaches "preach" in research. For instance, make sure to interpret what is collected in context if hermeneutics is chosen as a data analysis technique. Avoid saying one thing (e.g., grounded theory) while doing the other (e.g., phenomenology) in the process of theoretical approach research.

Third, do theoretical sampling genuinely to ensure all the ideas, concepts, themes, theories, and other entities are selected for data collection. Avoid confusing theoretical sampling with purposive sampling. Avoid doing sampling without prudent consideration of what is really needed in research. Similarly, determine data saturation after making certain that there is truly no new data emerging. Avoid simply taking a recurring term and so forth as a criterion for judging if data saturation is reached.

Last but not least, all the dos and don'ts outlined in the discussion of each major research tradition (e.g., phenomenology) in this chapter should be followed in conducing theoretical approach research. The hermeneutic guidelines proposed by Gilhus (2011) provide some good examples (see Section 6.2.3). Similar guidance can also be found for doing grounded theory and phenomenology research in Sections 6.2.1 and 6.2.2, respectively.

## 6.4 Case studies of theoretical approach research

Two case studies of theoretical approach research are to be made in this section. As data collection in theoretical approach research cannot be examined and discussed independently of data analysis, along with their unique presentation practice, case studies for this chapter will cover both report in their entirety, which differs from case studies done in other chapters of this book that mainly focus on data collection.

### *6.4.1 Case Study 1*

Lewicki, Roy J., Edward C. Tomlinson, and Nicole Gillespie. 2006. "Models of Interpersonal Trust Development: Theoretical Approaches, Empirical Evidence, and Future Directions." *Journal of Management* 32 (6): 991–1022. https://doi.org/10.1177/0149206306294405.

Lewicki, Tomlinson, and Gillespie (2006) attempt to formulate directions for future research on interpersonal trust development by examining selected existing work on the same topic. Although the authors did not openly indicate that theoretical sampling is used for choosing what to be collected, they describe their data selection as the scope of research: (1) The study focuses on trust development in interpersonal interactions only, excluding research on trust in teams, groups, and organizations. (2) The study does not cover trust repair (Lewicki, Tomlinson, and Gillespie 2006, 993). It would be highly desirable for the authors to describe how the reviewed studies are selected for the current research.

In Section 6.2, we have discussed grounded theory, phenomenology, and hermeneutics as three major techniques for theoretical approach research. Two of the three techniques are applied in Lewicki, Tomlinson, and Gillespie (2006) even though the authors do not explicitly include grounded theory and phenomenology in the title, abstract, or other major parts of their report. Specifically, the application of grounded theory in their study is reflected (in boldface) at least in the following two quotes:

- "Behavioral approaches to trust are **grounded** in observable choices made by an actor in an interpersonal context (Lewicki, Tomlinson, and Gillespie 2006, 993)".
- "Collectively, the models within the unidimensional approach tend to suggest that (a) expectations are **grounded** in perceptions of another's trustworthiness, which leads to a willingness to be vulnerable; ... (Lewicki, Tomlinson, and Gillespie 2006, 999)".

As to be explained below, the three authors identify four approaches to interpersonal trust development. The above two quotes indicate that in what kinds of data two of the four approaches (i.e., behavioral and unidimensional) are grounded. Besides, the word "grounded" appears nine more times in the rest of the research report when the authors illustrate how trust is grounded in various factors and other variables in related research. The phenomenological approach, in comparison, is mentioned just once as in "... and to understand whether 'deep' trust in close relationships is **phenomenologically** different from transactional trust" (Lewicki, Tomlinson, and Gillespie 2006, 1006). The seemingly casual references to grounded theory, phenomenology, and/or hermeneutics in theoretical approach research appear to be quite common in practice.

The authors categorize all the research approaches to interpersonal trust development into four kinds but at two levels. Of the four approaches, the behavioral approach and the psychological approach are in parallel at the first level. The psychological approach is further divided into three models or sub-approaches (i.e., unidimensional, two-dimensional, transformational) at the second level: (1) the unidimensional model, which treats trust and distrust as bipolar opposites; (2) the two-dimensional model, which argues that trust and distrust are two distinctly differentiable dimensions that can vary independently; and (3) the transformational model, which asserts that trust has different forms that develop and emerge over time. While the behavioral approach and the psychological approach are categorized according to one criterion, the same does not hold true regarding how the three sub-approaches under the psychological approach are grouped. We can see that the unidimensional model and the two-dimensional model are classified based on the number of dimensions. Nevertheless, it is apparent that the third model, the transformational model, is formed using a criterion other than the number of dimensions. When more than one criterion is adopted for categorization – a main data analysis technique for theoretical approach research, the principle of mutual exclusiveness cannot be observed. This inevitably undermines the validity and reliability of this theoretical approach study where categorization is utilized for data analysis.

Table 1 in Lewicki, Tomlinson, and Gillespie (2006, 994) encapsulates the four different approaches to interpersonal trust development, which encourages readers to compare and understand the various attributes of the four approaches to the three key questions listed in the most left column of Table 1. However, similar elements are shared among the three second-level approaches with regard to the last key question listed in Table 1: What causes the level of trust (distrust) to change over time? Namely, both "relationship history" and "trust declines when positive expectations are disconfirmed" appear under the unidimensional model and the transformational model; "increased knowledge about the other" is repeated under the two-dimensional model and the transformational model even though their wordings are slightly different. This degree of overlapping or non-mutual exclusiveness should have been avoided in the tabular presentation when the authors intend to demonstrate how the three models (i.e., unidimensional, two-dimensional, transformational) differ from each other instead of showing their similarities. This unintended outcome perhaps results from the practice that more than one criterion is adopted in categorizing the three models of the psychological approach.

Lewicki, Tomlinson, and Gillespie create a total of four figures in their report to visualize results. Figures are particularly helpful in enhancing the quality of research using the theoretical approach when most of the report is narrative. Of the four figures in Lewicki, Tomlinson, and Gillespie (2006), Figures 1, 3, and 4 are all effective presentations that help fully realize the potential of what a figure can offer. By contrast, the value of Figure 2 is in question because (1) it is actually a table rather than a figure; and (2) it merely repeats the cell headings listed in the most left column of Table 2 without any added value of a figure.

In concluding the report, the authors raise four questions for future research regarding trust development: (1) What are the defining features of trust? (2) What are the gaps in trust measurement? (3) How can we capture changes in trust over time? (4) How should researchers talk about the optimal level of trust? Do note that the present study focuses on interpersonal trust development only. Nevertheless, Lewicki, Tomlinson, and Gillespie in their report pose the above four questions for further research on trust development which is broader than the scope of their current study – interpersonal trust development. Although this study does not involve any external validity issue given its qualitative nature, the authors' suggestions do overstep the boundary they set for themselves in their current project and are not grounded in the data they have collected and analyzed. We should always stay within the boundary of our research by knowing that any recommendations beyond the current research will not be complying with grounded theory and other techniques of qualitative data analysis. Such practice may also make the validity of a study suffer.

### 6.4.2 *Case Study 2*

Saarijärvi, Hannu, P. K. Kannan, and Hannu Kuusela. 2013. "Value Co-creation: Theoretical Approaches and Practical Implications." *European Business Review* 25 (1): 6–19. https://doi.org/10.1108/09555341311287718.

Saarijärvi, Kannan, and Kuusela (2013) explore the different theoretical approaches to value co-creation to determine what kind of value is co-created for whom, using what resources, and through what mechanism in the process of value co-creation. A business-oriented analytical framework is then developed to help practitioners assess and approach the opportunities presented by value co-creation after the conceptual complexity involved in the process is identified and examined. Practitioners in this study refer to business practitioners who are interested in value co-creation.

It is not uncommon that theoretical approach research makes no particular references to any theoretical traditions and origins (e.g., grounded theory, phenomenology, hermeneutics) in its research report because qualitative research possesses various other tools such as comparison and reasoning for data analysis. This is exactly the case with Saarijärvi, Kannan, and Kuusela's study except that the word "grounded" appears once when the authors review a prior study. Theoretical sampling likewise is not alluded to in the report even though they should have at least indicated how the studies included in the literature review are selected. Same as Lewicki, Tomlinson, and Gillespie (2006), Saarijärvi, Kannan, and Kuusela do not present their research following the general format depicted in Table 2.1. There are no sections such as literature review, methodology, and results in the reports of theoretical approach research. Instead each study using the theoretical approach normally chooses a presentation format that suits its research objective. This characteristic is outstanding in most, if not all, theoretical approach research as we briefly mentioned in Section 6.1.

Based on a brief review of related literature, the authors examine and clarify different approaches to value co-creation that include the service-dominant (S-D) logic, service science, service logic, many-to-many marketing, social constructionist, new product development, and post-modernism approaches. These approaches together provide diverse perspectives on the complex nature of value co-creation.

The authors then decompose the concept of value co-creation into three individual questions (see Figure 1 in Saarijärvi, Kannan, and Kuusela 2013, 10): "Value" is concerned with the question "What kind of value for whom?"; "Co" is related to the question "By what kind of resources?"; and "Creation" is about the question "Through what kind of mechanism?". According to the authors, "value" in value co-creation can be customer value, firm value, or both, which also explains the "for whom" part of the question raised. The "co" in value co-creation refers to actors involved in the process employing resources from either the firm or the customer. However, linking "co" to "type of resources" does not seem natural in this context. The linkage is somewhat far-fetched because "co" in value co-recreation should be about who (i.e., the firm and the customer) jointly create the value. Yet, the question currently linked to "co" is "By what kind of resources?". Resources in this study are of two kinds: firm resources and customer resources. The "creation" in value co-creation pertains to the mechanisms through which value co-creation is made. Firm-, customer-, or community-led activities are the three mechanism options depending on who leads the value co-creation. Co-production, co-design, and co-development serve as another type of mechanism for value co-creation, which would perhaps make more sense if they were listed as options for the "co" in co-creation. This decomposition analysis explains as well as helps unravel the complexity involved in value co-creation.

Table 1 in Saarijärvi, Kannan, and Kuusela (2013) presents an analytical framework of value co-creation for practitioners. It consists of the three elements of "value", "co", and "creation" along with "customer" and "firm" representing parties involved in value co-creation. Due to some of the difficulties we have observed above regarding the authors' explanation and analysis, the framework in its current form runs into some problems when we take a closer look at it. For example, the "co" in value co-creation defines the actors involved in the value co-creation process but instead is supposed to address the question "By what kind of resources?". Furthermore, the authors specify that "… the additional resources - whether deployed by individual customers, firms, groups of customers, brand communities, or other configurations of actor roles – that are put together for the purpose of enhanced value creation" (Saarijärvi, Kannan, and Kuusela 2013, 11). Then, why are there only two types of actors (i.e., customer and firm) in Table 1, while groups of customers, brand communities, or other configurations of actor roles are listed in the text? Which group of actors will "brand communities" belong to, for instance, in the framework presented in Table 1?

Table 2 in Saarijärvi, Kannan, and Kuusela (2013) illustrates the framework presented in Table 1 with three real-life examples, which is a good practice in assisting readers to better understand the framework conceptually and practically when the framework ultimately is intended for business practitioners to assess and facilitate value co-creation. A more up-to-date term for co-creation is customer-centered or customer-empowered as in customer-centered system development. Given the fact that this report was published in 2013 and their research must have been done even earlier, there is nothing surprising in their usage of co-creation because it takes time for newer terminology to get utilized in research presentation.

The authors are successful in analyzing prior research on value co-creation with regard to various approaches as well as their complexity. The business-oriented analytical framework subsequently created, however, can certainly employ more systematic and rigorous analyses so that the inconsistency and other issues in the report can be addressed.

**Chapter Summary**

- Theoretical approach is a technique for gathering data through conceptual analysis, theoretical examination, or other similar activities. Some theoretical approach studies aim to develop a model, framework, or theory in a respective domain, while others intend to address their research questions by performing theoretical or conceptual analysis of collected data.
- Grounded theory, phenomenology, and hermeneutics are the three major traditions for research adopting the theoretical approach. Grounded theory emphasizes that research outcome can only be obtained by grounding its analysis on data instead of an existing theory, model, or framework. Phenomenology by contrast focuses on analytical description of participants' pure and direct experience using interview as the chief data collection technique. Hermeneutics, on the other hand, concentrates on interpretation of source material and collected data in related contexts.
- Constructivism in theoretical approach research means to obtain findings and draw conclusions by piecing together what is collected and analyzed.
- Data saturation in theoretical approach research refers to the point where no new data emerge in the data collection process, signaling the end of data gathering.
- Do prepare a solid plan for doing theoretical approach research in advance as well as observe other dos and don'ts when using the theoretical approach in research.

## References

Breckenridge, Jenna, and Derek Jones. 2009. "Demystifying Theoretical Sampling in Grounded Theory Research." *The Grounded Theory Review* 8 (2): 113–26.

Byrne, Michelle. 2001. "Hermeneutics as a Methodology for Textual Analysis." *AORN Journal* 73 (5): 968–70. https://doi.org/10.1016/S0001-2092(06)61749-3.

Charmaz, Kathy. 2000. "Grounded Theory: Objectivist and Constructivist Methods." In *Handbook of Qualitative Research*, edited by Norman K. Denzin and Yvonna S. Lincoln, 509–35. 2nd ed. Thousand Oaks: Sage.

Cibangu, Sylvain K., and Mark Hepworth. 2016. "The Uses of Phenomenology and Phenomenography: A Critical Review." *Library & Information Science Research* 38 (2): 148–60. http://dx.doi.org/10.1016/j.lisr.2016.05.001.

Corbin, Juliet M., and Anselm L. Strauss. 2008. *Basics of Qualitative Research: Techniques and Procedures for Developing Grounded Theory*. 3rd ed. Thousand Oaks: Sage.

Coyne, Imelda T. 1997. "Sampling in Qualitative Research. Purposeful and Theoretical Sampling; Merging or Clear Boundaries? *Journal of Advanced Nursing* 26 (3): 623–30. https://doi.org/10.1046/j.1365-2648.1997.t01-25-00999.x.

Croucher, Stephen M., and Daniel Cronn-Mills. 2015. *Understanding Communication Research Methods: A Theoretical and Practical Approach*. New York: Routledge.

Dowling, Maura. 2004. "Hermeneutics: An Exploration." *Nurse Researcher* 11 (4): 30–39. http://dx.doi.org/10.7748/nr2004.07.11.4.30.c6213.

Draucker, Claire B., Donna S. Martsolf, Ratchneewan Ross, and Thomas B. Rusk. 2007. "Theoretical Sampling and Category Development in Grounded Theory." *Qualitative Health Research* 17 (8): 1137–48. https://doi.org/10.1177/1049732307308450.

Educational Broadcasting Corporation. 2004. "What is constructionism?" *Concept to Classroom*. Accessed July 6, 2023. https://www.thirteen.org/edonline/concept2class/constructivism/.

Engler, Steven. 2011. "Grounded Theory." In *The Routledge Handbook of Research Methods in the Study of Religion*, edited by Michael Stausberg, and Steven Engler, 256–74. New York: Routledge.

Gadamer, Hans-Georg. 1989. *Truth and Method*. 2nd ed. Translated by Joel Weinsheimer and Donald G. Marshall. London: Sheed & Ward.

Gentles, Stephen J., Cathy Charles, Jenny Ploeg, and K. Ann McKibbon. 2015. "Sampling in Qualitative Research: Insights from an Overview of the Methods Literature." *The Qualitative Report* 20 (11): 1772–89. 10.4135/9781412950589.n885.

Gilhus, Ingvild Sælid. 2011. "Hermeneutics." In *The Routledge Handbook of Research Methods in the Study of Religion*, edited by Michael Stausberg and Steven Engler, 275–84. New York: Routledge.

Giorgi, Amedeo. 2012. "The Descriptive Phenomenological Psychological Method." *Journal of Phenomenological Psychology* 43 (1): 3–12. https://doi.org/10.1163/156916212X632934.

Giorgi, Amedeo, and Barbro Giorgi. 2003. "The Descriptive Phenomenological Psychological Method." In *Qualitative Research in Psychology: Expanding Perspectives in Methodology and Design*, edited by Paul M. Camic, Jean E. Rhodes, and Lucy Yardley, 243–73. Washington: American Psychological Association.

Glaser, Barney G. 1978. *Theoretical Sensitivity: Advances in the Methodology of Grounded Theory*. Mill Valley: Sociology Press.

Glaser, Barney G., and Anselm L. Strauss. 1967. *The Discovery of Grounded Theory: Strategies for Qualitative Research*. Chicago: Aldine.

Green, Frederick P., and Stuart J. Schleien. 1991. "Understanding Friendship and Recreation: A Theoretical Sampling." *Therapeutic Recreation Journal* 25 (4): 29–40.

Gubrium, Jaber F., and James A. Holstein. 2000. "Analyzing Interpretive Practice." In *Handbook of Qualitative Research*, edited by Norman K. Denzin, and Yvonna S. Lincoln, 487–508. 2nd ed. Thousand Oaks: Sage.

Lewicki, Roy J., Edward C. Tomlinson, and Nicole Gillespie. 2006. "Models of Interpersonal Trust Development: Theoretical Approaches, Empirical Evidence, and Future Directions." *Journal of Management* 32 (6): 991–1022. https://doi.org/10.1177/0149206306294405.

Merton, Robert King. 1968. *Social Theory and Social Structure*. Enlarged ed. New York: Free Press.

Moss, Simon. 2016. "Constructionism." Last modified June 1, 2016. Accessed July 6, 2023. https://www.sicotests.com/newpsyarticle/Constructionism.

Patton, Michael Quinn. 2002. *Qualitative Research and Evaluation Methods*. 3rd ed. Thousand Oaks: Sage.

Relph, Edward Charles. 2014. "Phenomenology." In *Themes in Geographic Thought*, edited by Milton E. Harvey, and Brian P. Holly, 99–114. 3rd ed. New York: Routledge.

Saarijärvi, Hannu, P. K. Kannan, and Hannu Kuusela. 2013. "Value Co-Creation: Theoretical Approaches and Practical Implications." *European Business Review* 25 (1): 6–19. https://doi.org/10.1108/09555341311287718.

Schwandt, Thomas A. 1998. "Constructivist, Interpretivist Approaches to Human Inquiry." In *The Landscape of Qualitative Research Theories and Issues*, edited by Norman K. Denzin and Yvonna S. Lincoln, 221–59. Thousand Oaks: Sage.

Schwandt, Thomas A. 2000. "Three Epistemological Stances for Qualitative Inquiry: Interpretivism, Hermeneutics, and Social Constructionism." In *Handbook of Qualitative Research*, edited by Norman K. Denzin and Yvonna S. Lincoln, 189–213. 2nd ed. Thousand Oaks: Sage.

Spickard, James V. 2011. "Phenomenology." In *The Routledge Handbook of Research Methods in the Study of Religion*, edited by Michael Stausberg and Steven Engler, 333–45. New York: Routledge.

Strauss, L. Anselm, and Juliet M. Corbin. 1990. *Basics of Qualitative Research: Techniques and Procedures for Developing Grounded Theory*. Thousand Oaks: Sage.

Tuckett, Anthony G. 2004. "Qualitative Research Sampling – The Very Real Complexities." *Nurse Researcher* 12 (1): 47–61. https://doi.org/10.7748/nr2004.07.12.1.47.c5930.

Wikipedia Contributors. 2022. "Hermeneutic Circle." In *Wikipedia, the Free Encyclopedia*. Last modified May 17, 2022. Accessed July 6, 2023. https://en.wikipedia.org/wiki/Hermeneutic_circle.

Wikipedia Contributors. 2023. "Hermeneutics." In *Wikipedia, the Free Encyclopedia*. Last modified July 4, 2023. Accessed July 6, 2023. https://en.wikipedia.org/wiki/Hermeneutics.

# 7 Metrics-Based Research Methods

## From Bibliometrics to Altmetrics

**Chapter Overview Questions**

- What is a metrics-based research method?
- What are the major types of metrics-based research method?
- What are the main features of the bibliometric method?
- What are the chief bibliometric techniques for data collection?
- What are the dos and don'ts in bibliometric research?
- What are the principal characteristics of the altmetric method?
- What are the dos and don'ts in altmetric research?

Metrics-based research methods refer to a range of techniques for collecting metric data (e.g., publication numbers, citation counts, tweet mentions) about publications, citations, social interaction, and more. Such techniques consist of bibliometrics, informetrics, scientometrics, webometrics, and altmetrics. They all focus mainly on quantitative data that can be processed to yield metrics such as journal impact factors (JIFs), popularity scores on social media, and the like.

Chronologically speaking, bibliometrics is established first and has subsequently been applied to the scholarly world for forming scientometrics and to the information realm for developing informetrics. Similarly, webometrics is the outcome of applying bibliometric methods to the web although webometrics has been on its way out as a research method after the emergence of social media applications like Facebook and Twitter. Gathering metric data from those social media platforms along with other sources having identical features enables researchers to address a variety of questions (e.g., the public's reactions to the COVID-19 outbreak), leading to the emergence of the newest metrics-based research method: alternative metrics or altmetrics in short.

Metrics-based research methods, as a unique category of research methods, have been utilized for evaluating research, tracing the development of a discipline, understanding the intellectual structure of a field or domain, exploring scholarly exchanges among researchers, gauging the sentiment of social media users, and more. The characteristics of these metrics-based methods vary from one to another, and they will thus be discussed as each is presented in this chapter.

DOI: 10.4324/9781003021711-9

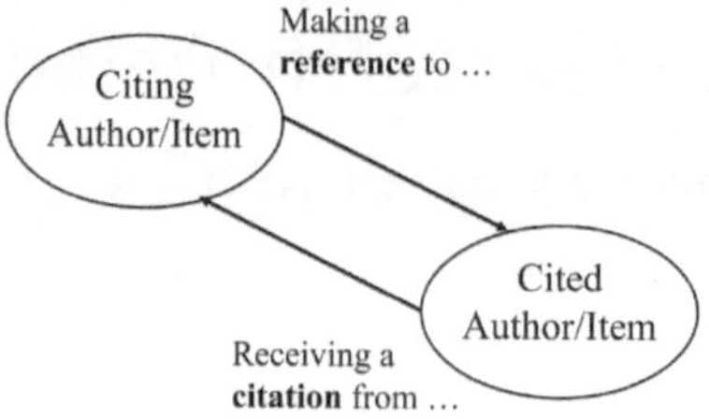


*Figure 7.1* References Versus Citations

## 7.1 Bibliometrics, informetrics, and scientometrics

Bibliometrics is a method used for collecting publication and reference/citation data from written publications. References and citations are two terms that represent a same concept from two different angles (see Figure 7.1). If this concept is looked at from the perspective of the citing author, it is called "reference" when the citing author makes a reference to a publication. When the same activity is examined from the cited author's viewpoint, we would say that the cited author receives a "citation". References and citations are thus synonymous. Nevertheless, citations will be used for representing both citations and references throughout this book for brevity purposes.

Bibliometrics is often applied in research with an evaluative intention. In this book, bibliometrics also includes citation analysis, informetrics, and scientometrics because they all share the same type of data – publication and citation data. It is the research focus that actually differentiates one from the other. When the research focus is on citations, it is called citation analysis. When the research focus is on information, it is called informetrics. If the research emphasis is on sciences, it is called scientometrics. In addition to publication and citation data, bylines, acknowledgments, and other similar components of publications can also serve as source data for this research method.

Besides its connection with bibliometrics, scientometrics (aka science of science) studies the history and sociology of science using primarily citation and publication data as well as other kinds of data (e.g., national expenditure on research and development, scientific manpower). Informetrics, on the other hand, explores the quantitative aspects of information in any form, not just records or bibliographies (Rousseau, Egghe, and Guns 2018, 3). Data sources for informetrics range from publications and citations to data on information growth and obsolescence (e.g., half-life or the median age of the citations produced or received). Although both scientometrics and informetrics have their unique orientations as research methods, what to be discussed in this book, however, will mainly cover the common areas scientometrics and informetrics share with bibliometrics in their usage of publication and citation data.

## 7.2 Major bibliometric data sources

There are two major sources when bibliometrics is adopted as the data collection techniques: Web of Science (WoS) and Scopus. Google Scholar (GS) and other data sources are all supplemental when compared with WoS and Scopus although the magnitude of GS is indeed impressive.

### *7.2.1 ISI's citation products*

The Institute for Scientific Information (ISI), part of Clarivate in 2021 after having several different parent companies in the past two decades, was founded by Eugene Garfield in Philadelphia in 1956. It got settled on ISI as its name in 1960 after going through several name changes (Small 2017). The ISI is known particularly for two of its citation products: citation indexes (predecessors of the Web of Science Core Collection, abbreviated as Web of Science or WoS in this book) and Journal Citation Reports (JCR).

ISI's citation indexes started with the publication of the Science Citation Index (SCI) in 1961 (Garfield 1963), followed by the Social Science Citation Index (SSCI) and the Arts & Humanities Citation Index (A&HCI or AHCI). The Conference Proceedings Citation Index (CPCI), the Book Citation Index (BCI), and the Emerging Sources Citation Index (ESCI), which cover scientific literature of regional importance and in emerging scientific fields, were all developed much later. The CPCI, BCI, and ESCI are also much less influential than the first trio as sources for citation data. Among the first three citation indexes, the SCI has been used the most often as a source for citation data collection, and the SSCI comes the next while the A&HCI ranks the last in terms of usage frequency. The Science Citation Index Expanded (SCIE) is an online enhancement of the SCI which implies enlarged coverage compared to the printed volumes of the SCI.

All the citation indexes are created on the same essential assumption that the references authors make when writing an article serve as indexing to the cited documents. When an author cites others' publications, a subject relationship is created between the citing document and cited document. The citing author will not make the citation if the cited publication is of no relevance. Different from subject indexing through which the indexer assigns descriptors and/or subject headings to target documents, citation indexing offers an innovative form of indexing from the citing author's perspective and can easily be automated for retrieval purpose. In other words, citation indexes were initially constructed to facilitate searching documents related to citing articles (Garfield 1963). Later the same citation indexes such as the SCI and its counterparts in the social sciences, arts, and humanities are eventually used for research evaluation. The rationale is, in its simplest and naïve form, that the more citations a publication received, the better its research quality (Rousseau, Egghe, and Guns 2018, 35). It is for the purpose of research evaluation that citation indexes become a source for data collection in bibliometric research. Therefore, the present book approaches citation indexes primarily from the perspective of research evaluation.

The journals indexed in the SCI, SSCI, and A&HCI are selected by an editorial board to ensure the production feasibility and quality of the final citation index products. Due to the journal selectivity, journal editors commonly take great pride when the journals they edit are indexed in ISI's citation indexes. Only major and top journals in a field are included in the SCI, SSCI, and A&HCI.

In addition to the citation indexes, ISI has been publishing its JCR annually since 1976. The JCR presents a number of journal metrics such as the impact factor (IF), the immediacy index, and half-life statistics. The latter two each deal with the time distribution of citations received or produced, respectively. These metrics, particularly the IF score, are used to guide journal selection by authors when submitting their research manuscripts. They are also adopted by journal editors and publishers, as well as by some librarians as measures of a journal's performance (Small 2017). Further explanation and elaboration on IF will be covered in Section 7.3.2.

One more unique service that WoS currently provides is the linkages to citation databases in other countries and regions at its website. They include Russian Science Citation Index, Chinese Science Citation Index, SciELO Citation Index (covering journals in Latin America, Spain, Portugal, the Caribbean, and South Africa), and Arabic Citation Index.

### *7.2.2 Scopus*

Elsevier, a major publisher then and a company of information and analytics now, launched Scopus in 2004 which has steadily been becoming a contender of WoS. Like WoS, Scopus is a subscription-based database that includes citation data as well. In addition to journals, Scopus also covers trade publications, book series, conference proceedings, and books. The "in press" articles from close to 4,000 journals Scopus gathers offer researchers with more timely access to scholarly publications (Rousseau, Egghe, and Guns 2018, 151). According to Elsevier (2022), newly suggested titles to be included in Scopus are reviewed, using transparent criteria, by the independent and international Scopus Content Selection and Advisory Board (CSAB).

For each source journal included in Scopus, Scopus lists a set of metrics (e.g., CiteScore, Scimago Journal Rank – SJR) for measuring a journal's impact based on citation and publication data. Unlike its fee-based counterpart – the JCR, the SJR is a publicly available portal (https://www.scimagojr.com/journalrank.php) developed jointly by Elsevier and Scimago Lab in Spain. The same portal also provides ranking of countries involved in the Scopus database (https://www.scimagojr.com/countryrank.php).

Because Scopus was developed more than four decades later than WoS, it takes the advantage of the web platform and does not have any baggage WoS had with its initial mainframe system. The old platform of WoS' predecessors (i.e., SCI, SSCI, and A&HCI) featured poor interface and clumsy operations as far as its users are concerned. If purely serving as a source for citation data, Scopus offers friendlier interface to its users when compared with WoS. The characteristics of Scopus, however, have been studied less extensively than those of WoS due to Scopus' shorter history (Waltman 2016). Most of the research on Scopus has been done in contrast to WoS (e.g., Chadegani 2017; Goodman and Deis 2005; Harzing and Alakangas 2016). Table 7.1 compares WoS and Scopus in several aspects. All the numbers listed in Table 7.1 were derived from https://clarivate.com/webofsciencegroup/solutions/web-of-science and https://www.elsevier.com/solutions/scopus, respectively in January 2021 unless indicated otherwise. Changes in those numbers are inevitable as time passes. Similarly, the URLs referred to in this chapter are likely to change or be removed for one reason or another.

As shown in Table 7.1, the coverage of Scopus is wider than that of WoS although WoS currently has a lead in the total number of records due to its much longer history. Their difference in coverage mainly results from their selection/evaluation criteria and subject orientation. While WoS was initially created to facilitate citation searching, Scopus was introduced by Elsevier who has been a major publisher and information provider in STEM (M in this context stands for "medicine" rather than the usual "mathematics") since 1880 (Groen 2007, 217). Scopus contents are thus derived from Elsevier's numerous publications and databases whose quantity far exceeds that of WoS. Furthermore, WoS appears to be more selective in determining which titles are included in its data (Clarivate 2023), while Scopus seems to be more inclusive but does not mean less selective (Elsevier 2023a).

As far as subject orientation is concerned, Scopus' strengths lie in the sciences. In most science fields, it has significantly greater coverage of third world and other non-English

*Table 7.1* Comparison of WoS and Scopus

| | *WoS* | *Scopus* |
|---|---|---|
| Coverage[1] | Over 171 million records from 14,000+ titles | 75+ million records from 24,600+ titles |
| Selection/evaluation criteria | 24 quality and 4 impact criteria[2] | 5 categories of criteria[3]: journal policy, content, journal standing, publishing regularity, online availability |
| Subject orientation | SCI – 178 (67%) sciences; SSCI – 58 (22%) social sciences; A&HCI – 28 (11%) arts and humanities | Health sciences – 25%; physical sciences – 27%; life sciences – 16%; social sciences – 32% |
| Data provided | Publication data (e.g., citing author, citing journal), citation data (e.g., cited author, cited journal), IF, immediacy index, half-life statistics, h-index | Publication data (e.g., citing author, citing journal), citation data (e.g., cited author, cited journal), h-index, CiteScore, Scimago Journal Rank (SJR) |
| Number of cited authors indexed | First author | Up to first eight authors |
| Cost[4] | $100,000 per year | 85–95% of WoS cost |

language publications than WoS with regard to recent publications (Goodman and Deis 2005). However, its weakness in covering the humanities is apparent while WoS has the A&HCI devoted to this domain. In the years since its creation in 2004, Scopus has expanded its titles in the social sciences that accounted for 32% of its total content in 2019. In comparison, the percentage of the social sciences titles in WoS only reached 22 (see Table 7.1) in 2020.

When WoS or Scopus is chosen as the source for bibliometric data collection, what each can provide becomes the determining selection factor if the researcher does not have access to or does not want to use both sources. Both sources offer publication and citation data such as publication and citation counts. In addition, they each provide a variety of ways to analyze and visualize search results. Table 7.2 summarizes the common and unique options of both WoS and Scopus in presenting search results at their own respective site.

Furthermore, WoS only indexes the first author of each cited reference, while Scopus indexes up to the first eight cited authors, which means that citation data on any author after the first are not available from WoS (Zhao and Strotmann 2010). This indexing policy would exclude the possibility of collecting data on any coauthors and thus eliminate the potential of studying coauthorship, collaboration and more in research when using WoS as a bibliometric data source.

*Table 7.2* Common and Unique Data Options in Search Results

| *Data Source* | *Common Data Options* | *Unique Data Options* |
|---|---|---|
| WoS | Author,[5] publication year, affiliation, country/region, source, document type, subject category, funding agency | Organization-enhanced, book series title, editor, group authors, language, research area, KeyWords Plus, grant number |
| Scopus | | Source by year, author keywords and indexed keywords (including subject terms assigned by professional indexers) |

The search results from WoS or Scopus can be printed, emailed, or downloaded according to one's needs. Such results can also be exported in multiple formats for future data analysis. Specifically, WoS supports CSV and other unspecified formats, whereas Scopus offers CSV, BibTeX, and plain text in addition to formats for other purposes (e.g., reference management).

### *7.2.3 Google Scholar*

While WoS and Scopus are the two major fee-based sources for bibliometric data collection, GS can serve as additional sources for gathering basic bibliometric data. In fact, GS is primarily a search tool for locating scholarly publications. Offering data for bibliometric research is its secondary feature only. Microsoft Academic was a source similar to GS until the end of 2021 and it has been retired since then (Microsoft Academic 2021).

Google launched a beta version of GS in 2004 which focuses exclusively on scholarly contents available from the public web (Rousseau, Egghe, and Guns 2018, 105). GS in 2007 announced to expand its coverage to include abstract and indexing information from main publishers who would be more than willing to share their bibliographic data in order to publicize their fee-based full-text publications. According to Khabsa and Giles (2014), it is estimated that at least 114 million English-language scholarly documents are accessible on the public web, of which GS has nearly 100 million. One can search GS as in regular Google searching. Search by title in GS can yield incredibly accurate results than search by author or keyword. Along with each search result, GS lists corresponding citation counts. If an author has set up a profile at GS, GS would provide additional citation data (e.g., total citation counts, h-index) for that author.

GS can provide meaningful bibliometric data to complement the selective coverage of both WoS and Scopus. Although some researchers may rely solely on GS for collecting bibliometric data, it is not recommended if you have access to WoS and/or Scopus because the latter offer much more data than citation counts and the h-index. In addition, GS cannot be utilized directly for collecting massive bibliometric data as can be done with WoS or Scopus. Instead, an app or algorithm must be created specifically before substantial bibliometric data can be collected. Publish or Perish (PoP) (https://harzing.com/resources/publish-or-perish), an OA (open access) program developed by Anne-Wil Harzing, is such an example for gathering bibliometric data from GS and more. This program can produce the following metrics for a given set of retrieved articles:

- The total number of publications
- The total number of received citations
- The average number of citations per publication
- The average number of citations per author
- The average number of articles per author
- The average number of citations per year
- The h-index, some of its variants, and more

## 7.3 Major bibliometric techniques

Several major bibliometric techniques have been developed over the decades for gathering bibliometric data in research. As noted in Section 7.1, bibliometric data consist of both publication and citation data. But we in general only use citation data to cover both types of them. Citation analysis is perhaps the best known data collection technique for bibliometric research.

### *7.3.1 Citation analysis*

Swales (1986, 39) once commented that "[c]itation analysis is something of a misnomer, in that it usually describes studies of bibliographical references, rather than studies of citations embedded in continuous academic text". Indeed, citations as a term in bibliometric research refer to not only citations but also references as depicted in Figure 7.1. We customarily call it citation analysis chiefly for reasons of brevity. Furthermore, the word "analysis" in citation analysis denotes that some analysis must be performed in order to obtain citation data, including the simple raw counts of citations, authors, journals, and so forth.

Even though citations have broader connotations to cover publication and reference data, any citation analysis is made on the assumption that there must be some subject relationship between the two publications involved when an author (i.e., citing author) cites another one (cited author). Another assumption regarding citations is that the citing author makes use of the cited document through the act of citation (Smith 1981). Both assumptions as well as the so-called reasons for citation (i.e., for what purpose is a citation made), however, are open to debate. The topic of reasons for citation will be covered in Section 7.4.1.

There are three measures commonly used for collecting citation data:

- Direct counting – A technique that determines how many citations a given document, author, journal, etc. receive over a period of time. For example, Vannevar Bush, whose concept of the Memex device motivating research and development on hypertext and eventually on the web, published in 1945 a paper titled "As we may think". This paper has been cited at least 10,444 times by June 2023 according to GS.
- Bibliographic coupling – A technique, proposed by Kessler (1963) of MIT, for grouping documents according to a measure called bibliographic coupling. As shown in Figure 7.2-I, two documents are bibliographically coupled if their reference lists share one or more cited documents. The more times the two documents cite the same other documents, the stronger the relationship (i.e., coupling strength) is between the two citing or bibliographically coupled documents.
- Co-citation counting – A technique, simultaneously proposed by Henry Small in the United States and Irina Marshakova from the former Soviet Union in 1973 (Smith 1981), for gathering data about two documents that are cited jointly (i.e., co-cited) in one or more subsequently published documents (see Figure 7.2-II). The more times the two documents are cited together, the stronger their relationship, measured in co-citation frequency, is (Small 1973).

Among the three citation data measures, direct counting is the simplest and most straightforward technique for gathering citation data. But simple citation counts cannot be employed for exploring the "assumed" subject relationships among citing and cited documents. Bibliographic coupling can be used as a technique for grouping documents to reflect the current research activities in a domain or field (Zhao and Strotmann 2008). Co-citation analysis, on the other hand, has been frequently used in the past decades to illustrate the intellectual structure of a discipline (White and Griffith 1981). Both bibliographic coupling and co-citation analysis are able to achieve what direct counting of

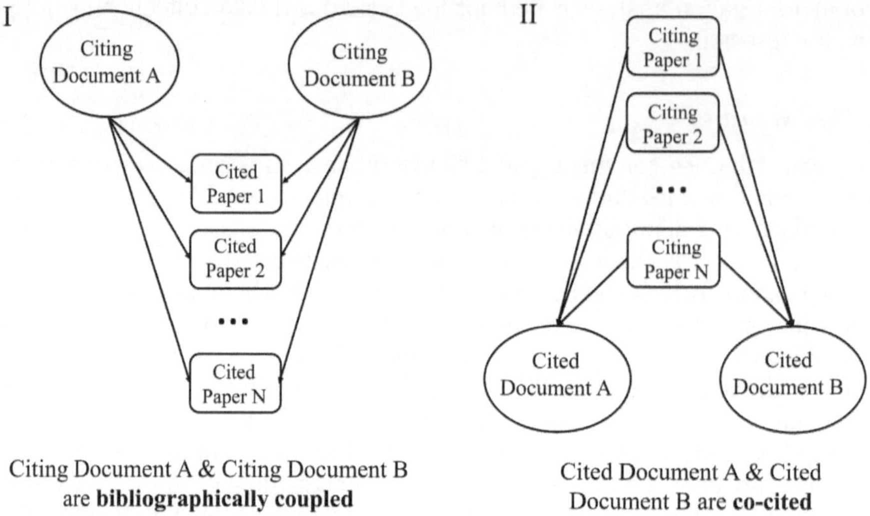


*Figure 7.2* I. Bibliographic Coupling; II. Co-Citation

citations cannot: revealing the subject relationships that otherwise underlie the citing and cited documents. However, bibliographic coupling is adopted much less often than co-citations as a technique of data collection in bibliometric research (Rousseau, Egghe, and Guns 2018, 139).

### *7.3.2 Impact factor, the h-index, and more variations*

Besides citation, bibliographic coupling, and co-citation counts presented in Section 7.3.1, additional measures are devised as bibliometric indicators to facilitate research. The IF is one such measure that has been long employed in bibliometric studies.

#### *7.3.2.1 Journal impact factor and CiteScore*

IF scores are usually computed for journals (including conference proceedings, book series, and other similar titles) and thus called JIF scores. The JCR, as briefly described in Section 7.2.1, present JIF among other bibliometric statistics such as the immediacy index (i.e., the average number of times an article is cited in the year it is published). The formula for calculating JIF is as follows:

$$\text{JIF} = \frac{\text{Citations(A)}}{\text{Citable Items(B)}}$$

where

- A – Number of times the "citable items" (e.g., articles, reviews) published in a journal in two years (e.g., 2015–2016) were cited in all indexed journals during the third year (e.g., 2017)
- B – Number of citable items published in the same two preceding years (e.g., 2015–2016)

As seen, JIF is the ratio between the number of citations received in a given year for articles of a journal that were published in the two preceding years and the total number of citable items published in that journal during the two preceding years. Take *Nature* as an example, this world's leading multidisciplinary science journal published 902 and 880 citable items, respectively, in 2015 and 2016 while those 1,782 citable items (i.e., B) together received 74,090 citations (i.e., A) in 2017. Hence, *Nature*'s JIF score in 2017 is A/B = 74,090/1,782 = 41.577. The JCR customarily reports three decimal points of a JIF in order to allow differentiation among the large number of journals it covers. A JIF of 41.577 implies that all the citable items published between 2015 and 2016 in *Nature* on average received more than 41 citations in 2017.

The JCR is published annually, which means that the JIF score for a same journal varies from one year to another. JIF scores are commonly regarded as a quality indicator of individual journals. However, we need to take any JIF's value in perspective rather than in isolation because citation counts used in computing JIF are influenced by many factors. For instance, journals in sciences and medicine generally have higher JIF scores with values in two or three digits (e.g., *Nature* – 41.577 in 2017; *CA: A Cancer Journal for Clinicians* – 292.278 in 2019), whereas journals in the humanities and social sciences receive JIF scores of a single digit or even a lower value (e.g., *Comparative Politics* – 1.481 and *Visual Communication* – 0.523 in 2019). Therefore, multiple other factors (e.g., reputation of the author or journal) should be taken into consideration if the quality of a journal is to be assessed fully and objectively.

CiteScore is a journal impact indicator that Elsevier, the parent company of Scopus, introduced in December 2016, serving the same purpose as the JCR's JIF but with some differences (Elsevier 2023b; Rousseau, Egghe, and Guns 2018, 188). First, CiteScore is computed for all Elsevier publications indexed in Scopus whose quantity is much larger than that of WoS (see Table 7.1). Second, CiteScore values are currently displayed to one decimal point to avoid an impression of superfluous precision and be in line with industry best practice while CiteScore values were previously displayed to two decimal points. Third, CiteScore is listed with every journal in Scopus, rather than in a separate publication as JIFs in the JCR. The full Scopus source list, along with CiteScore and related journal information, can be downloaded free from https://www.scopus.com. Fourth, the most essential difference between JIF and CiteScore is that CiteScore is calculated in a longer time window than JIF (see Figure 7.3). Instead of including citations made in one year to the two preceding years, the current version of CiteScore extends that two-year window to three preceding years.

For instance, when we calculate a CiteScore for *Nature* in 2017, the numerator A stands for all the citations *Nature* received in 2017 for the citable items it published in 2014, 2015, and 2016. The denominator B refers to all the citable items *Nature* published

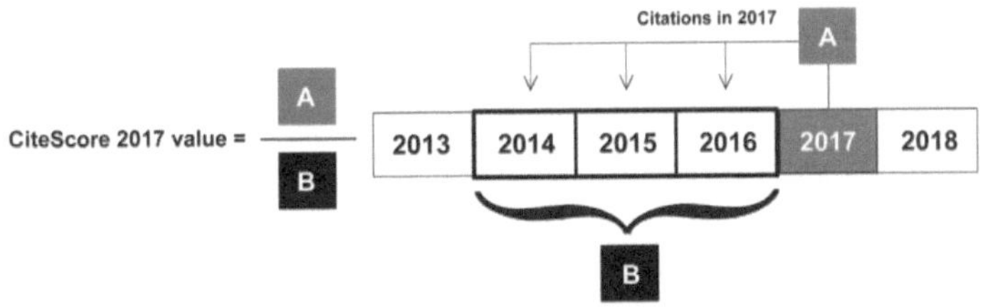


*Figure 7.3* Formula for Calculating CiteScore[6]

in 2014, 2015, and 2016. Figure 7.3 visualizes the calculation of CiteScore for any single title in 2017.

According to Figure 7.3, *Nature*'s CiteScore in 2017 was 53.7, higher than its JIF score of 41.577 in the JCR in the same year. That tendency remains true for all the CiteScore values due to the longer time window of CiteScore calculation, and the different data sources JCR and Scopus each use in computing the IF.

JIF and CiteScore are the two IFs dominating the bibliometrics domain for journal evaluation. In neither case should the IF be used as a proxy for journal quality because IF reflects only one aspect of journal quality (Rousseau, Egghe, and Guns 2018, 201–202). This indicates that IF scores should just be collected and utilized as one type of data for assessing the quality of a journal. Other relevant data also need to be gathered in order to gain a complete picture of any journal.

#### *7.3.2.2 H-index and its variations*

While JIF and CiteScore are bibliometric indicators for journals, the h-index serves a similar function in measuring individual authors' influence on a subject area or domain. This index is named after its creator, an American physicist Jorge E. Hirsch, who suggests that both the publication activity and citation impact of individual scientists should be considered in assessing their research performance (Hirsch 2005). Before Hirsch's proposal of the h-index, citations are often used alone in evaluating individual scientists' impact on the field. By integrating both the number of publications and citation counts to form the h-index, a new indicator for assessing individual scientists' research performance, some limitations in using only citations for evaluation purpose can be overcome in bibliometric research.

The calculation or, more accurately, identification of h-index is straightforward. First, rank an author's all publications by the number of citations each receives in descending order. Second, identify which publication has at least the same number of citations (i.e., h) as its rank (i.e., $h^{th}$) in the ordered list of all publications. Figure 7.4 illustrates one hypothetical example of the h-index.

The hypothetical author in Figure 7.4 has 17 publications that are ranked by the number of citations each receives in descending order. Figure 7.4 shows that the sixth of the

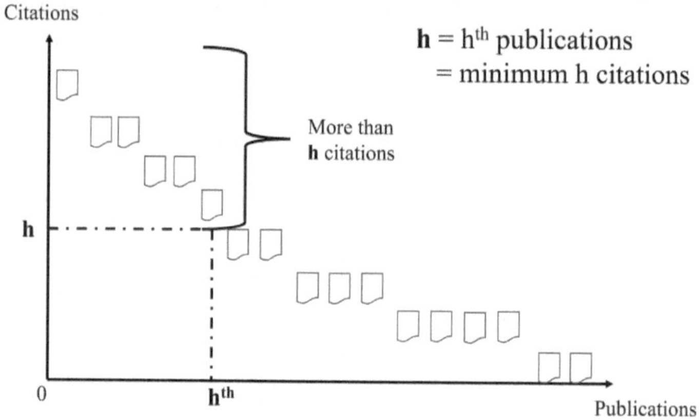


*Figure 7.4* A Hypothetical Example of the h-index[7]

author's publications is cited at least (i.e., a minimum of) six times. Thus, the h-index for this author is 6. In other words, the h-index is a metric that denotes the $h^{th}$ publication of an author receives at least h citations. The higher the h, the more influential the author is when compared with others in the same domain or field. Of the major citation sources covered in this chapter, WoS, Scopus, and GS all provide the h-index as one bibliometric measure.

The main advantage of the h-index is that it combines the assessment of both the number of publications (quantity) and their impact (quality) in terms of citations (Glänzel 2006). Other advantages of the h-index include its computation simplicity, encouraging highly visible works, and allowing any level of aggregation (e.g., at the level of individual authors, institutions, countries). Furthermore, single top publications have no influence on the h-index whose computation involves publication counts as well as citation counts (Rousseau, Egghe, and Guns 2018, 209). The h-index has therefore become one of the most widely used metrics in academia for measuring the productivity and impact of researchers.

When Hirsch proposed the h-index as an objective measure of scientific achievement in 2005, he did not think that it would be used outside his domain of interest – theoretical physics (Hirsch 2020). The h-index consequently has been mis- or over-interpreted partly because of its intrinsic characteristics. Like most citation measures, the h-index is field-dependent, database-dependent, and could be influenced by self-citations (i.e., authors cite their own publications). The number of coauthors in a publication may influence the number of citations received and subsequently inflate the h-index for people who coauthored many publications. It is also rather difficult to collect all data necessary for the determination of the h-index. Often an author's complete publication list is needed in order to discriminate between researchers with the same name and initial(s). This author name disambiguation problem is especially troublesome when only initials are given for authors' first and middle names in databases and other citation sources (Rousseau, Egghe, and Guns 2018, 209). The same disambiguation problem becomes more serious when Chinese authors' last names are recorded in *pinyin*, the Romanized spelling system for transliterating Chinese. There are only hundreds of Chinese last names in China's population of over 1.4 billion people. Yet the top three Chinese last names Wang, Li, and Zhang combined account for more than 270 million people (Wikipedia Contributors 2023). There could be thousands of authors in a large database whose names are, for example, J. Wang. Additionally, it is almost impossible to establish a standard of reference for addressing questions such as "What is a 'high' h-index in field X for a researcher who has been active in the past 10 years?" (Rousseau, Egghe, and Guns 2018, 210).

Moreover, Rousseau, Egghe, and Guns (2018, 210) point out that "[t]he h-index lacks sensitivity to performance changes: it can never decrease and is only weakly sensitive to the number of citations received. Concretely, once an article belongs to the h-core it does not matter anymore how many extra citations it receives. Conversely also for most articles in the tail the exact number of citations or increase in citations has no influence" on the h value. The h-core in the above quote refers to all the publications of an author that each receives h or more citations (e.g., the six publications with six or more citations in Figure 7.4). The lack of sensitivity is the main reason for the introduction of most variations of the h-index.

The i10-index is one alternative to the h-index, referring to the number of publications with ten or more citations. This measure considers any publication of an author as long as all of them are each cited at least ten times. The i10-index is always higher than the h-index if an author's h-index is greater than 10. GS lists the i10-index next to the h-index for any scholar who has a profile at its website. For example, the h-index for Albert Einstein is 113 (i.e., a total of 113 Einstein's publications are each cited at least 113 times), while his i10-index is as high as 385 (i.e., altogether 385 Einstein's publications are each cited at least 10 times). While the i10-index may not mean much for great scientists like Einstein, it could make novice researchers' academic profiles look a bit better.

Because Hirsch's h-index cannot be used to compare researchers who coauthor publications or are at different career stages, Harzing, Alakangas, and Adams (2014) develop a metric that corrects for these differences. It is called the hI,annual (or hIa for short) index. According to Harzing and her colleagues, authors in the life sciences and sciences typically publish more but shorter articles and also with a large number of coauthors. In comparison, authors in the social sciences and humanities normally publish fewer but longer articles (or even books) and with a smaller number of coauthors, if there is any. The hIa-index thus includes two additional pieces of information in its calculation, coauthorship and academic age, as follows (Harzing, Alakangas, and Adams 2014):

$$\text{hIa} = \frac{\text{hI,norm}}{\text{academic age}}$$

where

- hI, norm – Normalize (abbreviated as "norm" in the above formula) the number of citations for each publication by dividing the number of citations with the number of authors for that publication (i.e., fractional counting), and then calculate the h-index for a coauthor using the normalized citation counts. For instance, when a publication has four coauthors and receives 60 citations, each author would be considered having received 15 (i.e., 60/4) citations instead of the total of 60 citations to that publication.
- academic age – Number of years elapsed since an author's first publication.

The hIa-index enables the comparison of publications by the number of coauthors and academic age on top of what is already implied in the h-index. The software for computing the hIa-index, along with instructions, is available from https://harzing.com/resources/publish-or-perish.

The g-index is another variant of the h-index. It was proposed by Leo Egghe, a noted figure in metrics-based research, in 2006 (Egghe 2006), about one year after the h-index was published. It aims to improve the h-index by measuring the global citation performance of a set of publications and is thus named the g-index where g may stand for "global" although Egghe did not explicitly specify so. The g-index allows citations from the top cited publications to be used in its calculation in order to bolster publications that are cited fewer times or even without any citation in meeting this threshold as long as the top cited publications each on average receive at least g citations. Table 7.3 illustrates the difference between the h-index and the g-index using a set of hypothetical data.

In Table 7.3, the $g^2$ column needs some explanation. The $g^2$ value is a maximum number (i.e., g) in squared form ($g^2$) that can be derived from the citation total (i.e., g ≤ citation total). The g could be either equal to (e.g., A2, A3) or less than (e.g., A1, A4) the sum of all the citations an author receives for his/her publications. For example, the citation total for Author A1 is 33, the maximum number $g^2$ can take is 25 (i.e., $5^2$), smaller than the citation total of 33. The g-index for A1 is hence 5. The g-index, according to Egghe (2006), is intended to overcome the h-index's insensitivity to an author's top cited publications once

*Table 7.3* Comparison of the h-Index and the g-Index with Hypothetical Data

| *Author* | *Publications* | *Citations Received* | *h-index* | *Citation Total* | $g^2$ | *g-index* |
|---|---|---|---|---|---|---|
| A1 | 4 | 15, 10, 5, 3 | 3 | 33 | $5^2$ | 5 |
| A2 | 6 | 3, 3, 3, 0, 0, 0 | 3 | 9 | $3^2$ | 3 |
| A3 | 6 | 60, 4, 0, 0, 0, 0 | 2 | 64 | $8^2$ | 8 |
| A4 | 9 | 15, 15, 12, 12, 8, 7, 7, 3, 1 | 7 | 80 | $8^2$ | 8 |

they are among the h-core. However, as shown in Table 7.3, the g-index merely considers the average number of citations of the top cited publications in its computation. The number of publications an author has (e.g., A1, A3) and low or no citations to them (e.g., A2, A3) are not taken into account. In the cases of A1 and A3, the g-index value is larger than the number of publications while the g-index completely ignores the fact that there are no citations to several of A2 and A3's publications at all. Both points apparently deviate from Hirsch's original intention in creating the h-index, namely, measuring authors' research performance by considering both their publication activity (i.e., number of publications) and citations to their publications (Hirsch 2005).

As shown in Table 7.3, g is always at least equal to h in value; and in most cases, g is higher than h. If one believes that a good scientist is characterized by a reasonable number (not just a few) of highly cited publications, then the h-index is preferable to the g-index. If, however, one believes that even a single very highly cited publication makes a good scientist, the g-index is preferable (Bartolucci 2015). That is the recommendation from Bartolucci (2015) for helping one to decide whether the h-index or the g-index should be adopted as a metric for assessing an author's research performance.

### *7.3.3 Bibliometric laws*

Bibliometric laws are intended to help researchers better understand certain phenomena about publications, author productivity, and more. Bradford's Law, Lotka's Law, and Zipf's Law are the three most prominent bibliometric laws in studies using metrics-based research methods. They all seek to describe some aspects of scientific publications by mathematical means. While each of these laws applies to a specific phenomenon, they all have a common base as they each demonstrate that a small portion of journals, authors, or words, respectively, accounts for the many relevant contributions, publications, or terms used in scientific writings (Ikpaahindi 1985).

Bradford's Law depicts how the scholarly literature on a subject is distributed in journals. Samuel C. Bradford, the creator of this law, was a British mathematician and librarian at the Science Museum in London. When searching for articles on applied geophysics and lubrication in 1948, Bradford noticed that the scatter of such papers among journals had a common pattern. For this reason, Bradford's Law is also known as the Law of Scattering or Law of Concentration. Starting with the journals that contained the most relevant articles, Bradford divided all those articles he found on applied geophysics and lubrication from different journals into three roughly equal groups (see Table 7.4).

The mathematical expression of Bradford's Law is $k{:}kn{:}kn^2$, where $k = 9$, $n = 5$ in the above example of publications on applied geophysics and lubrication. When journals contributing to a subject area are ranked and then grouped in such a way that each group contributes approximately the same number of articles, the numbers of source journals

*Table 7.4* Application Example of Bradford's Law

| *Group* | *# of Articles* | *# of Source Journals* | *k:kn:kn²* |
|---|---|---|---|
| 1 | 429 | 9 | $k = 9$ |
| 2 | 499 | 59 | $kn = 9 \times 5 = 45$ |
| 3 | 404 | 258 | $kn^2 = 9 \times 5^2 = 225$ |

in each group increase geometrically. That is, the first group of 9 journals contribute 429 articles on applied geophysics and lubrication. The second group of 59 journals carry 499 articles, while the third group of 258 journals only contain 404 relevant articles, much fewer than what the first group of 9 journals can offer. Do note the difference between the actual number of source journals and the number calculated using the $k:kn:kn^2$ formula for the second and third groups of source journals (see Table 7.4). The difference is due to the fact that $k:kn:kn^2$ is an approximation rather than an exact calculation.

Over the decades, Bradford's Law has been found to be applicable in many other fields such as astronomy, information science, marine biology, tropical agriculture, and veterinary medicine (Ikpaahindi 1985). Implications of this law include how researchers should direct their attention to publications to stay current in their own respective domains. They may only focus on reading journals in the first group (i.e., k), scanning titles in the second group (i.e., kn), and can ignore journals in the third group (i.e., $kn^2$) if they do not have time. For any subject domains where Bradford's Law is observed, journals in the first group are considered core or key journals. Both authors and readers would select journals in the first group for publishing or perusing if possible. It must be pointed out, however, that the scattering or concentration of articles on a subject in journals is only an approximation rather than an exact calculation in almost all the domains examined.

Lotka's Law posits that there are a few researchers who publish a lot, while many publish very little or nothing at all in a given field according to Alfred James Lotka, an American physical chemist and statistician. Lotka did a study on authors in chemistry and physics in the 1920s and observed that the number of authors (k) publishing a certain number of articles (n) in a given field can be expressed in the formula $k/n^2$. Table 7.5 displays an application example of Lotka's Law. For instance, when there are 100 authors (k) who publish one article in a given field, 25 authors would publish two articles; ... and only one author would publish eight articles in the same field based on the $k/n^2$ formula.

What Table 7.5 presents, as seen, is an empirical observation instead of an exact calculation using the $k/n^2$ formula. When the number of articles published increases, the number of authors producing that many publications becomes fewer. This law hence is

*Table 7.5* Application Example of Lotka's Law

| *k (# of Authors)* | *n (# of Articles Published)* | *k/n²* |
|---|---|---|
| 100 | 1 | $100/1^2 = 100/1$ |
| 25 | 2 | $100/2^2 = 100/4$ |
| 11 | 3 | $100/3^2 = 100/9$ |
| 6 | 4 | $100/4^2 = 100/16$ |
| 4 | 5 | $100/5^2 = 100/25$ |
| 3 | 6 | $100/6^2 = 100/36$ |
| 2 | 7 | $100/7^2 = 100/49$ |
| 1 | 8 | $100/8^2 = 100/64$ |

also known as the Inverse Square Law of Authors' Productivity. In addition, Lotka's Law has been tested and applied in anthropology, computer science, geography, information science, and library science (Ikpaahindi 1985). Lotka's Law can help people identify top or key authors who publish many articles in their respective fields.

George Kingsley Zipf, a Harvard professor of linguistics and philology, studied statistical occurrences in different languages and published in 1949 a book titled *Human Behavior and the Principle of Least Effort: An Introduction to Human Ecology*. The principle of least effort proposed in the book suggests that people find it easier to choose and use familiar words rather than unfamiliar ones. Therefore, the probabilities of occurrence of familiar words in writing and speaking are higher than those of unfamiliar ones. This principle explains a statistical law of linguistic distribution which states that if words are ranked (r) according to their frequency of occurrence (f), the product of r and its corresponding f is equal to a constant (C), namely, $r \times f = C$. In other words, the more frequently a word is used, the higher it would be ranked. The expression $r \times f = C$ is called Zipf's Law. It is named after Zipf because he worked on it and subsequently popularized it. The regularity expressed in this statistical law though was noted by other researchers before Zipf (Rousseau 2002).

In order to demonstrate the law, Zipf arranged the 29,899 different words found in Irish writer James Joyce's *Ulysses* in descending order of their frequency of occurrence. To each word he assigned a rank (r), from 1 to 29,899 with 1 being the highest. He then multiplied the numerical value of each rank by its frequency (f) and obtained a product C which was more or less a constant throughout the entire list (Zipf 1949). Table 7.6 shows the values of selected sets of r, f, and C ($r \times f$) in the distribution of words in Joyce's *Ulysses* illustrating Zipf's Law. This table[8] is taken from Table 2-1 in Zipf's book *Human Behavior and the Principle of Least Effort: An Introduction to Human Ecology*.

*Table 7.6* Distribution of Words in Joyce's *Ulysses* Illustrating Zipf's Law

| *I* *Rank* *(r)* | *II* *Frequency* *(f)* | *III* *Product of I and II* *(r × f = C)* |
|---:|---:|---|
| 10 | 2,653 | 26,530 |
| 20 | 1,311 | 26,220 |
| 30 | 926 | 27,780 |
| 40 | 717 | 28,680 |
| 50 | 556 | 27,800 |
| 100 | 265 | 26,500 |
| 200 | 133 | 26,600 |
| 300 | 84 | 25,200 |
| 400 | 62 | 24,800 |
| 500 | 50 | 25,000 |
| 1,000 | 26 | 26,000 |
| 2,000 | 12 | 24,000 |
| 3,000 | 8 | 24,000 |
| 4,000 | 6 | 24,000 |
| 5,000 | 5 | 25,000 |
| 10,000 | 2 | 20,000 |
| 20,000 | 1 | 20,000 |
| 29,899 | 1 | 29,899 |

Zipf's Law holds for all languages, even non-natural ones like Esperanto, which indicates that of all the words that are available to us in a language, only a small portion is frequently used partly because of the principle of least effort. Zipf's Law has been applied in multiple areas, ranging from language teaching to keyword indexing in search engines such as Google.

All the three bibliometric laws have provided some mathematical and theoretical enlightenment for bibliometrics as a research method. On the other hand, bibliometric data can also be collected to further enrich these three bibliometric laws.

## 7.4 Characteristics and related topics of the bibliometric method

In the previous three sections of this chapter, various aspects of the bibliometric method are presented. Unlike other research methods covered in this monograph, the bibliometric method for data collection has its unique features that are to be discussed shortly.

### *7.4.1 Features of the bibliometric method*

The bibliometric method for data collection features unobtrusiveness and, to some extent, ensures the objectiveness of what is collected when compared with other data collection techniques such as questionnaire and interview. Publication and citation data are natural byproducts of the publishing process of scientific research reports. Additional bibliometric data (e.g., co-citations, IFs, the h-index), as discussed in Section 7.3, can be derived from the basic publication and citation data.

Given the availability of major bibliometric data sources like WoS and Scopus plus special apps for data collection such as PoP, collecting bibliometric data is becoming easier than ever before. Long gone are the days when citation data could only be collected from printed citation indexes (e.g., SCI). Overall, the bibliometric method provides rich data for quantitative as well as qualitative analysis in bibliometric studies while the data collection itself is not as costly as in the case of, for example, using the interview method.

Despite the obvious advantages of the bibliometric method, the most outstanding issue with this data collection technique stems from how to find out why authors cite when writing their research reports or reasons for citation in short. Eugene Garfield, the creator of what is known as WoS today, enumerated 15 reasons for citations out of many more (Garfield 1965, 189).

1) Paying homage to pioneers
2) Giving credit for related work (homage to peers)
3) Identifying methodology, equipment, etc.
4) Providing background reading
5) Correcting one's own work
6) Correcting the work of others
7) Criticizing previous work
8) Substantiating claims
9) Alerting for forthcoming work
10) Providing leads to poorly disseminated, poorly indexed, or uncited work
11) Authenticating data and classes of fact – physical constants, etc.
12) Identifying original publications in which an idea or concept was discussed

13) Identifying original publications or other work describing an eponymic concept or term ....
14) Disclaiming work or ideas of others (negative claims)
15) Disputing priority claims of others (negative homage)

Although the list of reasons for citation is well versed, actual determination of an author's motive for citing others' work is almost impossible because no author would record such reasons at the time of citing. Yet asking authors to recall reasons for citation later would depend entirely on the notoriously unreliable memory, which inevitably leads to inaccurate data.

Superimposing on the issue of reasons for citation with the bibliometric method is the problem about how individual citations are valued or measured. As briefly indicated in Section 7.2, a citation supposedly represents a relationship between the cited and citing documents. But what kind of relationship? Is the relationship weak, moderate, or strong? The determination and assessment of such a relationship is at best a guess if the determination and evaluation can be performed at all. Other assumptions about citations include the following (Smith 1981): All publications referenced have been cited and nothing has been omitted; more frequently cited publications have greater merit than less frequently cited, i.e., the best work is cited; the content of the cited publications is highly related to the overall research topic; all citations are created equal, i.e., equally important. Such assumptions without doubt can hardly be validated by bibliometric researchers even if they had the time and resources. On the other hand, those assumptions are virtually underlying in all bibliometric studies.

There are other issues relating to the bibliometric method, especially citation analysis, as a technique for data collection. As two prominent critics of citation analysis, MacRoberts and MacRoberts (1989) identify a list of its problems. Some of the problems are obvious (e.g., self-citations, citations to coauthored publications), while others are obscure (e.g., biased citations, most important publications not cited). As the obscure problems of citations require rigorous research to address and thus are beyond the scope of this book, only the apparent problems associated with citation analysis are reviewed below.

Self-citations are among the most criticized problems of citation analysis. There are various kinds of self-citations: author self-citations, research group self-citations, institution self-citations, country self-citations as well as journal self-citations (Rousseau, Egghe, and Guns 2018, 113). Each type of self-citations, if not properly done, can raise questions regarding whether those citations are made for self-promotion rather than for scholarly exchange. From the research perspective, it is quite logical that authors need to make citations to publications of their own, their colleagues at the same institution or journals that publish their previous works because research has continuity and is built on prior studies. Later works need to refer to early ones. But self-citations, regardless of what type, may become a problem (e.g., inflated self-citations) when their proportion goes beyond a reasonable level even though it is challenging to determine what can be considered "reasonable".

The term "citation doping" is coined when Italian scientists allegedly cited publications authored by Italian researchers after the introduction of research impact metrics such as citation counts and journal IFs (Van Noorden 2019). Cronin (1981) found via his research, though at a much earlier time, that US authors cited American works for 95% of the time, a percentage much larger than the US share of publications. Both of the two instances are about country self-citation. Journal self-citation likewise emerges as an issue in bibliometric research after JIF is adopted as a quality indicator for journals. In order to boost their JIF values and subsequently journal rankings, some journal editors go as far as to demand those who submit their manuscripts to the journals to cite articles from the same journal (i.e., coercive citation). For this reason, Clarivate, the publisher of the JCR, has to take measure in curtailing such gaming behavior to suppress those journals with abnormal self-citations (Oransky 2020). Such suppressed journals would be re-evaluated in succeeding years or maybe removed from the JCR altogether if the same gaming behavior is not corrected.

Coauthorship is another problem that has often been criticized in bibliometric research. When the number of coauthors has been steadily increasing as is the case in most science disciplines, the coauthorship problem becomes more noticeable. This problem naturally leads to the question regarding how the coauthors should be counted. Fractional counting, briefly described in Section 7.3.2 when the hIa-index is discussed, is a technique widely adopted in mitigating the coauthorship problem. This fractional counting technique divides a publication or citation by the number of coauthors. For example, each author would be considered published 1/4 of an article when there are four coauthors. However, this technique, though simple, apparently does not consider the different amount of contribution each author makes to the publication. The first author in a coauthored publication usually should be given the largest share of credit, and the "average", fractional counting approach clearly is not appropriate for acknowledging the first author's contribution. An alternative approach is to assign tiered weights to each coauthor. A new problem surfaces, however, when using this "weighted" approach because how a weight can be objectively determined to each author is anything but feasible.

In addition to those major problems in using the bibliometric method for data collection, other issues that may also cause concerns in bibliometric research include the following:

- Review articles are cited more often than original research reports although review articles at least should not receive more credit or recognition than original research reports.
- The Matthew effect of cumulative advantage is observed in citation practices when well-known researchers receive more citations than those doing similar research due to their invisibility in the field or domain.
- Citation practices are different in different disciplines. Yet, people often apply the same criterion when choosing metrics based on publications and citations (e.g., citation counts, the h-index) across all disciplines.
- The majority of bibliometric indicators (e.g., publication counts, CiteScore) do not take into account the years an author is in the field with the exception of the hIa-index discussed in Section 7.3.2.

It is because of those problems collectively that we should pay particular attention to the dos and don'ts when conducting bibliometric research.

### 7.4.2 *Dos and don'ts in bibliometric research*

Bibliometric research relies on the techniques we discuss in Section 7.3 for data collection. Because of the bibliometric method's salient features as well as the many issues being brought up above, we must know the dos and don'ts to ensure the quality of research when choosing the bibliometric method for data collection. It is obviously impossible for us to exhaust all the dos and don'ts in bibliometric research in this brief subsection. Nevertheless, we attempt to group and summarize them in three categories: data sources, data collection, and data preparation for analysis.

The importance of choosing the right sources for collecting bibliometric data cannot be overstated. Luckily there are only two major fee-based bibliometric data sources, WoS and Scopus, and one leading OA source GS, which makes the task of data source selection easy. In general, we would at least use one fee-based bibliometric data source for large-scale research. If both WoS and Scopus are available, collect data from both because, as described in Sections 7.2.1 and 7.2.2, WoS and Scopus could be complementary to each other in many ways. Use GS as only data sources if neither WoS nor Scopus is accessible due to financial constraints. In those cases, it is recommended that software specifically designed for the bibliometric method such as PoP be utilized to facilitate data collection.

Data collection in bibliometric research requires searching in the chosen data sources. Even though WoS and Scopus are both user-friendly, some practices in citation searching are necessary and shall be helpful in gathering what is needed for research. Apply the sampling technique when dealing with a large set of bibliometric data. Do not use too narrow a time span (e.g., two to three years) in data collection because meaningful bibliometric data, especially citation data, can only be gathered in a reasonable time span. During the actual data collection process, do decide (1) whether bibliometric data would be collected on the first author, first X number of authors, or all authors; (2) how self-citations are to be handled; (3) how the coauthorship issue is to be addressed. In addition, take advantage of the data analysis functionality that WoS and Scopus each provide to obtain an overview of the data collected (e.g., top cited authors or institutions). Download or export the collected data by choosing a format (e.g., BibTeX, CSV) appropriate for data analysis later.

When preparing data for analysis, use citation counts as a rough indicator of quality because the quality of a publication is determined by many other factors and a publication could be cited for reasons other than for its good quality. As posited in Section 7.4.1, do not assume that all citations are equal, and do not over-interpret citation data. Always look at citation data in context rather than in isolation.

The bibliometric method is one of the few whose research objectives are clearly defined from the very beginning. Rousseau, Egghe, and Guns (2018, 102) recapitulate information retrieval, research evaluation, and social study of science as the main applications of the bibliometric method. If properly utilized, the bibliometric method could be an extremely valuable tool for researchers to have in their toolbox of research methods.

## 7.5 Case study of bibliometric research

There is only one case study for each research method if a chapter includes more than one method in its coverage. As this chapter presents both bibliometrics and altmetrics, one case study for bibliometric research will be performed in this section and another one for altmetric research in Section 7.6.5. The bibliographic information about the publication for this case study is as follows:

Harzing, Anne-Wil. 2013. "A Preliminary Test of Google Scholar as a Source for Citation Data: A Longitudinal Study of Nobel Prize Winners." *Scientometrics* 94 (3): 1057–75. https://doi.org/10.1007/s11192-012-0777-7.

The study by Harzing (2013) is intended to explore if GS is a reasonable source of citation data. Because most governmental research assessment exercises do not use citation data for the humanities and social sciences when WoS or Scopus coverage in these disciplines is considered to be insufficient, Harzing collects citation data (i.e., citation counts and the h-index) on 20 Noble Prize winners from both WoS and GS before comparing how these two citation sources perform in terms of coverage measured in citation counts and the h-index.

WoS is chosen over Scopus because the former has a longer history which is important when extensive bibliometric data should be retrieved for the 20 Noble Prize laureates who might have received the award for their research published several decades ago. However, Harzing should have briefly mentioned that Scopus is not selected for the study due to its shorter time coverage.

Unlike WoS, GS on its own cannot provide citation data as a data set for further research. PoP, a free software program Harzing provides, is utilized for gathering bibliometric data from GS. The PoP software is available at https://harzing.com/resources/publish-or-perish.

The 20 Nobel Prize winners are selected from the fields of chemistry, economics, medicine, and physics because Nobelists in literature and peace are not chosen based on their academic performance. In each of the four fields, five Nobel Prize laureates are selected from each of the five years: 1990, 2000, 2008, 2009, and 2010. Harzing's intention is to have three winners in recent years (i.e., 2008, 2009, 2010) and two from earlier years (i.e., 1990, 2000). But this sampling procedure for choosing Nobel Prize winners is entirely purposive and the resulting sample may not be representative. Perhaps 1990, 1995, 2000, 2005, 2010 could create a more presentative sampling frame for choosing the five Nobel Prize winners from each of the four fields of chemistry, economics, medicine, and physics.

Because WoS only indexes the first cited author in coauthored publications (see Table 7.1), the current study has to choose the first winner of the Nobel Prize regardless of if the prize is shared by one or two cowinners. Once the 20 Nobel Prize winners are determined, the top 20 most cited articles of each Nobelist are identified in WoS and then located in GS to assess the comprehensiveness of GS coverage (i.e., the second research question of the present study). Data on citation counts and the h-index are collected directly from WoS for each of the 20 Nobel Prize laureates. The same type of bibliometric data is obtained from GS via PoP in three separate time points, April 2011, September 2011, and January 2012, to test the stability of GS coverage over time (i.e., the first research question in Harzing's study). The three time points, however, are not equally distributed as the time span between April 2011 and September 2011 is one month longer than that from September 2011 to January 2012. Extending the last time point for data collection to February 2012 would eliminate the one-month difference. A quick note on the order of the first two research questions Harzing formulates for her study: The order of these two research questions perhaps should be swapped because the author cannot address the first research question (i.e., stability of GS) until she answers the second research question (i.e., comprehensiveness of GS).

The third research question of the current study is to compare the citation counts and the h-index computed using data from WoS and GS. Data analysis in this study, though not a focus in the current case study, consists of frequency and percentage only. The conclusion of the study is that "Google Scholar might provide a less biased comparison across disciplines than the Web of Science" (Harzing 2013, 1057).

## 7.6 Webometrics and altmetrics

The internet is a rapidly evolving infrastructure as well as a fertile platform for many applications. The web is probably the most widely adopted internet application while social media applications have been mushrooming on the web since the mid-2000s. A variety of social media applications are embraced enthusiastically by people young and old, and from all walks of life. The web and social media each would, respectively, serve as an incubator for the emergence of the two distinctive metrics-based research methods: webometrics and altmetrics.

### *7.6.1 From webometrics to altmetrics*

Webometrics, to a great extent, is bibliometrics in the web environment where webpages and websites are largely regarded as "publications", inlinks (i.e., hyperlinks or links a webpage/site receives) and outlinks (i.e., links a webpage/site makes to others) as "citations and references" correspondingly. Moreover, the web has its own "citation" or link indexes in the form of commercial search engines (e.g., predecessors of Google like AllTheWeb.com in the late 1990s), and so it is ready for researchers to exploit (Thelwall 2008). Webometrics, a term coined by Ingwersen (Almind and Ingwersen 1997), became a technique for collecting web data via search engine in the late 1990s and early 2000s. Link analysis or cybermetrics are sometimes used in place of webometrics. A study may be considered using the link analysis method if it collects only link data. The internet, of which the web is a part, was also known as the cyberspace in the last decade of the 20th century. In that sense, cybermetrics can be considered a synonym of webometrics although it began to fade away as a term in the vocabulary of metrics-based methods in the beginning of the 21st century.

Although the bibliometric method could be applied for studying websites in research using links and other types of webometric data (Almind and Ingwersen 1997), differences do exist between these two methods. First of all, Egghe (2000) argues that hyperlinks are not the same as references or citations commonly used in bibliometric research. Hyperlinks could be bi-directional, while references are always unidirectional because only later publications can cite earlier ones. Moreover, websites are not scholarly publications and links are created in a context entirely different from that for making citations. Hyperlinks are made for a variety of reasons (e.g., direction, subject content) by webmasters who may or may not be responsible for the content of the sites when marking up a document already written required some learning of the hypertext markup language (HTML) at that time. In contrast, citations are made, though also for different reasons (see Section 7.4.1), by authors who should be fully responsible for the documents they compose. Additionally, there is no counterpart in citations for directional links (e.g., "Go to next page") on websites (Chu, He, and Thelwall 2002).

Second, the web as the data source for webometric research is different from those for bibliometric studies (i.e., scholarly publications). The former has little quality control over its content and quality, while the latter implements a time-tested quality control mechanism (e.g., the referee system) to determine, for example, which articles should get published in a journal and which journals should be included in bibliometric sources such as WoS or Scopus. In addition, the data source for webometric research is not stable. One site could be available on the web one day and disappear the next day for some reason. The same does not hold true for data sources used in bibliometric studies.

Third, data collection for webometric research depends on web search engines (e.g., predecessors of Google). Yet search engines have their own set of problems such as coverage and search functionality (Chu, He and Thelwall 2002). According to the study conducted by Lawrence and Giles (1999), search engine coverage relative to the estimated size of the publicly indexable web has decreased substantially since December 1997, with no search engine indexing more than about 16% of its estimated total size. In other words, the coverage of search engines is limited and decreasing. Although Lawrence and

Giles (1999) reported their study before the turn of the 21st century, the coverage issue of search engines would only become direr while the web's size has been increasing dramatically since then. The search functionality provided by search engines is also of concern because search engines are not developed for collecting webometric data in the first place. In no aspect can search engines be compared with WoS, Scopus, and other similar tools in terms of their data collection capabilities.

All these issues surrounding webometrics contribute to the diminishing use of this method in research. Instead altmetrics is emerging around the same time as a new metrics-based technique for gathering data from various social media platforms (e.g., Facebook, Twitter), metric data produced at academic social media sites (e.g., Academia.edu, Mendeley), and scholarly databases (e.g., ACM Digital Library).

The predecessors of social media are the so-called Web 2.0 applications coming around in the early 2000s. Those applications collectively feature user-generated content, participatory culture, and ease of use. They include blogging, instant messaging (IM), information sharing (e.g., YouTube), RSS (Rich Site Summary), social bookmarking (along with tagging and folksonomies), social networks (e.g., Facebook), virtual communities (e.g., Second Life), and wikis (Xu, Ouyang, and Chu 2009). Some of those Web 2.0 applications have become popular social media sites (e.g., Facebook, YouTube); some have evolved to a more advanced stage (e.g., from wikis to Wikipedia), while some are disappearing as a stand-alone application and integrated into existing social media sites. For example, tagging or social bookmarking becomes a function available in many social media sites such as Instagram and Twitter instead of a stand-alone application.

### *7.6.2 Altmetric data sources*

Altmetric data include bookmarks, clicks, comments, downloads, likes, posts, recommends, saves, shares, tags, tweets, and views that can be gathered from social media and other similar sites. In comparison to bibliometric data, altmetric data encompass more varieties although that does not imply that altmetric data are cleaner (i.e., requiring less data preparation for data analysis) or more valuable. To the contrary, altmetric data on one hand need to be cleaned (i.e., data cleansing in professional terms) before they can be analyzed because altmetric data are generated by social media users and everything (including IP addresses, time stamps) is captured automatically at the host site. On the other hand, the quality of altmetric data is notoriously mixed because such data are logged at their host sites without any selection and organization. Furthermore, an app or algorithm specifically developed is needed in many cases for collecting altmetric data from individual social media and similar sites.

As briefly indicated in Section 7.6.1, there are several kinds of sources for gathering altmetric data. Table 7.7 lists some sample sites for each type along with corresponding altmetric data that can be collected for research.

Social media sites such as Facebook and Twitter are the most common source type for altmetric data collection. Each social media site provides a unique set of altmetric data in accordance to its function and features. It is thus necessary to choose an appropriate site for gathering data in light of one's altmetric research objective.

Academic social media sites include altmetric data for scholarly publications (e.g., journal articles, proceedings papers) in the academic world. For instance, Academia.edu offers views, downloads, saves, readers, mentions, visitors by university, city or country, and more. However, much of its altmetric data are available only to people with its fee-based

*Table 7.7* Major Altmetric Source Types and Data

| *Type* | *Sample Site* | *Altmetric Data* |
|---|---|---|
| Social media | Facebook | Posts, friends, likes, comments, groups, ... |
| | Twitter | Hashtags, tweets, retweets, likes, followers, followings, bookmarks, ... |
| | YouTube | Views, shares, saves, comments, thumb-ups, thumb-downs, ... |
| Academic social media | Academia.edu | Views, downloads, saves, readers, mentions, visitors, ... |
| | Mendeley | Downloads, readers, annotations, tags, groups, group members, ... |
| Scholarly database | ACM Digital Library | Downloads, top subject, top keyword, top affiliation, ... |

premium accounts. Those without an Academia premium account can only have access to very limited altmetric data in its database. Mendeley, becoming a subdivision of Elsevier in 2013, is better known as a reference management system although it also provides a good amount of altmetric data as well. Thanks to its association with Elsevier, Mendeley has much better organization and search functionalities than its counterparts such as Academia.edu or ResearchGate (https://www.researchgate.net). Various altmetric data are available at topic, group as well as individual article levels from Mendeley.

As bibliometric and altmetric indicators are increasingly adopted by institutions and organizations for research evaluation, grant awarding, and other related purposes, many databases of scholarly publications start supplying these two types of data. According to what is presented in Section 7.2, bibliometric data can be obtained from such established sources as WoS and Scopus. But the same does not hold true for altmetric data collection. That is why scholarly databases like ACM Digital Library become appreciated sources for gathering altmetric data. Scholarly databases nowadays also provide bibliometric data (e.g., publication counts, citation counts) besides altmetric data.

Of the three major types of altmetric sources, social media sites generate altmetric data for non-academic contents, while academic social media sites and scholarly databases focus on altmetric data for scholarly contents. As briefly mentioned earlier, an app or algorithm must be used for collecting altmetric data. More importantly, all the data collection must be done with the host site's permission and each host site has its own policy that must be complied with.

A fee-based altmetric service, similar to WoS and Scopus in nature but much less established in terms of coverage and quality, is PlumX Metrics (https://plumanalytics.com), acquired by Elsevier in 2017. The emphasis of PlumX is on altmetric data that include data on usage (e.g., clicks, downloads, views, library holdings, video plays), captures (e.g., bookmarks, favorites, readers, watchers), mentions (e.g., blog posts, comments, reviews, Wikipedia references, news media), and social media (e.g., shares, likes, comments, tweets). In addition, PlumX has data on citations that contain "both traditional citation indexes such as Scopus, as well as citations that help indicate societal impact such as Clinical or Policy Citations" (Plum Analytics 2023). However, researchers normally rely on WoS and/or Scopus for collecting bibliometric data. Altmetric.com is another altmetric service like PlumX and obtains sponsorship of major companies such as John Wiley & Sons and Taylor & Francis. Interested readers can visit its website which holds a lot of relevant information. Gumpenberger, Glänzel, and Gorraiz (2016) particularly explore the value and issues that associate with the Altmetric Score, a composite indicator Altmetric.

com computed based on its data. The Altmetric Score is also visualized as a multicolored "donut" badge. Although the Altmetric Score can serve as one more indicator of research quality and dissemination landscape, we should be cautious about its true value and related issues when using that composite indicator in altmetric research.

### *7.6.3 Using altmetrics in research*

Altmetrics is considered an interesting option for assessing the societal impact of research as it offers new ways to measure public or social engagement with research output (Bornmann 2014a). Like bibliometrics, altmetrics can be applied for evaluating a researcher, a publication, an institution but mostly as a supplemental means. Altmetric indicators utilized in research include but not limited to the following:

- Views – HTML or PDF views on a site, often this is a publisher's site but other sites (e.g., Academia.edu, Mendeley) may also provide view data. Viewed, though lacking an explicit definition in related literature, appears to refer to a publication that is accessed, scanned, or skimmed.
- Readers – According to Academia (2023), readers are those who "interact" with a publication at its site while Mendeley and other similar sites do not provide any specification about this altmetric indicator.
- Comments – This altmetric indicator includes comments and posts in blogs, academic social media sites, and general social media sites such as Facebook and Twitter.
- Saves – It means that a copy of the publication is saved at academic social media sites for later reference.
- Downloads – A copy of the publication is downloaded from academic social media and other similar sites to one's own computer or a similar device (e.g., iPad) possibly for later peruse.

Data on those altmetric indicators are normally measured in frequency (e.g., download counts) except the "comments" category which comprises comments, posts, or annotations, depending on how individual sites name this activity. Altmetric data gathering from this category are primarily texts or qualitative in nature. They are in general to be analyzed qualitatively in research.

A clear advantage of using the altmetric method is that altmetric data are quicker to accumulate than bibliometric data. By virtue of being sourced from the web and not from publications like journals, it is possible to monitor and collate, for instance, "mentions" of work online as soon as it is published. Mention in altmetric terminology is regarded as citations in the bibliometric vocabulary (Academia 2023). For example, the journal *Nature* can be immediately mentioned at altmetric sites after its publication and such mentions can then be counted as altmetric data for *Nature*.

The other advantage of the altmetric method is that it can capture more diverse impact data than the bibliometric approach. Altmetrics can complement bibliometrics in that altmetric data, in the form of views, reads, comments, bookmarks, and downloads, are able to help understand the many "flavors" (e.g., viewed and commented; downloaded and read; read, bookmarked, and shared) than the IFs based on publication and citation data (see Section 7.3.2).

A third advantage of the altmetric method is that altmetrics applies to more than publications. Researchers can also share their data, software, presentations, and other scholarly outputs via altmetric sites more conveniently than ever before. All these kinds of sharing would generate more altmetric data (e.g., mentions, downloads) for research.

All the advantages of the altmetric method can be summarized as speed, diversity, and broadness (Bornmann 2014a). On the other hand, altmetric data are created more casually than bibliometric data. Any person can comment on, download, and share a publication as long as the individual has internet access. Like any metric, there is a potential for gaming of altmetrics. Any people with enough time on their hands can artificially inflate the altmetrics for their research. Furthermore, altmetrics are relatively new, and how each metric means and could be interpreted in research requires additional exploration and examination. For example, can downloads be regarded as a measure of readership? Or would downloading ultimately lead to reading and use in one's own research? Are mentions indicators of impact? We do not have answers to such questions yet. Rousseau, Egghe, and Guns (2018, 201–2) believe that it is inappropriate to take the number of downloads as a proxy for readership. Bornmann (2014a) also points out that commercialism, data quality (including biases, targets, multiple versions, duplications), missing evidence, manipulation are all limitations of the altmetric method. Roemer and Borchardt (2015) illustrate some opportunities for altmetrics after a close look at its issues and controversies such as altmetrics' limited or negligible correlation with bibliometrics and the use of general social media sites (e.g., Twitter) in evaluating scholarly publications. What is called altmetrics today cannot actually be seen as alternatives to bibliometrics according to Rousseau, Egghe, and Guns (2018, 4). Altmetrics or social media metrics, as Haustein, Costas, and Larivière (2015) suggest, can only function as complements to bibliometrics because of its afore-discussed limitations.

In addition to its application in the scholarly world, altmetrics can also be adopted for gauging public opinions about various issues, societal reactions to events and happenings, and much more. Each of the sample altmetric indicators listed in Table 7.7 can be exploited to help learn more about social media users. For example, we can look at the list of friends at one's Facebook to see who are the individual's friends even though we understand a friend on social media is different from a friend in the real world. We also can gather data about one's followers and followings to learn the individual's influence by and on others, respectively. The likes from Facebook, the thumb-ups, and thumb-downs at YouTube all reflect others' reactions to individuals' posts, tweets, and so on at their own social media accounts. While such altmetric indicators are measured at the individual level, group level measures are also available. Sentiment analysis, done at the group level, is one application of the altmetric method to find out, for instance, the public's views on a product, a service or an event by collecting and analyzing the posts at corresponding sites. In this regard, altmetric research can be performed in areas far beyond scientific domains. Research using altmetric data from non-academic spheres is on the rise as social media evolve and penetrate in many sectors of our work and daily lives.

### *7.6.4 Dos and don'ts in altmetric research*

Altmetric research shares some dos and don'ts delineated for bibliometric research in Section 7.4.2. In general, the quality of altmetric data is lower than that of bibliometric data because the former is generated entirely on a crowd basis without any quality control mechanism. Given this very feature of altmetric data, we ought to observe the following dos and don'ts when using the altmetric method for data collection in research.

First, we should fully understand the nature of altmetric data and sources when using the altmetric method. Unlike bibliometric research, there is no comprehensive and time-honored sources for us to collect altmetric data. Instead, we have to gather what is needed in research from a number of altmetric data sources (e.g., Facebook, Mendeley) using either an app specifically created for data collection or an API (application programming interface) that some sites (e.g., Altmetric) provide. PlumX offers fee-based data collection for researchers but only 4.7% of its data come from social media (Plum Analytics 2016).

Second, do collect proper altmetric data based on the research questions you intend to address. For instance, if you plan to examine the degree of sharing and collaboration among researchers, you would gather such altmetric indicators as shares and types of data shared (e.g., raw data, software, analysis results) rather than other metrics. Furthermore, altmetric data collection should not be limited to indicators alone (e.g., views, saves, downloads). Contextual data (e.g., comments, annotations) relating to those indicators ought to be gathered as well because qualitative data are usually more illuminating than indicators alone. That is the very reason why we always emphasize the integration of both the quantitative and qualitative approaches in research, a topic to be fully discussed in Chapter 13, Techniques for Data Analysis and Their Selection.

Third, make certain to obtain permission from altmetric data host sites before data collection due to the fact that altmetric data often can only be gathered directly from target sites via an app or algorithm. This practice concerns not only with the data ownership issue but also about privacy and other related issues regarding social media users. Altmetric data are created at platforms of companies who have the complete data ownership. Thus, getting permission from such companies is a must instead of an option. On the other hand, altmetric data involve a lot of personal information about users which must be protected throughout the research process. In no circumstances shall users' personal information be revealed in all the steps of research ranging from data collection to research reporting. Anonymizing collected data is a recommended and widely adopted approach to protecting personal information in altmetric research.

Last, use altmetric data in conjunction with bibliometric and other kinds of data if needed because altmetric data alone have its limitations as discussed in Section 7.6.3. The most outstanding limit of altmetric data lies in how a metric can be meaningfully interpreted in answering research questions. Although the general consensus is that, for example, downloads are treated as an indicator of readership or usage, the actual linkage between downloads and readership/usage is hard to be verified or proven. Moreover, such verification becomes even harder when downloads are often made at academic social media platforms like Mendeley by different people with different purposes. Hence, treat altmetrics as a complement to, rather than a replacement for, other methods adopted for bibliometric and other kinds of research. Altmetrics is just one tool of many in your method toolbox for your research undertakings.

### *7.6.5 Case study of altmetric research*

Bornmann, Lutz. 2014b. "Validity of Altmetrics Data for Measuring Societal Impact: A Study Using Data from Altmetric and F1000Prime." *Journal of Informetrics* 8 (4): 935–50. http://dx.doi.org/10.1016/j.joi.2014.09.007.

Bornmann, a seasoned researcher using metrics-based methods, compares in his study altmetric counts and Twitter counts, both obtained from the Altmetric database, with citation counts from WoS to investigate the societal impact of publications in the Faculty of

1,000 (F1000) system. The subscription-based version of F1000 was called F1000Prime and in 2021 known as Faculty Opinions (https://facultyopinions.com) with different company parameters. Societal impact is defined by Bornmann as the impact of research publications on readers from the altmetric world, including the Twitter platform.

F1000 is a recommendation system of research publications in biology and medicine at the time of Bornmann's study. The number 1,000 in F1000 is symbolic as it involves more than 5,000 experts worldwide, assisted by 5,000 associates, to provide services (Bornmann 2014b, 936). Those F1000 members would identify, evaluate (using the good, very good, or exceptional rating), and tag publications that they read recently in more than 40 subject areas. F1000 members would do the tagging by choosing from a list of tags F1000 supplies (see Bornmann 2014b, 937). In January 2014, Bornmann obtained over 100,000 records from F1000 with unique DOIs (digital object identifiers). Bornmann then tried to locate those records in the following two sources:

- Altmetric.com – An altmetric service, often referred to as Altmetric, that offers various kinds of data including Twitter, Facebook, and other mentions for the F1000 records published after 2011. This time point is chosen because "Altmetric can only reliably provide data for papers published after 2011" (Bornmann 2014b, 938).
- WoS – A citation database available at the Max Planck Society (MPG) for locating F1000 records published before 2011. This time point is selected because papers published in 2011 and later do not have adequate time to accumulate citations.

After all the limits, Bornmann collected altmetric mentions and Twitter mentions for the qualified F1000 records as two measures of societal impact as opposed to the research impact assessed via citation counts from WoS. Two issues arise in this data collection procedure. First, altmetric counts and Twitter counts are both obtained from the Altmetric database, while Twitter counts are the largest part of all altmetric counts presented in Table 1 of Bornmann (2014b). In other words, Twitter counts are counted twice, once alone and once as part of total altmetric counts. It would be desirable if Bornmann had placed all altmetric counts into two categories: Twitter counts and non-Twitter accounts to avoid the double counting of Twitter mentions in the study. Second, how validly can altmetric mentions serve as a societal impact indicator? The author should have offered more explanations and justification about this usage in his research.

In addition to the mention counts in the Altmetric database, Bornmann also gathered data on the three types of ratings (e.g., good, very good, or exceptional) the F1000 members had made along with the tags they had assigned to the research publications in the F1000 system. The ratings and tags can be regarded as academic altmetric data for determining the research impact of the target F1000 records. Citation data were simply collected from WoS, one of the major sources for citations, in the study.

Because of the many limits (e.g., F1000 records published before or after 2011) applied in the data collection process, data sets of different sizes are gathered for the study. For instance, altmetric data are collected from the Altmetric database for 13,678 F1000 records, while tag data are collected from the F1000 system for 17,805 of its records whose total was over 100,000 in January 2014. It would be of great help if a flow chart were drawn to illustrate this data collection process with many inclusions and exclusions.

As can be seen from this research report (Bornmann 2014b), both altmetrics and citation analysis are utilized in the study to examine the societal impact as well as the research impact of F1000 publications rated and tagged by F1000 members. The results of this investigation show that publications in F1000 with the tag "good for teaching" do achieve higher altmetric counts (mainly Twitter counts) than those without this tag when the quality of the publications is controlled (i.e., publications receiving the same rating of good, very good, or exceptional). At the same time, a higher number of citation counts is received by publications with the tag "new finding" than those without it. These findings of the study indicate that publications that are rated and tagged by peers as done in the F1000 system would enhance both their research impact in terms of citation counts as well as societal impact in terms of altmetric counts.

**Chapter Summary**

- A metrics-based method is a technique for collecting metric data (e.g., publication numbers, citation counts, tweet mentions) about publications, citations, and more.
- There are two major types of metrics-based research methods: (1) bibliometrics that focuses on publication and citation data, and (2) altmetrics that involves various metrics generated by people of different ages and from all walks of life (e.g., mentions at social media sites like Facebook or Twitter; views, saves, downloads at academic social media sites like Mendeley).
- The bibliometric method features unobtrusiveness in its data collection which enhances its data credibility. It is less costly for researchers to use than other data collection techniques like questionnaire and interview. Bibliometric data can be gathered from established sources such as Web of Science and Scopus as well as from other sources like Google Scholar. This method also provides rich data for quantitative and qualitative analysis in bibliometric studies.
- The dominant bibliometric techniques for data collection include citation analysis, impact factors, the h-index, and bibliometric laws.
- The dos and don'ts in bibliometric research center around data source selection, data searching, and handling of bibliometric indicators. Always keep in mind that any bibliometric measure is no more than a rough indicator of research quality.
- The altmetric method can be used to gather data that are quicker to accumulate than bibliometric data. Altmetric data are richer than bibliometric data as they are created by the public from different perspectives at a variety of social media sites. But this method is prone to gaming due to its social nature.
- The altmetric method shares many dos and don'ts with the bibliometric technique for data collection. One must obtain permission from the host site before data collection in addition to paying particular attention to the privacy issue throughout the altmetric research process.

## Notes

1. According to Chadegani (2017, 19), the total number of titles WoS has is 12,311, while Scopus holds 19,809. Of these two different totals, WoS and Scopus share a common set of 11,377 titles. The unique titles are hence 934 for WoS and 8,432 for Scopus.
2. Detailed information about WoS' 24 quality criteria (e.g., scholarly content, editorial board composition) and 4 impact criteria (e.g., author citation analysis, content significance) can be found at https://clarivate.com/webofsciencegroup/journal-evaluation-process-and-selection-criteria/.

3 Detailed information about Scopus' 5 categories of criteria for journal selection and evaluation can be found at https://www.elsevier.com/solutions/scopus/how-scopus-works/content/content-policy-and-selection. For example, the journal policy category includes convincing editorial policy, type of peer review, diversity in geographical distribution of editors, and diversity in geographical distribution of authors.

4 The cost is estimated for large organizations (Goodman and Deis 2005). The general consensus is that WoS holds its pricing information tightly, while Scopus sets its pricing according to an annual subscription fee with unlimited usage, ranging from $20,000 to $120,000 per year based on the size of organizations (Chadegani 2017).

5 Author names are now fully spelled out in both WoS and Scopus. However, WoS recorded only authors' first name initials at least until Scopus was created in 2004, which makes it hard for differentiating authors with common last names in data collection. For example, J in Smith J. could be John, Jack, James, and more.

6 Figure 7.3 was derived with permission from: Elsevier. 2018. *CiteScore Metrics*. Page 8. Copyright ©2018 Elsevier BV. But this brochure is no longer available to the public from Elsevier's website.

7 Figure 7.4 is adapted from Wikipedia Contributors. 2023. "h-index." In *Wikipedia, The Free Encyclopedia*. Last modified August 10, 2023. Accessed September 11, 2023. https://en.wikipedia.org/wiki/H-index.

8 Zipf's book *Human Behavior and the Principle of Least Effort: An Introduction to Human Ecology* was published in 1949 and the author passed away in 1950. So, this book is in public domain after 2020.

## References


Academia. 2023. "What is Academia Premium?" Accessed July 8, 2023. https://support.academia.edu/hc/en-us/articles/360043383853-What-is-Academia-Premium-.

Almind, Tomas C., and Peter Ingwersen. 1997. "Informetric Analyses on the World Wide Web: Methodological Approaches to 'Webometrics'." *Journal of Documentation* 53 (4): 404–26. https://doi.org/10.1108/eum0000000007205.

Bartolucci, Francesco. (2015). "A Comparison Between the G-Index and the H-Index Based on Concentration." *Journal of the Association for Information Science and Technology* 66 (12): 2708–10. https://doi.org/10.1002/asi.23440.

Bornmann, Lutz. 2014a. "Do Altmetrics Point to Broader Impact of Research? An Overview of Benefits and Disadvantages." *Journal of Informetrics* 8 (4): 895–903. http://dx.doi.org/10.1016/j.joi.2014.09.005.

Bornmann, Lutz. 2014b. "Validity of Altmetrics Data for Measuring Societal Impact: A Study Using Data from Altmetric and F1000Prime." *Journal of Informetrics* 8 (4): 935–50. http://dx.doi.org/10.1016/j.joi.2014.09.007.

Chadegani, Arezoo Aghaei. 2017. "A Comparison Between Two Main Academic Literature Collections: Web of Science and Scopus Databases." *Asian Social Science* 9 (5): 18–26.

Chu, Heting, Shaoyi He, and Mike Thelwall. 2002. "Library and Information Science Schools in Canada and USA: A Webometric Perspective." *Journal of Education for Library and Information Science* 43 (2): 110–25. https://doi.org/10.2307/40323972.

Clarivate. 2023. "Web of Science Journal Evaluation Process and Evaluation Criteria." Accessed July 7, 2023. https://clarivate.com/products/scientific-and-academic-research/research-discovery-and-workflow-solutions/web-of-science/core-collection/editorial-selection-process/editorial-selection-process/.

Cronin, Blaise. 1981. "Transatlantic Citation Patterns in Educational Psychology." *Education Libraries Bulletin* 24 (2): 48–51.

Egghe, Leo. 2000. "Discussions on Informetrics of the Internet." In *Proceedings of the Second International Symposium on Qualitative Evaluation of Research Performances and Sixth National Conference on Scientometrics and Informatics*, Shanghai, China. 8–26.

Egghe, Leo. (2006). "Theory and Practise of the G-index." *Scientometrics* 69 (1): 131–52. https://doi.org/10.1007/s11192-006-0144-7.

Elsevier. 2022. "Scopus." Accessed July 6, 2023. https://www.elsevier.com/__data/assets/pdf_file/0017/114533/Scopus-fact-sheet-2022_WEB.pdf.

Elsevier. 2023a. "Content Policy and Selection." Accessed July 7, 2023. https://www.elsevier.com/solutions/scopus/how-scopus-works/content/content-policy-and-selection.

Elsevier. 2023b. "Measuring a Journal's Impact." Accessed July 7, 2023. http://www.elsevier.com/authors/journal-authors/measuring-a-journals-impact.

Garfield, Eugene. 1963. "Science Citation Index." *Science Citation Index 1961* 1: v–xvi.

Garfield, Eugene. 1965. "Can Citation Indexing Be Automated?" In *Statistical Association Methods for Mechanized Documentation* (NBS Misc. Pub. 269), edited by Stevens, Mary E., Vincent E. Giuliano and Lawrence B. Heilprin, 189–92. Washington: National Bureau of Standards.

Glänzel, Wolfgang. 2006. "On the H-index – A Mathematical Approach to a New Measure of Publication Activities and Citation Impact." *Scientometrics* 67 (2): 315–21. https://doi.org/10.1007/s11192-006-0102-4.

Goodman, David, and Louise Deis. 2005. "Web of Science (2004 Version) and Scopus." *The Charleston Advisor* 6 (3): 5–21.

Groen, Frances K. 2007. *Access to Medical Knowledge: Libraries, Digitization, and the Public Good*. Lanham: Scarecrow Press.

Gumpenberger, Christian, Wolfgang Glänzel, and Juan Gorraiz. 2016. "The Ecstasy and the Agony of the Altmetric Score." *Scientometrics* 108 (2): 977–82. https://doi.org/10.1007/s11192-016-1991-5.

Harzing, Anne-Wil. 2013. "A Preliminary Test of Google Scholar as a Source for Citation Data: A Longitudinal Study of Nobel Prize Winners." *Scientometrics* 94 (3): 1057–75. https://doi.org/10.1007/s11192-012-0777-7.

Harzing, Anne-Wil, and Satu Alakangas. 2016. "Google Scholar, Scopus and the Web of Science: A Longitudinal and Cross-Disciplinary Comparison." *Scientometrics* 106 (2): 787–804. https://doi.org/10.1007/s11192-015-1798-9.

Harzing, Anne-Wil, Satu Alakangas, and David Adams. (2014). "hIa: An Individual Annual H-index to Accommodate Disciplinary and Career Length Differences. *Scientometrics* 99 (3): 811–21. https://doi.org/10.1007/s11192-013-1208-0.

Haustein, Stefanie, Rodrigo Costas, and Vincent Larivière. 2015. "Characterizing Social Media Metrics of Scholarly Papers: The Effect of Document Properties and Collaboration Patterns." *PLoS ONE* 10 (3): e0120495. https://doi.org/10.1371/journal.pone.0120495.

Hirsch, Jorge E. 2005. "An Index to Quantify an Individual's Scientific Research Output." *Proceedings of the National Academy of Sciences of the United States of America* 102 (46): 16569–72. https://doi.org/10.1073/pnas.0507655102.

Hirsch, Jorge E. 2020. "Superconductivity, What the H? The Emperor Has No Clothes." *Physics and Society* 49 (1): 4–9. https://doi.org/10.1142/s0217979224300019.

Ikpaahindi, Linus. 1985. "An Overview of Bibliometrics: Its Measurements, Laws and Their Applications." *Libri* 35 (2): 163–77.

Kessler, Meyer Mike. 1963. "Bibliographic Coupling of Scientific Papers." *American Documentation* 14 (1): 10–25. https://doi.org/10.1002/asi.5090140103.

Khabsa, Madian, and C. Lee Giles. 2014. "The Number of Scholarly Documents on the Public Web." *PLoS ONE* 9 (5): e93949. https://doi.org/10.1371/journal.pone.0093949.

Lawrence, Steve, and C. Lee Giles. 1999. "Accessibility of Information on the Web." *Nature* 400: 107–9.

MacRoberts, Michael H., and Barbara R. MacRoberts. 1989. "Problems of Citation Analysis: A Critical Review." *Journal of the American Society for Information Science* 40 (5): 342–49. https://doi.org/10.1002/(SICI)1097-4571(198909)40:5<342::AID-ASI7>3.0.CO;2-U.

Microsoft Academic. 2021. "Next Steps for Microsoft Academic – Expanding into New Horizons." Accessed July 7, 2023. https://www.microsoft.com/en-us/research/project/academic/articles/microsoft-academic-to-expand-horizons-with-community-driven-approach/.

Oransky, Ivan. 2020. "Major Indexing Service Sounds Alarm on Self-Citations by Nearly 50 Journals." *Retraction Watch*, June 29. Accessed July 8, 2023. https://retractionwatch.com/2020/06/29/major-indexing-service-sounds-alarm-on-self-citations-by-nearly-50-journals/.

Plum Analytics. 2016. "Plum Analytics Coverage Infographic." Accessed July 8, 2023. https://plumanalytics.com/wp-content/uploads/2016/11/Plum-Analytics-Coverage-Infographic.pdf.

Plum Analytics. 2023. "About PlumX Analytics." Accessed July 8, 2023. https://plumanalytics.com/learn/about-metrics/.

Roemer, Robin Chin, and Rachel Borchardt. 2015. "Altmetrics." *Library Technology Reports* 51 (5): 3–37.

Rousseau, Donald. 2002. "George Kingsley Zipf: Life, Ideas, His Law and Informetrics." *Glottometrics* 3: 11–18.

Rousseau, Ronald, Leo Egghe, and Raf Guns. 2018. *Becoming Metric-Wise: A Bibliometric Guide for Researchers*. New York: Elsevier/Chandos Publishing. https://doi.org/10.1016/C2017-0-01828-1.

Small, Henry. 1973. "Co-Citation in the Scientific Literature: A New Measure of the Relationship Between Two Documents." *Journal of the American Society for Information Science* 24 (4): 265–69. https://doi.org/10.1002/asi.4630240406.

Small, Henry. 2017. "A Tribute to Eugene Garfield: Information Innovator and Idealist." *Journal of Informetrics* 11 (3): 599–612. https://doi.org/10.1016/j.joi.2017.04.006.

Smith, Linda. 1981. "Citation Analysis." *Library Trends* 30 (1): 83–106.

Swales, John. 1986. "Citation Analysis and Discourse Analysis." *Applied Linguistics* 7 (1): 39–56. https://doi.org/10.1093/applin/7.1.39.

Thelwall, Mike. 2008. "Bibliometrics to Webometrics. *Journal of Information Science* 34 (4): 605–21. https://doi.org/10.29085/9781856049986.018.

Van Noorden, Richard. 2019. "Italy's Rise in Research Impact Pinned on 'Citation Doping'." *Nature – News*, September 13. https://doi.org/10.1038/d41586-019-02725-y. Accessed January 4, 2021. https://www.nature.com/articles/d41586-019-02725-y.

Waltman, Ludo. 2016. A Review of the Literature on Citation Impact Indicators. *Journal of Informetrics* 10 (2): 365–91. http://dx.doi.org/10.1016/j.joi.2016.02.007.

White, Howard D., and Belver C. Griffith. 1981. "Author Cocitation: A Literature Measure of Intellectual Structure." *Journal of the American Society for Information Science* 32 (3): 163–71. https://doi.org/10.1002/asi.4630320302.

Wikipedia Contributors. 2023. "List of Chinese Common Surnames." *Wikipedia, The Free Encyclopedia*. Last modified July 6, 2023. Accessed July 7, 2023. https://en.wikipedia.org/wiki/List_of_common_Chinese_surnames

Xu, Chen, Fenfei Ouyang, and Heting Chu. 2009. "The Academic Library Meets Web 2.0: Applications and Implications." *The Journal of Academic Librarianship* 35 (4): 324–31. http://dx.doi.org/10.1016/j.acalib.2009.04.003.

Zhao, Dangzhi, and Andreas Strotmann. 2008. "Evolution of Research Activities and Intellectual Influences in Information Science 1996–2005: Introducing Author Bibliographic-Coupling Analysis". *Journal of the American Society for Information Science and Technology* 59 (13): 2070–86. http://dx.doi.org/10.1002/asi.20910.

Zhao, Dangzhi, and Andreas Strotmann. 2010. "Mapping the Highly Collaborative Stem Cell Research Field: Adding Last-Author-Based Analysis to the Author Co-Citation Analysis Family." In *Proceedings of the 73rd Annual Meeting of the American Society for Information Science and Technology* 47, Article No. 53. https://doi.org/10.1002/meet.14504701284.

Zipf, George Kingsley. 1949. *Human Behavior and the Principle of Least Effort: An Introduction to Human Ecology*. Cambridge: Addison-Wesley Press.

# 8 Other Stand-Alone Methods

Six research methods (i.e., questionnaire, interview, observation, theoretical approach, bibliometrics, and altmetrics) have been described and discussed from Chapters 3 to 7, one chapter for each method. In the current chapter, four stand-alone research methods (i.e., focus groups, transaction log analysis (TLA), research diaries, and historical methods) are to be presented as four separate sections in consideration of their more specialized functions and applications. In order to facilitate easy reading and comprehension, overview and summary points for each research method shall be provided in the corresponding section rather than in the beginning and at the end of the present chapter.

## 8.1 Focus groups

**Section Overview Questions**

- What are focus groups?
- What features do focus groups have as a research method?
- What are the dos and don'ts of research using focus groups for data collection?

### *8.1.1 What are focus groups?*

Focus groups are one research method used by social scientists, marketers, policy analysts, health and social services professionals, political consultants, and other scientists and decision makers to gather information (Steward, Shamdasani, and Rook 2009). Originally called "focused" interviews (e.g., Merton and Kendall 1946), this technique came into vogue after World War II and has been a part of researchers' tool kit ever since.

Focus groups refer to groups of individuals, selected and assembled by the researcher, to discuss and comment on a topic of the study from personal experience. Usually, more than one group of participants is invited for one research project, and that is why we have "groups" instead of "group" in the name of this data collection technique (Chu and Ke 2017). Much of focus group research, as one growing type of market research, does not accurately reflect the distinctive nature of this research method (Merton 1987). That is, focus groups are not a Q&A procedure as in interview. Rather, it is through participants' discussion and comments that the researcher collects data. In this sense, it is perhaps inappropriate to call focus groups group interviews (Madriz 2000), although some publications, including some textbooks, choose such usage.

DOI: 10.4324/9781003021711-10

The major differences between group interviews and focus groups are summarized in Bloor et al. (2001, 42–43). In group interviews, the group is asked a sequence of predetermined questions, just as if the interviewer were speaking to a single interviewee. The group format is simply a matter of convenience and/or economy, and the objective is to elicit the group's answers to the interview questions. Group interviews are rarely a preferred option over individual interviews in research as individual variations in viewpoint within the group will be blurred and under reported. In focus groups, pre-determined questions may also be asked, but the objective is not primarily to elicit the group's answers. Instead, such questions are asked to stimulate discussion and thereby understand, through subsequent analysis, the meanings and norms underlying those group answers. Hence, in group interviews, the interviewer seeks answers, whereas in focus groups, the facilitator seeks group member interaction. That is, focus groups involve not only "vertical interaction" or interaction between the moderator and participants, but also "horizontal interaction" among the participants (Madriz 2000, 840). Furthermore, there is typically only one group of participants in group interviews, while at least three or more groups of participants must be recruited in focus group research.

Focus groups are similar to some other research methods (e.g., interview) in that they all enable researchers to have access to the opinions, viewpoints, attitudes, and experiences of individuals. What makes focus groups unique is that they can be used to capture data on participants' collective viewpoints as well as on interaction among participants in the group, which is absent in other methods. For instance, no data on interaction with other participants can be collected via questionnaire and interview. Even though observation, as one data collection technique, can be adopted to study the interaction among participants, data on their opinions, viewpoints, attitudes, and experiences can hardly be observed and subsequently gathered. Breen (2006) summarizes that it is likely appropriate for researchers to use focus groups for data collection when they intend to get participants to (1) share and compare their experiences with each other, (2) develop and generate ideas, or (3) explore issues of shared importance.

There are three stages in the development of focus groups as a research method (Morgan and Krueger 1998, 37). In the 1920s, researchers used focus groups for a variety of purposes, one of the most important being for creating questionnaire questions. Such usage is echoed in Scheuren (2004, 33) when he discusses the focus group method. The second stage of focus groups' application spanned the time period between the 1940s and the 1970s, during which this technique was mainly utilized by market researchers to understand people's wants and needs. Researchers in sociology also made great use of this method in the period, even though it might often be called focused interview (e.g., Merton and Kendall 1946). The third stage of focus group research began in the 1980s when scholars from different fields (e.g., anthropology, information science, psychology) applied this method in their studies.

### *8.1.2 Major parameters in focus group research*

Compared with questionnaire, interview, and other major stand-alone methods, focus groups are used less frequently in research. However, they become the most viable option when the researcher intends to gather data from groups of participants, while individual views and input are not crucial.

Group size is one of the parameters that researchers must consider when choosing focus groups as a data collection technique. Yet little agreement exists among most sources about an optimal group size. For example, Morgan (1989, 43) and some authors recommend six to ten members in each group, while others (e.g., Wildemuth 2017, 262) indicate that most focus groups include 8–12 participants each. In addition, Krueger and Casey (2015, 22) suggest that for complex problems, the focus group size should be kept to no more than seven participants. Bloor et al. (2001, 92) in comparison think that "6 to 8 individuals seems to be a good thumbnail estimate of the optimum group size". Larger focus groups can be cost-saving but may not give each participant enough time to express their viewpoints and allow all the group members to adequately interact with each other, which nevertheless should be a prominent feature of the method. Smaller focus groups would function just the opposite of larger groups by comparison. The size of focus groups thus should range between 6 and 12 (Scheuren 2004, 35), and the researcher ought to ultimately determine the exact group size based on the project undertaken.

Identifying and recruiting participants is a task focus group researchers must take up in addition to group size determination. While it is neither possible nor feasible in selecting focus group participants with a random sampling plan, it is still recommended to choose participants that are representative of the target population. A common practice is to group participants who share a demographic attribute, such as age, gender, or race, based on the nature of the research topic. This selection practice would enable the researcher to focus on changes brought about by variables other than the demographic variable used to form a focus group. Further discussion of this topic will be provided immediately below with regard to focus group formation. Another recommendation for selecting participants for focus group research is to recruit participants of strangers in order to avoid unintended political or social effects.

The second parameter in focus group research, closely related to group size determination, is to decide the number of focus groups one study should have. As briefly indicated above, focus groups are usually formed by choosing participants who are similar or homogeneous on a demographic variable (e.g., gender, age, race, education level) in one group. For example, the college student participants in each of the seven focus groups in Gullifer and Tyson (2010), the case study of focus group research chosen for this section, were homogeneous in terms of discipline (e.g., psychology, public relations) and college year (i.e., first year or third year) of undergraduates. Such homogeneity by and large serves as not only a guideline for selecting focus group members but also a criterion for determining focus group numbers. If a study aims to explore similarities and differences on one or two variables (i.e., specific topics), fewer focus groups should be formed. Otherwise, more focus groups are needed in order to carry out a thorough examination of similarities and differences caused by each variable involved in research. With the homogeneity guideline in mind, focus group researchers normally create four to six groups in one study (Morgan 1996). Morgan (1996) further denotes that smaller groups are more appropriate for emotionally charged topics (e.g., domestic violence) that would generate high levels of participant involvement, while larger groups work better with more neutral topics (e.g., foreign policy) that lead to lower levels of involvement. Berg (2009, 161) advises against too few groups mostly because comparison among groups, an important characteristic of focus groups, would be impossible for researchers to make when the number of groups is lower than three.

A third parameter that focus group researchers must consider is who should serve as the moderator. The moderator plays a crucial role in any focus group research in that the moderator is the person who actually convenes each focus group, asks questions, and, more importantly, facilitates the focus group discussion. However, past practices show that several decision points must be made regarding the selection of a moderator.

- Should the researcher or someone else serve as the moderator? – There are apparently two and only two options to this question. If the researcher functions as the moderator, s/he knows what is expected to be gathered via focus groups, which excludes the need to help a moderator get familiar with the research topic. The downside is that the researcher may be so occupied in moderating the focus group session that s/he loses the focus on the research itself. If someone other than the researcher serves as the moderator, s/he can focus on moderating focus group discussion without bringing any assumption to the data collection process, while the researcher can focus on the discussion and make notes as another recording technique. The downside of this option is that the moderator needs to know enough about the research topic before focus group sessions start, which may lead to misunderstanding or misinterpretation of the topic.
- Should a professional moderator or someone in-house be hired to conduct the focus groups? – The advantage of hiring a professional moderator is that this moderator not only is trained for conducting the focus group sessions but also can be responsible for developing the focus group guide and writing up the report. Besides, the professional moderator option can alleviate the burden of data collection on the researcher's shoulders, although the price tag associated with the hiring should not be overlooked. For example, it costed several thousand dollars for a research project with five focus groups, each lasting 70–80 minutes (Shoaf 2003). Hiring in-house help would be much cheaper. However, moderators of this kind only conduct the focus group sessions without offering other assistance, as the professional moderator does. This is the option Gullifer and Tyson (2010) selected "to ensure students were not familiar with the moderator, thereby reducing any possible power dynamics within the focus groups" (468).
- How many moderators should be hired for one single study using focus groups? – Although published sources have no mention of this question, it is relatively easy for the researcher to determine a number according to the research topic, the number of focus groups involved, and the time frame. It is possible to have one moderator to conduct all the focus group sessions (e.g., Shoaf 2003), while it is also common for several moderators to hold different focus group sessions (e.g., Gullifer and Tyson 2010).

The focus group guide, a must in focus group research, is what researchers need to create before data collection so that focus group discussion will be conducted as planned. When the moderator is not the researcher himself or herself, the focus group guide can help make certain that all the focus group sessions are carried out consistently. The researcher should be responsible for the creation of a focus group guide even if s/he is not the moderator. When a professional moderator is hired for conducing focus groups, the researcher still should get involved in the development of a focus group guide because only the researcher knows the best about what should be included in the guide. Similar to the composition of an interview guide, a focus group guide contains main questions, probes, and follow-up questions. All focus group questions should be formulated according to a study's research questions. However, the questions in a focus group guide ought to be formulated to encourage discussion among focus group participants (e.g., "What do others in this room think about the gun control issue?") instead of interview questions for a conversation between the interviewer and interviewee (e.g., "What do you think about the gun control issue?"). In addition, all the dos and don'ts in formulating interview questions (see Section 4.2.5) are applicable in the creation of focus group questions.

The actual composition of a focus group guide can vary from study to study, depending on the research objective of each project. Berg (2009, 173–75), nevertheless, outlines the following five components of a typical focus group guide.

- Introduction – The beginning part of a focus group guide serves two functions: (1) A brief reiteration of the research project objective. Since the participants should have been informed of what the project is about during the recruitment process, a reiteration at this point can be short and concise. Moreover, it would work better if the research project's purpose is presented using plain, colloquial English rather than reading from a prepared text word-by-word. (2) Provide an opportunity for the moderator and participants to get acquainted with each other. The moderator should think about what techniques can be used to facilitate the introductory activities (e.g., each participant describing something unique about her/himself) ahead of the focus group sessions.
- Instructions for the focus groups – This part in the focus group guide is for the moderator to provide instructions to participants. In any focus group research, it might be the first time for most, if not all, participants to take part in a focus group session. The moderator thus needs to explain what is expected from the participants and general guidelines in taking part in focus group activities (e.g., one person speaking at any time, respecting each other). Again, such instructions should be presented clearly and succinctly. By all means, avoid reading from a list of dos and don'ts in focus group sessions.
- Questions for focus group sessions – This is the most crucial component of any focus group guide. When another person is invited to moderate the focus group session, the researcher and moderator should work together to develop a list of questions that would be asked in focus group sessions. Make certain that all the questions are created according to the research problem and questions of the study. Keep in mind that the questions should be good for stimulating discussion instead of purely Q&A or conversation as in interview.
- Focusing exercises – Some focus group research may incorporate focusing exercises, namely, activities specifically designed for gathering data from participants in focus group sessions. If one or more focusing exercises are to be used in focus group sessions, details about them should be included in the focus group guide. There are four major types of focusing exercises according to Bloor et al. (2001, 42–48). The first type is to ask participants to rank a list of statements in the order of importance, which is identical to the Likert-scale questions discussed in Section 3.2.1. Then discussion is carried out about the rankings to discover any deep differences and important similarities in the participants' tacit understandings of the statements and rankings. Vignettes, a second type of focusing exercise, are hypothetical cases or scenarios with particular features, which make suggestive real-life situations to participants in focus group research. The participants are then asked about what course of action should follow (see Box 3.2 in Bloor et al. 2001, 45). A third kind of focusing exercise is called the news bulletin exercise. The exercise comprises the distribution of a series of still photographs to the participants, asking them to use the photographs to compile their own news bulletin on a given topic before the focus group discusses what they have assembled. This should be an enjoyable exercise for the participants. The fourth type of focusing exercise is related to the new bulletin exercise: participants are shown some photographs and then asked to describe what they think was going on in the photograph in the focus group discussion. Additional types of focusing exercises are suggested by Colucci

(2007) such as label generation and role playing. Any of the focusing exercises should be helpful in facilitating data collection in focus group research.
- Guidance for dealing with sensitive issues – Focus group research often deals with sensitive issues such as LGBTQ+ movement, censorship, and gun control. Guidance on how to moderate focus group discussion on such issues constitutes the fifth component in a focus group guide. Berg (2009, 175) suggests to ask a question that may open the door for discussion in the group (e.g., "what do you think of legalizing marijuana?") instead of directly asking a question such as "tell me about the benefits or harms in legalizing marijuana". In other words, always approach sensitive issues with sensitivity.

The focus group guide is to some extent similar to an interview guide except that the emphasis in the former should always be placed on facilitating and encouraging discussion about the research topic among the participants.

### *8.1.3 Using focus groups in research*

After the major parameters in focus group research (e.g., focus group size and number) are determined, the researcher is ready to employ focus groups for data collection. The data collection process via focus groups generally consists of the following steps:

- Making appropriate arrangements before each focus group session: Arrangements for focus groups are specific and detail-oriented to ensure good attendance of focus group participants. For example, send a gentle reminder before the scheduled date to make sure that each recruited participant knows the date, time, and location of the focus group session. Provide participants with proper incentives (e.g., Amazon coupons for the general public or research report summary for academics) to show your encouragement and appreciation of their participation in the study. The guidelines for incentive offering presented when using other research methods in this book (e.g., Section 3.3.3 for questionnaire) are applicable in focus group studies.
- Making the setting inviting and comfortable for focus group discussion: The setting for focus group discussion is perhaps more central than anticipated for the success of this data collection process. The setup of the venue, the temperature in the room, and the arrangement of the table and chairs all play a part in having a proper setting for the focus group session. For instance, participants would not be able to focus on the discussion if the room temperature is too high or too low. An ideal furniture arrangement for focus group sessions is that participants can sit comfortably around a round or oval table so that they can clearly see each other and no one is hidden even partially from the view of other participants. This arrangement also can avoid the unintended separation between the moderator and participants if a rectangular table is used.
- Putting the focus in your focus groups: Plan ahead regarding how to structure your focus group sessions as this is the most crucial step in any focus group research – asking questions and facilitating discussion. Proven techniques in this step of focus group sessions include having warm-up discussion topics and making the flow of discussion from more general topics to more specific ones. Warm-up discussion topics could be something all the participants are able to take part in while relating to a study research question in a certain manner. For example, ask about a recent weather event (e.g., a heavy snowstorm, a hurricane) in a focus group discussion on global climate change as a warm-up exercise. Furthermore, the logic in asking questions and the emphasis on

facilitating discussion should be maintained throughout the focus group session. What is depicted in Section 4.4.2 regarding how to ask questions can also be of relevance in this step of focus group research.

- Moderating focus group sessions: One of the most serious challenges a moderator should prepare for is to handle group silence when no participant says anything in a focus group session. Multiple factors can attribute to this dreadful group silence. For example, participants are strangers to each other and no one wants to break the ice. The initial self-introduction before the focus group discussion and presenting some warm-up topics can be helpful for breaking the silence. Another possible common factor causing silence in focus groups is a pre-existing hostility or suspicion between some or all the group participants (Bloor et al. 2001, 51). Such an issue can usually be avoided by carefully recruiting participants who are at least not hostile to each other in the first place. Bloor et al. (2001, 51–52) and Krueger and Casey (2015, 47–49) suggest a number of additional tactics to help avoid silence in focus groups. For instance, start a conversation relating to the research topic before the focus group begins. Prepare a handful of conversational topics for breaking ice. Avoid using a long, rectangular table or seating arrangement, with the moderator alone at one end, while all participants at the other. As indicated before, a round or oval table should always be selected for focus group discussion. While a participant talks, use head nodding, short verbal responses (e.g., a good point, interesting), or similar techniques to keep the discussion flowing. Comments on nonverbal cues (e.g., Ann, you look as if you want to say something?) can also facilitate focus group discussion.
- Capturing the data collected in focus group sessions: With easy access to video recording devices such as smartphones and video cameras, recording the entire session of each focus group discussion becomes a norm for data capture in focus group discussion. Use multiple video cameras for recording a focus group session, if necessary and feasible, to attain video data from different angles. If focus groups are conducted through video conference software such as Zoom, the recording function coming with the application should be utilized. Audio-taping can still be used if video-taping is not available, but then note-taking must be used to record the missed body languages and other non-textual data (e.g., seating of participants) in this context. If neither video nor audio-taping is possible, the researcher has to rely on note-taking to record focus group data. In such cases, the researcher may need to designate a note-taker to help capture focus group data. The note-taking in observation research (see Section 5.3.3) is of great relevance to focus group studies. Regardless of which format is chosen for capturing data during the focus group session, consent must be obtained from all participants before the session begins.
- Concluding the focus group session: A focus group session will end after all the questions, both prepared and impromptu, are asked and discussed. It is recommended that the moderator briefly summarize what is discussed in the focus group session before thanking all participants for their time and contribution to the discussion.

The steps illustrated above, nevertheless, should be carried out coherently and with some flexibility. In reality, neither a starting point nor an ending point for each step is clearly marked in the data collection process. Instead, the researcher should decide when to move to the next step in a focus group session as well as do so seamlessly and naturally.

### *8.1.4 Dos and don'ts in focus group research*

Although interview and focus groups are two data collection techniques with different features, many dos and don'ts for the interview research method, depicted in Section 4.4.4, are applicable to focus groups as well. Additional dos and don'ts for focus group research are elucidated as follows:

- Do encourage every participant to talk. This is quintessential for focus groups due to the fact that there are usually at least several participants in each group. The focus group session cannot be considered a discussion session if only one or two participants take part in the discussion. At the same time, don't let a few participants dominate and influence the rest of the group in the discussion.
- Do facilitate, but don't take part in the discussion to make certain that what is gathered is from the participants rather than from the moderator.
- Do stay focused on the topic under discussion. It is quite easy for a group member to wander off from the discussion topic in a focus group session. The moderator at this point should politely but firmly bring the discussion back to track. In the meantime, don't overreach during any given focus group session or have too little influence in the process as a moderator. In addition, don't lead the group to the outcome the researcher wishes to have, as this would damage the credibility of the study if ever attempted.
- Do listen and observe attentively. In any focus group session, multiple activities are taking place simultaneously because multiple participants take part in the process. Theoretically, each participant can generate many activities (e.g., facial expression, gestures, body posture) even without speaking out. The moderator must pay full attention to each participant as well as all the participants as a group. In this sense, do keep the group size reasonable as discussed in Section 8.1.2.
- Do record and take notes of the discussion. Video recording is recommended as one data capture technique. But don't overlook the value of note-taking during a focus group session, especially when the researcher does not serve as the moderator. Taking notes can record what might be missed in the video recording because a video camera is always stationary and cannot capture all the dynamics (e.g., Participant A threw a glance at Participant D) in a focus group session even if more than one video camera is used. Meanwhile, setting up multiple cameras for a focus group session may cause uneasiness to the participants. Issues of this kind should not be overlooked.
- Do keep in mind that what has been gathered is from a group instead of an individual. This fact determines how the collected data will be analyzed and interpreted. Don't treat focus group data as interview data.
- Do determine the proper number of focus groups as well as appropriate group size. Using too few/many focus groups or having too few/many individuals in one group can both lead to undesirable outcome. For example, Bloor et al. (2001, 91) indicate that a single 90-minute focus group session, which is the typical length for one session, can easily generate 100 pages of transcript. Thus, only have as many focus groups as necessary based on the study's research questions.

It is always true that we are unable to exhaust all the dos and don'ts for any research method application, and focus groups are not an exception. What we can do is to follow the established practices, learn from previous research that uses the same method, and avoid repeating others' mistakes in our own studies.

We briefly discuss in Section 8.1.1 that focus groups can be used for gathering data from groups of participants when individual views or input are not crucial as well as for soliciting a wide range of responses from groups of people. Another usage of the focus group technique is for gathering data from temporary or transient populations (e.g., homeless people, people living in mobile homes) when other research methods are not feasible. In addition, focus groups are frequently used in conjunction with other methods of data collection (e.g., questionnaire, observation, experiment).

### *8.1.5 Case study of focus group research*

Gullifer, Judith, and Graham A. Tyson. 2010. "Exploring University Students' Perceptions of Plagiarism: A Focus Group Study." *Studies in Higher Education* 35 (4): 463–81. https://doi.org/10.1080/03075070903096508

Plagiarism is an issue often visited by researchers. But with digital technology advancement when electronic publications can be easily copied and pasted, the plagiarism issue continues to attract considerable attention from the media, academics, administrators, and students. It is in the rising concerns about plagiarism that Gullifer and Tyson (2010) conducted this study. Their research purpose is that "universities can benefit from learning about their own students' perceptions of plagiarism in order to develop appropriate strategies to promote academic integrity" (465).

The participants of the study by Gullifer and Tyson (2010) consisted of 41 students in total, 25 women and 16 men, who were either in their first or third year of study and from four different disciplines (i.e., psychology, policing, public relations, and advertising) in an Australian university. The participants were all recruited via subject coordinators, similar to academic advisors in the US higher education system, in the afore-listed departments. Gullifer and Tyson (2010) state that "[e]ach focus group was homogeneous with regard to discipline and year, as issues pertinent to understanding plagiarism may be discipline or faculty specific" (467). Therefore, college year and discipline are the two demographic variables used to form focus groups, which closely follow the established practices of creating focus groups in a study (see Section 8.1.2). Gender is not considered in forming focus groups in the current research, so it is not an issue when there are nine more female participants than male students in this study.

All the 41 participants were assigned to seven groups, ranging from 3 to 9 students per group. Gullifer and Tyson argue that, according to Wilkinson (2008), "... a focus group can involve as few as two or as many as a dozen participants, with the norm being between four and eight" (Gullifer and Tyson 2010, 467). The authors further reason that two participants per focus group is kind of low. They hence decide that three can be the minimum number of participants in a focus group to ensure adequate interaction among themselves. Of the seven focus groups in the study, Groups 1, 6, and 7 all have only three participants. This low number of participants in the three focus groups deviates from the recommendation by many authors (Bloor et al. 2001, 92; Scheuren 2004, 35) that six to eight as the minimum number of participants per focus group so that the group can truly have interaction and discussion. In addition, Group 1 was designated as the pilot group to test out the research design of the study under examination. However, Gullifer and Tyson did not exclude what they collected from Group 1 in their data analysis as generally done in research practices, that is, participants and data collected in the pilot phase should not be included in the full study. Otherwise, any revision made to the initial research design is not reflected in the pilot.

Groups 4 and 5 each, respectively, had eight and seven students, but the authors did not indicate which year, first or third, they were from, except just listing them as "1$^{st}$ session police students" in Table 1 of Gullifer and Tyson (2010). Is "1$^{st}$ session" the same as first year? The researchers should have explained. Furthermore, no participants came from STEM fields when those disciplines constitute the core in any university in the 21st century. The authors perhaps recruited the participants in four subject areas (i.e., psychology, policing, public relations, and advertising) for the sake of convenience because the participants were from the researchers' own affiliation: School of Social Sciences and Liberal Studies. The representativeness of the study participants by discipline thus is in question.

An interview guide was developed but then used as a focus group guide. The researchers, strictly speaking, treat focus groups as group interviews when we know by now that these two data collection techniques are different (see Section 8.1.1). The guide has in total five questions (Gullifer and Tyson 2010, 467) without any follow-up or probe questions and without using any focusing exercise as Bloor et al. (2001, 42–48) suggest to facilitate focus group discussion. Therefore, the data collected in the focus group sessions are answers to the five questions rather than from focus group discussion as anticipated. Gullifer and Tyson (2010) also indicate that "[q]uestions that were more important to the research agenda were presented early in the session" (468). However, the order of question presentation really should be determined by the nature of the research instead of by the significance of questions.

Four moderators, other than the two authors and all being employed by the university, conducted the seven focus groups "to ensure students were not familiar with the moderator, thereby reducing any possible power dynamics within the focus groups" (Gullifer and Tyson 2010, 468). But the authors did not specify how many focus groups each moderator did. Nor did they denote whether the moderators know how to moderate focus groups or receive any moderation training if they are new to this role.

Each focus group session lasted no more than 75 minutes, which is within the recommended norm of 90 minutes at maximum. Each session was digitally recorded, although Gullifer and Tyson did not specify if the recording is audio or video.

The study we just examined has the essential elements of focus groups (e.g., 41 participants in seven groups). The focus group sessions nevertheless were done basically as group interviews, which in reality fails to take advantage of what the true focus group method can offer. As indicated by the authors, their research is a pilot study for a large-scale project (Gullifer and Tyson 2014). Based on that project report, it becomes clear that Gullifer and Tyson used focus groups to develop a questionnaire for their large-scale project. This practice thus represents one of the several implementations of the focus group method depicted in Sections 8.1.1 and 8.1.4.

**Section Summary**

- Focus groups are a technique for collecting data via groups of individuals, selected and assembled by the researcher, to discuss and comment on a topic of the study from personal experience. The focus group method is primarily used for gathering data from groups of participants when individual views and input are not crucial.
- The characteristics of focus group research include dividing participants into three or more homogeneous groups on demographic variables, such as age, race, or gender, focusing on participants' discussion and comments as a group, and comparing differences and similarities among focus groups.
- Of all the dos and don'ts for focus group research, choosing the right kind of participants to form the right number of focus groups constitutes the most important "do", while avoiding turning focus group sessions into group interviews represents what focus group researchers must observe in their data collection.

## 8.2 Transaction log analysis

**Section Overview Questions**

- What is transaction log analysis?
- What features does transaction log analysis have as a research method?
- What are the major steps in collecting transaction log data and preparing them for analysis?
- What are the dos and don'ts of research using focus groups for data collection?

### *8.2.1 What is transaction log analysis?*

Transaction log analysis (TLA) is a technique for collecting transaction logs – data automatically captured and recorded by a system (e.g., search engines, e-commerce sites) that include the type, content, and time of transactions or interactions between the system and users of that system (e.g., searchers, online shoppers). Transaction logs were called computer-monitored data in the early days (e.g., Rice and Borgman 1983) before the current name was coined.

There are two major kinds of transaction logs people commonly gather from systems. One type, also the primary one, relates to transaction logs from information retrieval (IR) and other information systems. Common IR systems comprise databases, search engines, and OPACs (online public access catalogs), while other information systems (e.g., social media systems) are too many to be summarized. Research using transaction logs from IR and other information systems is often conducted and published for improving system performance or learn about system users' behavior and more. The other kind of transaction logs concerns with those from e-commerce sites, which is smaller in quantity when compared with that from IR and other information systems. Transaction logs from e-commerce sites are closely studied to gauge customers' needs and wants. However, research of this type is mainly for internal use to help with marketing or service improvement. Its findings are rarely published or available to the public. Consumer big data contain such transactional data, and they are transforming marketing decision making (Erevelles, Fukawa, and Swayne 2016). The focus of this section is on transaction logs from IR and other information systems because they are not meant to be used internally and have wide implications in the research arena. Google Trends (https://trends.google.com) is one example of TLA using transaction log data from the search engine Google itself.

TLA, as one data collection technique, possesses some distinctive features (Jansen 2006; Jansen, Taksa, and Spink 2009; Kurth 1993; Wildemuth 2017, 186–88). First, transaction log data can be gathered unobtrusively and in large quantity once permission for data collection is obtained from the owner of transaction logs. Second, transaction logs are captured literally in full and at the time when transactions or events (e.g., searching, clicking, browsing, posting) took place. In this sense, transaction logs are also called frozen data as they do not change once being recorded. Collecting transaction log data additionally involves no recall, biases, low responses, and incorrect memory that might occur with research participants in comparison with data gathering using other techniques (e.g., questionnaire, interview). Therefore, transaction logs serve as one of the few sources that can produce exact data as captured and be free of any interpretation from users or research participants. Transaction log data record precisely what system users have done, not what they say they might or would do; not what they were prompted to say; and not what they thought they did (Nicholas and Huntington 2003). Third, TLA is an efficient way of gathering longitudinal usage data. There is no time limit for data collection as long as the log files exist. Fourth, transaction logs can be analyzed quantitatively as well as qualitatively because they capture both numbers and texts in the interactions between the target system and users.

However, TLA has its weaknesses too as a data collection technique. It is extremely difficult in gaining access to transaction logs due to the proprietary nature of this kind of data and privacy concerns about system users. Transaction logs, as they are captured automatically, are messy and need cleansing after they are gathered for a research project. In addition, no contextual data on motivations, feelings, and other similar data about the system users can be recorded in transaction logs. Similarly, nor can system users' demographic data be collected via transaction logs. It is for these reasons that researchers are recommended to choose at least another data collection technique (e.g., questionnaire) to be used together with the TLA method. One more limitation of the TLA method is the difficulty in identifying users relying on their IP (internet protocol) addresses because a user may move from one computer to another while using a system or two users may alternatively use a single computer when accessing the same system. This is very likely to happen with public computers located in libraries, computer labs, and the like. The usage data from these computers thus represent a group of users rather than single users. This problem is exacerbated by the difficulties of distinguishing among sessions (i.e., periods of online time for such activities as searching or reading) when system users do not officially log off after each activity. Jamali, Nicholas, and Huntington (2005, 558–59) further depict such difficulties in their review of TLA research.

Transaction log data are captured in two ways, namely, at server-side or at client-side. A brief explanation of the client/server concept seems necessary at this point. The client/server concept is essential in any network environment. Networks could be centralized or distributed, but past experience shows that centralized networks are not efficient because the network center would easily be overloaded or overwhelmed. In contrast, distributed networks using the client/server structure are widely adopted and implemented. A client is an entity that works on the client's behalf to extract a service (e.g., performing a search, getting a file, printing a report) from another entity (i.e., server) on the network. Similarly, a server is an entity that provides service to another entity (i.e., client) on the network upon request. An analogy of the client/server structure could be what we experience at the restaurant. We as customers (i.e., clients) order food from a waiter (i.e., server) who would later serve us with what we have ordered. Today, all systems (e.g., search engines, social media sites, e-commerce sites) operate on the client/server model. In other words, the information systems people use for various purposes function as servers, while the computer or terminal users employ in interacting with the information system are called clients.

Server-side transaction logs refer to those data that are recorded on the system server (e.g., servers at Google or for library databases), whereas client-side transaction logs are obtained by using specific software installed on computers or terminals (i.e., clients) users utilize for accessing the system. Server-side transaction logs are automatically kept as long as the system is in operation, and they account for the majority of such data collected in TLA research. By contrast, client-side transaction logs are only gathered when researchers have a need to capture them. The case study chosen for this research method by Logan, Barrett, and Pagotto (2019) is a TLA study using client-side transaction logs.

Client-side transaction logs can contain such detailed data as keystrokes because these kinds of data are generated from the client side.

### *8.2.2 Essentials in obtaining and processing transaction logs*

Unlike the data collection techniques (e.g., questionnaire, observation) we have discussed before this chapter, gathering transaction log data is more about gaining access to transaction logs and preparing them for analysis because the data to be "gathered" are already captured by the system or the client-side software. Bernard J. Jansen, a noted researcher in TLA, details the procedure for how to do web search log analysis in one of his many publications on TLA (Jansen 2006). The procedure Jansen (2006) describes is for web search log collection, preparation, and analysis. Nevertheless, it is also applicable to other kinds of transaction logs (e.g., database usage logs, social media site logs) used in TLA research. The major steps presented below are for obtaining and processing transaction logs in general.

- Obtaining permission for data collection – This is the first step researchers should take in conducting any TLA study. As indicated earlier, transaction logs are usually proprietary, and their owners (e.g., database publishers, search engine companies) are not willing to share them with outside researchers. Thus, one feasible approach for researchers to gain access to transaction logs is through personal or other connections in order to get in touch with the system administrator who is able to grant access. It does not seem appropriate or effective to contact the system administrator directly with a request for obtaining transaction logs because the system administrator has every reason to dismiss such a request. Another factor that prevents researchers from obtaining transaction logs relates to concerns about personal information captured in the log data. For example, the IP address of any computer used for accessing a system is always recorded. If people use their own computers when accessing a system, their computers' IP addresses and subsequently their identities can be found without much difficulty. Therefore, all researchers who intend to use transaction logs for their studies must make certain that any information relating to individuals' identities in transaction logs should be anonymized even after they are given access permission.
- Collecting transaction logs – The ideal situation is that the owner of transaction logs allows researchers to download the log files. Most, if not all, TLA research has a specified time span. For instance, Logan, Barrett, and Pagotto (2019) gained access to the transaction log data between June 1 and December 1, 2016. The authors then gathered what they planned to collect from the log data (i.e., factors and behaviors associated with user dissatisfaction with a chat reference) for that time period. Another option is to use an API (application programming interface), a software intermediary either provided by the system or developed by the researcher, for data collection purpose. How to develop an API nevertheless is beyond the scope of the current book. Collecting data through an API is often known as data scraping, which mostly aims to gather real-time or streaming data from social media sites such as Facebook and Twitter.
- Cleansing collected transaction logs – Transaction log data are notorious for their messiness. It is thus absolutely necessary for researchers to clean the log data collected. Jansen (2006, 414–15) provides some instructions for data cleansing. (1) Export or download the log files into a spreadsheet (more common) or a database (less common) with necessary fields (e.g., user identification, time, search terms). (2) Sort each field in the exported

or downloaded log file in sequence to detect errors. Experience shows that this sorting method is efficient and effective to identify most errors "[s]ince the erroneous data will not fit the pattern of the other data in the field, these errors will usually appear at the top of, bottom of, or grouped together in each sorted field" (Jansen 2006, 415). (3) Delete all records with erroneous data in the sorted fields from the transaction log file.

- Preparing transaction log data for analysis – This step is done by labeling the sorted fields of log data as variables based on one's research questions. Common variables in TLA research include user identification (i.e., users' IP addresses), date (as recorded in the transaction log), time length, activities, or interactions (e.g., searches conducted, search results viewed, sites visited, messages posted, purchasing orders placed) with the system. Activities or interactions constitute the largest category of variables researchers want to examine in data analysis later. Activities commonly examined in TLA research on IR systems include commands/queries entered, response time, search lengths, errors, no results, failures, and their causes (Peters 1993).

Transaction logs, after all the above steps, are ready for researchers to analyze and answer any research question they formulate in their studies. How to analyze transaction log data quantitatively and qualitatively is to be discussed in Chapter 13 together with all other kinds of data collected using techniques the present book depicts.

### *8.2.3 Using transaction log analysis in research*

Transaction logs have been used for a variety of purposes since the 1960s. Peters (1993) divides TLA research into three phases when reviewing its history from the information science perspective:

1) From the mid-1960s to the late 1970s, the emphasis of TLA research was placed on evaluating system performance;
2) From the late 1970s to the mid-1980s, TLA was applied to online library catalog systems with equal attention to system usage and user behavior;
3) From the mid-1980s and on, diversified TLA research was conducted on multiple topics ranging from search types, systems of different types (e.g., database systems) to specific user groups (e.g., dial-in users before the internet era).

Researchers also start replicating previous TLA studies and conducting longitudinal studies as transaction logs can be obtained from the same system longitudinally with proper permission. Most TLA studies focus on operational systems accessible by public users, which reflects the implications of such research – improving operational systems and enhancing users' experience in utilizing those systems.

Not long after the advent of the internet and particularly its star application – the web – in the mid-1990s, TLA researchers set out to explore transaction logs generated by numerous web systems. The web has been becoming the default platform for various kinds of systems, from search engines to e-commerce websites. Hence, a fourth phase in the history of TLA research could be added to the three phases Peters (1993) already identified. Active TLA researchers (e.g., Jansen, Spink, and Saracevic 2000; Jansen and Spink 2006) mainly carried out studies using web transaction logs in this phase. When social media emerge as one major type of transaction log data, TLA research is entering the fifth phase in its history of development. Furthermore, this phase also shows some relevance to big data analytics because streaming data, as a data source, sit on the borderline between transaction logs and big data. Big data analytics itself is a book-size topic, which deals with gigantic data sets using artificial intelligence (AI) and statistical techniques. This topic, nevertheless, is beyond the scope of the present book and thus will not be mentioned again hereafter.

TLA, as a data collection technique, keeps evolving as its data sources change. Additional phases are likely to appear when transaction logs will come from new sources that we might be unable to perceive at present. The potential likelihood of data source change, however, would not lead to dramatic changes in TLA as a research method. Instead, all the fundamentals underlying this research method will remain the same, regardless of what the data source will be.

As more and more resources become available electronically (e.g., ebooks, journal articles, webpages, websites), log data for them (e.g., downloads, views, browsing time, links clicked, number of webpages visited) are often collected and analyzed to address research questions relating to observable user behaviors of systems with electronic resources (Oard and Kim 2001). This kind of TLA research further extends the application of the TLA technique beyond the monitoring and evaluating system performance. Understanding user-system interaction has also become a common objective in TLA research.

Sampling is a must rather than an option when TLA is chosen as a data collection technique in research because of the enormous size of transaction logs. No study can and should use all the transaction logs available from a given system, even if the system is small in size in consideration of practicability and efficiency in data collection. In general, TLA researchers would collect data via a sampling frame by, for instance, time of the day, day, week, or month. The sampling can be random or otherwise. If random sampling is selected, make sure to follow the established sampling practices described in Section 2.5.2.

Transaction logs are one type of trace data that we discuss in Section 5.2. Trace data include two major categories, one for erosion and the other for accretion (Berg 2009, 287–89). Data of erosion trace refer to, for example, the wear-tear of printed books, furniture, and staircases. Data of accretion trace include dust on printed books or furniture, fliers on walls, and transaction logs. Transaction logs additionally have the unique characteristic that they are "frozen" in time and space once captured. Unlike other data of accretion trace, the targeted transaction logs would not increase in quantity as time passes or if being moved from one storage place to another. Because of this distinctive feature of trace data, it is important to consider the following six questions when conducting TLA research (Jansen, Taksa, and Spink 2009, 7–8):

- How is this data defined?
- What is the population from which the researcher has drawn the data?
- Which data are analyzed?
- What is the context in which the researcher analyzed the data?
- What are the boundaries of the analysis?
- What is the target of the inferences?

The afore-listed six questions obviously go beyond data collection using the TLA method. In other words, TLA research must be done in context when there is no opportunity for researchers to contact system users for clarification or further exploration as in questionnaire or interview studies. The transaction logs are all what the researcher gets, and such data do not include any data about, for example, why users choose a particular system, what motivates online shoppers to continue completing a purchase even after a power outage, and whether users are satisfied with their interaction with the system. That is why researchers are highly recommended to adopt another data collection technique (e.g., questionnaire, interview) to complement the TLA technique.

### *8.2.4 Dos and don'ts in transaction log analysis research*

Although collecting transaction logs seems straightforward once permission for doing so is obtained, TLA research in reality can only be properly conducted when the following dos and don'ts are observed.

First, do follow the ethical guidelines in TLA research. Always obtain permission from the owner of transaction logs (e.g., system manager) before data collection. In addition, transaction logs contain more privacy information than many people realize, which can be illustrated by the following two infamous examples. One is about the Bush administration's subpoena to the major internet search engine companies, Google, America Online (AOL), Microsoft and Yahoo! in August 2005, requesting each to provide all search queries and associated result URLs from June 1 to July 31, 2005 as part of the government's effort to uphold the Child Online Protection Act (Fry 2006). While AOL, Microsoft and Yahoo! had complied with the subpoena, Google had resisted on the ground that, if complied, this could expose identifying information about its users. The final settlement is that Google submitted a list of 50,000 web addresses randomly selected from its logs of one week, and the Bush administration's original request for user search queries was denied (Fry 2006). In the other case, which concerns AOL's "accidental" release of more than 650,000 of its users' search histories, the 21 million search queries from those users have exposed an immeasurable number of life stories ranging from the mundane to the illicit and bizarre (McCullagh 2006). The damage has been done even though AOL issued an apology soon after. These two examples demonstrate how much privacy information can be gleaned from web search logs.

Other kinds of transaction logs likewise record privacy information. Take the Target store pregnant girl case in 2012 as an example (Hill 2012). An angry father went into a Target store outside of Minneapolis, demanding to know why his daughter received Target coupons for baby clothes and cribs when she was still in high school. It turns out that the girl bought a pregnancy test kit from Target and got a positive result. In other words, Target, like all for profit companies, analyzes all its customers' shopping records (i.e., transaction logs from Target's system) and noticed this purchase. Target thus wanted to sell baby items to this pregnant girl. Duhigg (2012) explains in great detail how companies like Target analyze or mine customers' shopping records for marketing purpose. Therefore, the privacy issues and ethical concerns should never be overlooked in TLA research.

Second, do use an appropriate sampling method for collecting transaction logs to make data size manageable. Strictly speaking, no researcher should or is able to collect all the transaction logs a system captures because that practice is counterproductive and not efficient. Choose a random sampling technique if you intend to generalize your research findings to the population representing the collected data. Otherwise, a nonrandom sample can be used, but researchers must know that they do not have any opportunity to generalize their research findings. Detailed description on sampling and related topics can be found in Section 2.5.2.

Third, be aware of the limitations regarding what transaction log data can record. Transaction log data are not captured by a system with any particular research questions in mind. Nor are they as rich as data gathered with some other techniques (e.g., questionnaire, observation). Data extracted from transaction logs are often only loosely linked to the concepts they are alleged to measure. For example, the time a user was online may or may not mean that s/he interacted with the system every minute during that period. The user might take a break or do something else instead of using the system. With transaction logs, the users may be conscious of the fact that they are being recorded, which may alter their actions. Their behaviors hence may not be altogether natural as a result (Jansen, Taksa, and Spink 2009, 10).

Fourth, do not over-interpret transaction log data because they only record what can be captured (e.g., search queries, download frequencies) and traced (e.g., time interacting with the system). The whys (e.g., why does one want to search for X) and many other data pieces (e.g., users' satisfaction or frustration with the system) cannot be captured due to the nature of transaction logs. Do not attempt to infer anything that is not captured in transaction logs at all.

Fifth, do analyze transaction log data with the appropriate context. In general, transaction logs are captured for certain purposes (e.g., evaluating system performance, studying user behavior) and on a certain platform (e.g., search engine logs). If transaction logs are analyzed without reference to why and how they are recorded in the first place, any results obtained from such analyses would have validity issues. For example, Google captures its transaction logs mainly for understanding how users search it and what they search for. Google then uses such data for improving its ranking algorithm of search results. That is, which Google search results should be presented, for instance, as the top ten for a given search query out of thousands of possible relevant ones for a given query because users normally do not view results beyond the first ten. This is the context we should keep in mind when using Google transaction logs for our TLA research. We can use Google transaction logs to evaluate its usability, its users' search behavior, and the like. If we intend to judge the usability of a library online catalog with Google transaction logs, for example, that would be totally out of the implied context of Google's log data. In the peak time of COVID-19, people naturally searched Google for information about how to deal with the pandemic. Transaction logs of Google from that period of time (e.g., the entire year of 2020) should constitute a rich source for TLA research relating to the pandemic (i.e., the appropriate context) as opposed to other topics. Analyzing data in the proper context is not unique to TLA research only. Instead, we always need to analyze our collected data in the context of our research questions.

### 8.2.5 Case study of transaction log analysis research

Logan, Judith, Kathryn Barrett, and Sabina Pagotto. 2019. "Dissatisfaction in Chat Reference Users: A Transcript Analysis Study." *College & Research Libraries* 80 (7): 925–44. https://doi.org/10.5860/crl.80.7.925

Logan, Barrett, and Pagotto (2019) conducted a TLA study with the intension to identify factors and behaviors associated with user dissatisfaction in chat reference of academic libraries (i.e., college and university libraries) so that suggestions can be provided to chat operators (i.e., reference librarians and other staff members) on what behaviors to avoid. We all use library reference services at one point or another. Since the 2010s, many libraries have been offering reference services using a chat application (e.g., LiveEngage used in Logan, Barrett, and Pagotto 2019), which enables libraries to provide reference services to their users regardless of geographical locations.

In their study, Logan, Barrett, and Pagotto (2019) refer to each text-based, synchronous chat reference session as one interaction. Transcripts for a total of 9,424 chat interactions from June 1 to December 1, 2016 were obtained from a chat reference consortium of academic libraries in Ontario, Canada. Those chat reference transcripts are transaction logs that recorded the interactions between the reference staff and users at the consortium's system but without any illegible data pieces commonly found in other transaction logs (e.g., search engine logs). The researchers in this study thus do not need to clean and process their log data after their data collection.

The researchers took six-month worth of transaction logs in 2016 as their initial data set with no use of any random sampling technique, which means that they could only discuss their study findings within the boundary of the 9,424 chat reference interactions. As part of their study design, Logan, Barrett, and Pagotto (2019) also administered a pre-chat survey as well as an optional exit survey in order to gather additional data on both reference staff and users' behaviors during the chat reference sessions.

Three comments are due about the pre-chat survey and optional exit survey employed in this study: (1) Both surveys change this study from entirely unobtrusive to partially obtrusive, although the exit survey is administered after the chat reference is completed. (2) The exit survey is optional, so less than 15% (i.e., 1,395) of the 9,424 chat reference interactions included in this study have a completed exit survey. (3) Transaction logs alone cannot yield much behavioral data, if any. So using both surveys in combination with chat reference transcripts in this study is a recommended practice for gathering data (e.g., dissatisfaction with chat reference) not available from the transcripts otherwise. As shown, four of the eight questions in the exit survey are "designed to gauge a user's satisfaction with the interaction" (Logan, Barrett, and Pagotto 2019, 929).

The responses to the four questions about users' satisfaction with the interactions from the 1,395 completed exit surveys are categorized as "satisfactory" or "neutral or dissatisfactory". The researchers labeled the following three answer options in three of the four multiple-choice questions as "neutral": "satisfactory", "too much assistance", and "a satisfactory way of getting library help" (see the italics on page 929 of Logan, Barrett, and Pagotto 2019). However, none of the three answer options appear to be "neutral or dissatisfactory" answers to the questions. They all to the contrary should be categorized as "satisfactory". The authors perhaps should have simply removed "neutral" from the "neutral or dissatisfactory" category and move those three so-called neutral responses to the "satisfactory" category, mitigating any confusion about the "neutral or dissatisfactory" category when "neutral" is apparently not a synonym of "dissatisfactory".

A total of 256 interactions, about 18% of the 1,395 interactions with a completed exit survey, were randomly selected from those that are categorized as "satisfactory". Another set of 217 interactions, approximately 16% of the qualified 1395 interactions, were purposively selected from those that are categorized as "neutral or dissatisfactory". Two notes seem necessary on the sampling processes of this study. One is that the authors should indicate which random sampling technique is used for selecting chat reference interactions from the satisfactory category. Whenever a sampling technique is applied in research, it is an established practice that the specific sampling technique (e.g., simple random sampling) should be given because there are multiple techniques in both random and nonrandom sampling methods (see Section 2.5.2). The other note is actually a question. Namely, why did the researchers use purposive sampling for chat reference interactions that fall into the dissatisfactory category? Logan, Barrett, and Pagotto (2019, 930) offer an explanation as follows: "Homogeneous purposive sampling was determined to be appropriate, as very few of the interactions with completed exit surveys displayed neutral or dissatisfied (16%) sentiments, so all available interactions would provide valuable data". The authors could still have directly indicated that all the 217 dissatisfactory interactions are included in the study due to its smaller number in comparison with the satisfactory interactions because, as discussed in Section 2.5.2.3.2, purposive sampling denotes that researchers choose sampling units subjectively rather than using all available cases or sampling units as done in this study. Another possibility is that Logan, Barrett, and Pagotto (2019) could have selected the same number of satisfactory interactions (i.e., 217 rather than 256) as dissatisfactory ones to make the comparison between the two more meaningful.

When the emphasis of this study is on identifying factors and behaviors associated with user dissatisfaction in chat reference, having more cases of satisfactory interactions puts the study out of focus to some extent. In addition, as mentioned earlier, all the "neutral" interactions in the study essentially belong to the "satisfactory" category. Although the authors did not separately list the number of "neutral" interactions in their report, we know that the total number of dissatisfactory chat reference interactions should be lower than 217 when "neutral" interactions, as pointed out earlier, are actually satisfactory ones by connotation in this study.

**Section Summary**

- TLA is a technique for collecting data automatically captured and recorded by a system (e.g., search engines, e-commerce sites) that includes the type, content, and time of transactions or interactions between the system and users of that system (e.g., searchers, online shoppers).
- Transaction logs, once permission for data collection is obtained from the appropriate party, can be gathered unobtrusively, in large quantity, and at the time when transactions or events (e.g., searching, clicking, browsing, posting) happened. Transaction logs involve no recall, biases, low responses, and incorrect memory from research participants that might occur when compared to other techniques such as questionnaire and interview.
- The steps in collecting transaction log data and preparing them for analysis include (1) obtaining permission for data collection, (2) gathering data, (3) cleansing collected log data, and (4) preparing data for analysis.
- Observe all the ethical guidelines (e.g., respecting people's privacy) while keeping in mind the rest of the dos and don'ts in TLA research.

## 8.3 Research diaries

**Section Overview Questions**

- What are research diaries?
- What types of research diaries are there, and what are participant-kept diaries?
- What can be done to ensure that participants will write and complete diaries as asked?
- What are the dos and don'ts in conducting diary research?

### *8.3.1 What are research diaries?*

Diaries are a common type of records that people write for various purposes. Many people may have been keeping a diary for years or even decades to record, for example, what happened during the day or week. Some may include reflections in their diaries. When diaries are used as a data collection technique in research, they are created and maintained by individuals on a regular, personal, and contemporaneous basis (Alaszewski 2006,[1] chap. 1, 2). Research diaries can be kept in different media formats (e.g., notebook, audio, video) with their contents ranging from descriptive event logs to narrative personal accounts (Wildemuth 2017, 228).

Diaries are adopted for data collection to facilitate the research process through recording events, activities, observations, thoughts, and questions as they occur at a specific interval. Researchers use diaries to gather data from participants or rely on self-kept diaries later to stimulate reflective thinking about the study (Newbury 2001). Common applications of research diaries include: (1) Seeking to understand hard to reach populations (e.g., victims of domestic violence) or private spheres (e.g., drug addicts) that can pose practical and ethical issues to observe or record; (2) asking participants to log the events of their daily lives or providing a space for participants to reflect on experiences in their lives, particularly those linked to health and other personal issues. Diaries can specifically be used as a tool for written reflections on experiences and feelings, as a memory aid for participants to recall mundane and daily moments, as a prompt in interviews, as one way for participants to control self-disclosure during an interview, and as one technique for both participants and researchers to explore ambivalence in their intimate daily lives (Harvey 2011).

There are other kinds of records that can be considered cousins of research diaries: the intimate journal, the memoir, the log, and the research log or field notes (Allport 1942; Toms and Duff 2002). The intimate journal contains chronologically ordered, free-flowing entries, without any censorship, often of a highly personal and intimate nature. The intimate journal perhaps is the closest "cousin" of research diaries, except that research diaries typically contain more entries over a limited time period (e.g., one or two weeks) and are kept at the solicitation of the researcher. The memoir is written after the event or events and with the intent to get published for an audience. By comparison, research diaries are kept just for one researcher or one research team. The memoir might be the farthest "cousin" of diaries based on its function. The log records an impersonal account of a series of structured and usually chronological activities. No explanation or reflection is offered about what is recorded in the log. The research log or field notes are often created in ethnography or field studies (see Chapter 11) in order to help the researcher to record what already happened while in the field but not necessarily at a regular frequency (e.g., daily) as in the case of research diaries. Researchers usually make field notes whenever they can before the events or activities fade away in their memories. This section only focuses on research diaries as a technique for data collection.

Research diaries possess several salient features when compared with other data collection techniques we have discussed so far (Toms and Duff 2002; Zimmerman and Wieder 1977). First, data are captured at or close to the time of occurrence, enabling the tight coupling of the record and event. As a result, diarists can minimize recall or memory errors that often occur when questionnaire or interview is the selected data collection technique. Second, diarists are familiar with what is recorded. Unlike in studies using other data collection techniques such as focus groups, participants might be asked to discuss topics that they are unfamiliar with. Third, no transcription is needed for dairy data as in interview or think-aloud research. The completed diaries can be directly analyzed without any additional data preparation, which saves researchers' time. Fourth, research diaries are unobtrusive in that diarists write them in their own space, at home or another location, without the presence of the researcher. This can reduce any uneasiness participants may experience when another data collection technique (e.g., interview or obtrusive observation) is used.

On the other hand, scholars do have some concerns about research diaries. For instance, how much can one capture in a diary? It is rare that the diarist is able to record what happens immediately. Usually, diarists would write later when they take a break or at the end of the day. By that time, many specifics about the event or activity may have become vague or even entirely lost in the diarist's memory. Similarly, how accurate or truthful is the captured content in a diary? According to Harvey (2011), diarists are constantly engaged in a process of self-editing or selection although they may do so subconsciously and unintentionally. Furthermore, the literacy issue might be a barrier to diary keeping. Meth (2003), for example, indicates that some of her prospective participants, black African women, are illiterate and hence cannot keep diaries about domestic and public violence she intends to study. Even if when literacy is not an issue, keeping a diary can appear challenging to participants who may have not written anything since grade schools. Training the participants on diary keeping thus becomes necessary but can be costly, putting additional load to researchers using the diary method for data collection.

The diary method can be used as a primary technique of data collection like what is done in Välimäki, Vehviläinen-Julkunen, and Pietilä (2007), the case study chosen for this section, for exploring how to extend home care of people with Alzheimer's disease by means of supportive practices. Many studies adopt the diary method in conjunction with another data collection technique such as interview. The diary-interview method, proposed by Zimmerman and Wieder (1977), is widely utilized by many researchers who ask participants to keep diaries for a period of time and conduct interviews with detailed questions about the diary entries. The diary-interview method can check the internal consistency of diarists' accounts, fill in omissions and move beyond events recorded in diaries into more general experience and attitudes. Another application of the diary method is to use research diaries as one of the multiple data collection techniques selected for a single study. This kind of application is frequently implemented in ethnography or field studies, one of two meta-methods discussed in this book (see Chapter 11, Meta-Method: Ethnography/Field Study).

### *8.3.2 Taxonomy of research diaries*

In addition to media formats and coverage detail, research diaries can also be categorized by diary writer, solicitation, structure, and writing prompt. Table 8.1 presents a taxonomy of diaries using these criteria.

As seen from Table 8.1, this taxonomy is not mutually exclusive across the four criteria listed (i.e., diary author, solicitation, structure, and writing prompt), which means that more than one criterion can be applied to research diary categorization simultaneously. For instance, participants write a solicited, structured diary that is signal-contingent. When diaries are chosen as a data collection technique, they are most likely solicited and kept by participants. Due to this reason, unsolicited diaries written by researchers are usually

*Table 8.1* Taxonomy of Research Diaries

| *Criterion* | *Category* | *Feature* |
|---|---|---|
| Diary writer | Participant | Participants write diaries at the request of researchers. |
| | Researcher | Researchers in ethnography, field study, or other investigations may write diaries for their own use. |
| Solicitation | Solicited | Researchers ask participants to write diaries. Participant-kept diaries are usually solicited. |
| | Unsolicited | Researchers keep diaries for their own studies. |
| Structure | Structured | Diarists are instructed to keep diaries using a specified form. |
| | Semi-structured | Part of the diaries are kept using a specified form and part of them can be written freely. |
| | Unstructured | Diarists are free to write without using a specific form. |
| Writing prompt | Interval-contingent | Diaries are kept in a specified interval (e.g., daily, three times per day). |
| | Signal-contingent | Diaries are kept according to a given signal (e.g., clock alarm). |
| | Event-contingent | Diaries are kept whenever a specific event (e.g., a need for information) occurs. |

not examined in studies other than in ethnography. However, both types are discussed in Section 8.3.3 because dairies kept by researchers belong to the diary method after all.

There are also unsolicited diaries kept by individuals who are not researchers. Such diaries are in essence data sources for historical research and therefore cannot be regarded as a technique for data collection. This kind of diaries is depicted as one type of data sources for historical research in Section 8.4.

Structured, semi-structured, and unstructured diaries are, to some extent, analogous to the three interview types we discuss in Section 4.3: structured, semi-structured, and unstructured interviews. As is known, interview types are mainly determined by whether interview questions are prepared beforehand and whether follow-up and probe questions are allowed. The categorization of diaries by structure in comparison is simply decided by if diaries must be kept according to a specified form. A structured diary means that the researcher creates a diary form to specify what kinds of data (e.g., date, time, location, who, and what) must be recorded in the diary. Generally speaking, structured diaries are easier for participants to keep because they only need to complete the diary form or record what is specified. Unstructured diaries, in contrast, are harder for participants to write, although participants on the surface are free to write anything. But, in reality, participants may write little when they are not sure what should be included in the diary or write something that is beyond the scope of the study and not what the researcher anticipates. Therefore, clear and detailed instructions must be provided when unstructured diaries are chosen for a study. The topic of how to prepare instructions for diarists is to be discussed in Section 8.3.3.

Researchers must decide when their participants should keep a diary if diaries are selected for data collection purpose. Interval, signal, or event can each be used as a prompt for diary keeping. The interval-contingent prompt means that diarists keep diaries at a specified interval such as daily or three times a day. The signal-contingent prompt denotes that participants write their diaries according to an external prompt such as clock alarm. For example, some diaries are kept using an alarm set at semi-random intervals to avoid repetition at set times of day (Burton, Weller, and Sharpe 2007). This specific type of prompt apparently is only feasible when researchers have ready access to their participants (e.g., in-patients in a hospital). The event-contingent prompt implies that participants keep their diaries when they experience an event (e.g., a seizure, a need for information). For instance, Sun, Sharples, and Makri (2011) conduct a study with doctoral students working on their dissertation research and ask them to record in their dairies whenever they find any resource by serendipity.

### *8.3.3 Using research diaries in studies*

As mentioned earlier, research diaries can be written by either participants or researchers. When dairies are chosen as a data collection technique, it usually implies that diaries are written by participants. In comparison, researcher-kept diaries are generally mentioned in ethnography or field study when researchers work in the field and need to keep a diary to stay informed and make reflections.

#### *8.3.3.1 Solicited diaries kept by participants*

Because keeping research diaries for a study may require weeks or even months of participants' time, it is quintessential for investigators to recruit candidates who can and are willing to make such a commitment. For the same reason, it is rare for researchers to recruit participants via random sampling in studies using the diary method. Instead, purposive or convenient sampling is often adopted in choosing participants for research on non-sensitive topics (e.g., people with heart disease). Snowball sampling could be a preferred technique for recruiting participants to keep diaries on sensitive topics (e.g., condom use). In the latter cases, the number of participants might be very small. For example, only five people were willing to keep a diary about their condom use out of all the potential candidates who met the recruitment criteria (Harvey 2011).

Alaszewski (2006, chap. 3, 6) enumerates three criteria for recruiting participants in diary research. Namely, those who can take part in the research must have (1) the competence to maintain a diary, (2) an understanding of the purposes of keeping the diary, and (3) the motivation to keep an accurate record. All these three criteria are necessary because only participants meeting them can properly undertake the task of keeping diaries as required. In addition, researchers should also factor in the huge time commitment participants are expected to make in diary research, as other data collection techniques (e.g., focus groups, interview) typically require a total of one to two hours of participants' time.

Since keeping a research diary is new to most, if not all, participants, researchers should provide guidelines and instructions to assist participants to write diaries as asked. Such guidelines should specify all the details participants may need in their diary writing. In general, the following points ought to be spelled out in the guidelines:

- Media format – In what media format shall the diaries be kept, in a notebook, with a computer (e.g., laptop), or using an audio/video recording device? Hand-written diaries used to be the most common media format in the past because participants can maintain a diary as long as they know how to write. Computers nowadays are increasingly adopted as a media format of diary writing for those participants who are computer literate and have ready access to a computer. When diaries are kept on computers, participants can easily submit their complete diaries via email or other similar means. Researchers would no longer need to physically gather diaries from their participants, which might mean individual visits to participants' homes. Diaries created using a computer also eliminate the legibility issue of hand-written diaries. Audio diaries are usually created if diarists are unable to write for one reason or another, whereas video diaries are made when the environment and other visuals must be included in the diary (e.g., rehabilitation at home). Audio and video dairies on the whole are rarely used in research, although today's smartphones can produce both rather easily. If notebooks are chosen as the diary recording medium, researchers should provide one to participants before diary keeping begins. Researchers always try to do anything they can to facilitate participants to take part in studies.

- What to write – Instructions on what needs to be recorded in a diary should be clearly spelled out according to the research questions a study attempts to answer. It is helpful and in fact mandatory for researchers to first look at their research questions to see what diary data should be gathered in order to answer them. Then develop instructions accordingly as to what to put down in the diary. For example, the case study for this section by Välimäki, Vehviläinen-Julkunen, and Pietilä (2007) aims to find out "[w]hat kind of data can unstructured diaries produce in a study of family caregivers of people with Alzheimer's disease" (69). The researchers hence ask their participants (i.e., family caregivers of people with Alzheimer's disease) to write in their diaries "a wide range of topics about daily life, the impact of Alzheimer's disease on this, emotions and events experienced in the different stages of the disease, ways of solving problems, and moments of contentment" (71). If a structured diary is selected, researchers must provide a form or a list of items to participants. If an open or unstructured dairy is chosen as in Välimäki, Vehviläinen-Julkunen, and Pietilä (2007), researchers ought to specify exactly what they expect their participants to write in the diary. For instance, shall participants merely log what happened or do they also need to comment and reflect on the events and activities? Researchers must be as specific as possible in their instructions to participants about what to write in unstructured diaries. Otherwise, participants might feel confused or at loss about what should and should not be recorded in the diary.
- Diary-keeping frequency and time length – Instructions for diarists need to include details about how often and for how long they are asked to write diaries. Participants are usually instructed to keep diaries once a day. However, other diary-keeping frequencies can also be chosen depending on the diary method design. Researchers may ask participants, for instance, to write diaries according to the signal- or event-contingent prompt. No matter which diary-keeping prompt is chosen, researchers should explicitly state so. The same is true with the diary-keeping time length, which might lead to more choice options. Many diary studies select a time length from two weeks to one month. Any other option shorter or longer than that time range may need the researcher to justify because a shorter time span may cause concern about if adequate diary data are collected, while a longer time period may discourage participants from completing all their diary entries as requested. It is therefore imperative for researchers to choose a reasonable time length so that adequate diary data can be collected and participants are likely to complete all their diary entries. A short diary-keeping time span (e.g., 24 hours, several days) is sometimes selected when diaries are a secondary or supplemental technique for data collection in research.

The guidelines and instructions for diarists are usually printed on the first page of the notebook if notebook is the chosen media format or included in the introduction note if another diary media format (e.g., computer, audio recording) is selected. The rule of thumb for developing diary-keeping guidelines is to be clear, specific, and comprehensive. Those guidelines can help participants understand what they are expected to do and how they can do well in diary keeping.

Nevertheless, diary keeping is a minority habit, and most of the participants in a diary study do not usually keep a diary (Alaszewski 2006, chap. 4, 14). Researchers may want to offer training and assistance to recruited participants via such approaches as explanation of related terminology (e.g., what serendipity is in Sun, Sharples, and Makri 2011) and development of a list of FAQs on diary-keeping. "Researchers will avoid detailed prescriptive instructions that restrict diarists' freedom of expression. Instead, the researcher will seek to influence the diary keeping process by developing a relationship with the diarist through regular contact initiated before the diary period, maintained during the diary keeping period with perhaps visits to look at sample entries or to collect initial diaries, ..." (Alaszewski 2006, chap. 4, 14).

Rapport building with diarists is another task that researchers must accomplish during the data collection process. Unlike in studies using some other data collection techniques (e.g., interview, focus groups), gathering data via the diary method is not a one-time activity for researchers. Rather, researchers have to get involved and maintain contact with participants throughout the diary keeping process, which could take weeks or even months. Researchers ordinally provide participants with their own contact information (e.g., email address, phone number) so that participants can ask questions when needed. Many researchers choose to check with diarists three times during the diary keeping period: once in the beginning, once in the middle, and once at the end of the diary keeping period. These contacts, when properly done, can help ensure that participants write diaries according to the provided guidelines as well as address any issues that may emerge during the diary keeping period. Contacts fewer than three times in the diary keep period may not be adequate for researchers to stay informed with participants' diary writing, while contacts more than three times may appear intrusive to diarists to some extent. It is also a good practice for researchers to ask for diarists' own comments on and clarifications of any peculiarities relating to their entries at the end of the diary keeping period.

#### *8.3.3.2 Researcher-kept diaries*

Diaries kept by researchers are different from participant-kept diaries in that the former are to facilitate the research process through recording observations, thoughts and questions as they happen for later use and to stimulate reflective thinking about the study. In contrast, the latter are basically a technique for gathering data from participants. Researcher-kept diaries "... can be seen as a melting pot for all of the different ingredients of a research project – prior experience, observations, readings, ideas – and a means of capturing the resulting interplay of elements" (Newbury 2001, 3).

Furthermore, researcher-kept diaries are not a daily record of all research activities. Instead, researchers would write in their diaries about other elements in their investigations that include the following (Newbury 2001, 8):

- observations
- thoughts and reflections
- a record of reading (with comments/summaries/quotes)
- a record of phone calls and meetings

- notes on methodology
- unresolved problems, issues or questions
- plans for action
- keywords for events and activities relating to the ongoing study. It is an effective way for researchers to use keywords to quickly record what happens in the field (Yin 2016, 167).

It is generally accepted and even expected that research-kept diaries should encompass some analytical commentary on how the research develops. Some researchers indicate that the act of diary writing can be considered cathartic in so far as their deepest frustrations, fears, and anxieties can be channeled from the internal to the external in the form of writing, especially when they work alone in the field – an environment that is often times unfamiliar or volatile to them (Browne 2013).

As noted earlier, researchers keep diaries mainly in ethnography or field studies. The reason why researcher-kept diaries are covered in this section instead of in Chapter 11, Ethnography/Field Study, is because they share more commonalities with participant-kept diaries as an individual data collection technique. Ethnography or field studies, known for their meta-method property, generally utilize more than one technique for data collection. Those research methods (e.g., observation, interview, focus group) are each discussed in respective chapters of this book and so should diaries. In addition, many of the fundamentals we discuss for participant-kept diaries are essentially applicable to researcher-kept ones as well.

### *8.3.4 Dos and don'ts in research diary studies*

Research diaries are a unique technique for data collection in comparison with such methods as interview and questionnaire in that diarists are asked to engage in the data gathering process for days, weeks, or occasionally months by writing diaries. This and other features of research diaries determine that the following dos and don'ts should be observed when using diaries for collecting data in research.

First, clearly and explicitly describe in your instructions what participants should record in their diaries. If you want to collect data not only about a log of items, events, or activities but also participants' observation and reflections on what they record, indicate so in your instructions. However, your request or anticipation of the diarists should be reasonable and appropriate with respect to their background and qualifications. Anything beyond the reasonable level would discourage or disengage the diarists from taking part in your research.

Second, pretest the diary form and instructions if the structured diary format is chosen. As discussed in other chapters of this book, we pretest questionnaire, interview guide, and other data collection instruments. The same is applicable to the diary form and instructions. Revise them if needed after the pretest and before actual usage in data collection.

Third, provide lucid guidelines about how to keep diaries, how often, and for how long. Always choose a reasonable prompt or interval for diary keeping based on your research objective, feasibility, and participants' availability. Avoid having too few or too many prompts without thinking realistically and practically. If participants are given freedom in naming their activities that are to be coded later, it is important to give precise guidelines on what type of behavior to include, what definitely to exclude, and the level of detail required (Corti 1993).

Fourth, stay in touch with participants throughout the data collection process to encourage and support them to complete their diaries as instructed. This will help reduce the occurrence of the so-called participant conditioning and first-day or first-week effects in diary keeping. Participant conditioning refers to the phenomenon that participants change their behavior as a result of diary keeping when they consciously know what they record in the diary would be read and used in research. Similarly, first-day or first-week effects mean that participants record more content initially and trail off toward the end of the data collection period when they become tired of diary keeping (Toms and Duff 2002). Issues relating to diarist retention and diary-keeping completion can at least be partially addressed if researchers maintain their contact with diarists.

Research diaries are mostly used for collecting qualitative data, yet participants are asked to produce them in writing. This salient characteristic of research diaries, along with others, requires researchers to be extremely mindful when conducting studies using this data collection technique.

### *8.3.5 Case study of investigation using research diary*

Välimäki, Tarja, Katri Vehviläinen-Julkunen, and Anna-Maija Pietilä. 2007. "Diaries as Research Data in a Study on Family Caregivers of People with Alzheimer's Disease: Methodological Issues." *Journal of Advanced Nursing* 59 (1): 68–76. https://doi.org/10.1111/j.1365-2648.2007.04273.x.

Välimäki, Vehviläinen-Julkunen, and Pietilä (2007) report the use of unstructured diaries as a primary technique for gathering research data in a study of family caregivers of people with Alzheimer's disease to find out what works in increasing the duration of home care by means of supportive practices. The authors also identify the benefits and limitations related to the use of diaries as a main data collection technique in nursing research.

The study participants were recruited from memory clinics serving three hospital districts in Finland based on interest and willingness. The criteria for recruiting participants for this study were: (1) patients being diagnosed with Alzheimer's disease but in the mild dementia stage, and (2) patients' family caregivers in close, preferably daily, contact with the patients. A total of 241 people with dementia and the same number of their family caregivers were "evaluated once a year during the project by means of several instruments for 5 years from the initial visit (2002–2011)" (Välimäki, Vehviläinen-Julkunen, and Pietilä 2007, 69). Those two sets of patients and family caregivers constitute the population of the current study. One quick comment is due on the time span for the investigation. The authors stated that they would do their evaluation of the patients once a year for five years. Yet the time span listed is 2002–2011, far more than five years. As this research report is published in 2007 and the initial (hospital) visit was in 2002, the correct ending year for evaluation should be 2006 instead of 2011.

All the 241 pairs (one patient with Alzheimer's disease and one family caregiver) in the study were first randomly assigned to two groups: (1) the intervention group who got usual care and intensive psychoeducational courses during the first two years after diagnosis; and (2) the control group who just received usual care. This makes up the experiment method we discuss in Chapter 10. We briefly mention this experiment in the current case study because those who participated in the diary portion of the current study were all from the intervention group. That is, for all the participants who kept diaries in this study, their family members with Alzheimer's disease received not only usual care but also intensive psychoeducational courses from 2002 to 2004. At the end of 2004, 132 out of the 241 family caregivers were recruited from the intervention group by the researchers. The number of 132 family caregivers who participated in the diary phase of the study suggests that all the 241 pairs of participants were randomly assigned to but not equally divided into two groups in the experiment phase of the study (see above). In other words, there were 132 participants, instead of 120 or 121 if equally divided, assigned to the intervention group in the afore-depicted experiment. The researchers perhaps should have explained in their report why the two groups in the experiment were not assigned equally in number as common practices do.

As explained in Section 2.9.2, informed consent must be obtained from human participants in all research projects. Research participation in general is always voluntary rather than mandatory. All the 132 family caregivers in the intervention group of the experiment phase in this project consented to write diaries. At the first visit of the 132 family caregivers, Tarja Välimäki, the first author of this research report who was working on her doctoral dissertation, asked them to keep a diary for two weeks. Two weeks are a reasonable time length in consideration of the nature of the current research, which lasted for five years.

Diary writing was discussed with all the consented participants, who were also given written instructions to take home with them. It was emphasized that the diary content, i.e., the description of their own experiences, was more important than formal literary expression. The family caregivers were also given brief instructions on how to write in a way that would enable the use of their diaries as research data. The basic question for the participants to address in their diaries was: "What is your life like now that a member of your family has Alzheimer's disease?" More specific topics that can be covered in the diary include "daily life, the impact of Alzheimer's disease on this, emotions and events experienced in the different stages of the disease, ways of solving problems, and moments of contentment" (Välimäki, Vehviläinen-Julkunen, and Pietilä 2007, 71). Those instructions are detailed enough for guiding the participants to keep diaries for the research project. However, one note is due about "emotions and events experienced in the different stages of the disease". The diarists in the study would be unable to write about anything "in the different stages of the disease" when they only kept the diary for two weeks.

Furthermore, the participants were given an A5-sized notebook (i.e., 5-7/8 × 8-1/4 inch) containing 40 pages and a prepaid envelope for returning the diary. This is in accordance with the recommended practice when notebook is the chosen media format for participants to keep diaries. The principal researcher – Tarja Välimäki's contact details were provided in case the participants had any questions. Välimäki also made a telephone call to each participant a month later to ask about the writing and to encourage the caregivers to return the diary. However, it would be better (1) if Välimäki had contacted the participants one more time in the middle of the two-week period to check their diary keeping and encourage them to continue their writing as instructed; and (2) if Välimäki had made the final phone call immediately after the two-week period was over. Unless Välimäki instructed her participants to keep the diary for any two weeks in one month of time, she should not call the diarists a month later after the first visit to avoid any possible missed or delayed diary entries. A total of 83 family caregivers or 63% of the 132 consented participants completed a diary. The diary completion rate could have been higher if Välimäki had stayed in close contact with her participants in the two weeks of time, although, honestly speaking, 83 participants are a very decent number for diary research. Välimäki, Vehviläinen-Julkunen, and Pietilä (2007) should be proud of this outcome.

As for ethical concerns, this study follows the recommended practice that the participants were informed that they could withdraw from the study at any point without any effect on their future treatment. The consent form was signed by both the patient with dementia and the family caregiver. Any questions about the study from participants could be discussed if they so wish in the follow-up phone call and hospital visits in the five-year time period. The privacy of the participating families was guaranteed by changing the names in the research report and by not publishing their home localities.

Overall, the 83 participants in the research found diary-writing in general personally rewarding. They were motivated to write by the idea that they could contribute to improving the treatment of Alzheimer's disease by keeping a diary and sharing their feelings and thoughts. This is a positive indicator that Välimäki, Vehviläinen-Julkunen, and Pietilä (2007) successfully engaged their participants by convincing them to appreciate their value in the study. From this perspective, the current study sets a fine example for other researchers in participant recruitment and retention.

**Section Summary**

- Research diaries are entries created and maintained by individuals to record what happened in a specified period of time and/or their reflections on a regular, personal, and contemporaneous basis for research purposes.
- Research diaries are commonly categorized by diary author, solicitation, structure, and writing prompt. Diaries can be kept by both participants and researchers. Participant-kept diaries are typically solicited, and account for the majority of diary data collected. Researcher-kept diaries in comparison are unsolicited and often created to facilitate the research process.
- Clear and explicit guidelines should be provided to participants on what should be recorded in the diary and how to do so. Researchers should also make reasonable decisions on how often and for how long participants should keep the diary. Similarly, researchers ought to maintain contacts with participants during the dairy-keeping period to check if the diary is kept as instructed and to encourage diarists to complete all the entries as requested.
- Be fully aware that diary-keeping is a long-time commitment by participants, and researchers should make every effort to help and support diarists in accomplishing this challenging task while observing all the dos and don'ts discussed for conducting diary research.

## 8.4 Historical method

**Section Overview Questions**

- What is the historical method?
- What features does the historical method have?
- What are the steps in historical research?
- What are the dos and don'ts in conducting historical research?

### *8.4.1 What is the historical method?*

The historical method refers to studying a research problem in history by critically collecting, assessing, analyzing, interpreting, and presenting evidence from both published and unpublished materials. Unlike most other research methods discussed in this book, the historical method integrates all the elements, including data collection and analysis, of a project into one inseparable research process. When conducting historical research (i.e., research using the historical method), we do not single out and consider data collection or data analysis separately. Instead, these two key steps in research are done integrally with the rest of steps in the whole process. The historical method therefore is not considered only for data collection but also for all other steps in historical research.

Historical research reports or narratives are products of investigations using the historical method. They are called histories that provide accounts of past events and analyses of various sources in order to explain what happened at that time and what might imply at present. Some histories are written to forecast the future by identifying repeated patterns because seeing the past can help envision the future. Simply put, histories enable us to learn about the past, understand our own origins and present, and map out a better future by learning from the past as well as previous mistakes (Porra, Hirshcheim, and Parks 2014).

History is also a discipline at academic institutions, while the same does not hold true for any other research methods in usage (e.g., experiment, questionnaire). A study of history offers a backdrop from which to determine what is novel in the current situation and which factors serve to distinguish the present situation from any others in the past. It helps understand the sources of contemporary problems, how they arose, and how their characteristics unfolded through time. It ascertains the solutions that worked in the past and those that did not. It reminds us of the richness of the human experience and of the broad degree of complexity, intricacy, and unpredictability that surrounds any real circumstance (Mason, McKenney, and Copeland 1997). Hence, an understanding of history prevents us from viewing our present circumstances in idiosyncratic, traditional, narrow, or shallow ways. Reading history also stimulates the imagination and reveals new possibilities.

Conducting historical research relies on the historical method, which indeed facilitates researchers to address problems from perspectives that are unavailable from any other methodological means. Nevertheless, it has certain limitations as a research method. Lack of mathematical tractability, no possibility for generalizing research findings, and being prone to reduction in interpreting evidence are just a few examples of its limitations. The historical method is even questioned by some as to whether it is truly a scientific research method. For instance, Wildemuth (2017) writes: "[f]or a variety of reasons, we could say that historical inquiry is not really research in the scientific sense at all" (156). The reasons listed for supporting this argument include: (1) There is no controlling for variables nor any attempt to build a strictly scientific case as in experiments. (2) No rigorous techniques for data collection and analysis are available to historical research as in, for example, questionnaire research. (3) It relies on imperfect and incomplete sources to form conclusions. The afore-listed limitations and concerns about the historical method do not mean that it cannot be adopted and utilized in research. Rather, it only demonstrates that extra caution is needed when conducing historical studies.

### *8.4.2 Features of the historical method*

The most prominent feature of the historical method is that researchers adopting it do not have a discrete phase for data collection, as done in most other investigations using methods such as experiment and focus groups. This is because research question formulation, data collection and evaluation, and evidence interpretation in historical research are all integrated with one another, although we do acknowledge in Section 8.4.4 that there are still different steps in historical research. In countering the assumptions that historians first collect all their data and then start writing, Carr (1961) pointed out the necessary intertwining of critical thinking and writing in historical research: "I am convinced that, for any historian worth the name, the two processes of what economists call 'input' and 'output' go on simultaneously and are, in practice, parts of a single process" (28–29).

The second feature of historical research is that the process is iterative, as one's hypothesis or working theory in the study is constantly in need of revision. We denote in the beginning of Chapter 2 that the research process is spiral, cyclical, yet progressive. This iterative feature gets amplified in historical research mainly because, unlike research adopting other methods (e.g., experiment, interview), researchers cannot collect all the data needed for a project before moving to evaluate, analyze, and interpret what is gathered all at once. Instead, researchers locate some sources for a given problem, perform some analysis, and then notice that additional data are available or that the interpretation done early needs modification after additional evidence shows up in the process.

Third, the historical method is employed primarily for gathering qualitative data. A research question that requires the use of the historical method is usually answered with qualitative data embedded in, for example, government records, diaries, memoirs, private correspondence, and oral testimonies. These sources are made of words and texts that can be gathered via the historical method.

The fourth feature of the historical method is that it is always utilized in research concerning the past, even though historians can never recreate that actual past (Gorman and Clayton 2005, 162). This feature defines when and where the historical method can be applied. Additional features of the historical method are to be depicted and discussed in the remaining subsections of Section 8.4.

### *8.4.3 Data sources and evaluation in historical research*

As indicated earlier, research using the historical method is considered historical research, the process of which, like in other kinds of investigations, consists of several steps. Since data sources for historical research and evaluation of them occupy essential positions in the research process, this subsection is devoted to them and presented first and separately from other steps in historical research.

#### *8.4.3.1 Primary sources*

Data sources in historical studies are universally categorized as primary sources and secondary sources based on their potential value in research. Primary sources are original materials created at the time under study. They present information in its original form, neither evaluated nor interpreted by others (Meroño-Peñuela et al. 2015). They include oral, written, and recorded eyewitness accounts of events in the past in the form of diaries, oral histories, personal letters, reports, photos, recordings, and the like. The word "eyewitness" should be emphasized here in that only first-hand accounts, instead of second- or third-hand ones, are qualified to be primary sources. Any materials, for example, from the survivors and those who witnessed the 9/11 terrorist attacks can serve as primary sources for research on this subject. If a person born after September 11, 2001 writes a document about the 9/11 attacks, that document is not a primary source because that individual came to this world after the horrific date. Primary sources are generally regarded as the core concept of historical method applied in historical research (Grigg 1991).

Primary sources can be further divided into the following two kinds (Meroño-Peñuela et al. 2015). One is called administrative sources that comprise records of some administration, such as census data, birth certificates, marriage certificates, administrative accounts of taxes and expenses, minutes of administrative bodies, deeds, and contracts. The other type of primary sources is referred to as narrative sources. They are full text documents containing a description of the past and made by an individual being an eyewitness of the past event. Examples of narrative sources include autobiographies, biographies, diaries, chronicles, newspaper articles, oral histories, and more. Do note that diaries in historical research are not the same as research diaries discussed in Section 8.3 of the current book. Diaries in this context were kept by people in the time under study for their own specific purpose and without any solicitation from the researcher. Oral history is another primary source that needs some attention. It refers to the collection and study of historical information using sound or video recordings of interviews with people having personal knowledge of past events. It might be recorded previously (e.g., interview recordings of Michael Jackson) or the researcher conducts interviews with participants to create it (e.g., interview recordings of Jackson's family and friends). When interview is adopted to record oral history, all the guidelines for conducting interview should be observed (see Chapter 4). Elliott and Davies (2006), for instance, describe how to use oral history in consumer research.

#### *8.4.3.2 Secondary sources and electronic sources*

Secondary sources are materials that others authored about the past on given topics or events. They describe, evaluate, analyze, and interpret the primary sources to aid people in learning history or conducting historical research (Meroño-Peñuela et al. 2015). They consist of oral, written, and recorded accounts by people not immediately present at the time of past events but through gathering, evaluating, and interpreting information from primary sources for a specific purpose or audience. Examples of secondary sources are bibliographies, encyclopedias, books, journal articles, literature reviews, obituaries, and so on. Most writings and publications about WWI put out at present belong to the secondary sources because few eyewitnesses of that war are still alive today.

More and more sources nowadays are available electronically, and they constitute another type of sources for historical research. Email messages, blogs, and posts from other social media (e.g., Facebook, Twitter, TikTok) are some examples of electronic sources, although they may not be clearly labeled as primary or secondary sources. Electronic sources can serve as an important reservoir for historical studies as their magnitude has been increasing exponentially since the 2010s. While their impact on historical research is to be determined, researchers should pay serious attention to this kind of sources whenever appropriate.

At this point, it is necessary for us to ask what sources we should use for data collection after learning all the sources available for historical research. The answer is to draw on as many sources of evidence as possible for two principal reasons. First, it allows the study to address a wider range of issues in historical research. Second, it provides different data of the same phenomenon as well as permits cross-examination and development of converging lines of inquiry through triangulation (Mason, McKenney, and Copeland 1997). In addition to the number of sources gathered for the study, historical researchers are generally in favor of primary sources over secondary ones, and unsolicited sources (e.g., personal letters) over solicited sources (e.g., reviews). The preference of primary sources results from the fact that such sources are original and without any interpretation. Unsolicited sources are preferred because they record past events more truthfully. Davie and Wyatt (2011) furthermore summarize in a tabular form the pros and cons of using unsolicited document types. Although the authors do not create the summary exclusively for historical sources, their summary is helpful for historical researchers to get a better understanding of unsolicited sources.

#### *8.4.3.3 Source evaluation*

There is striking unanimity among researchers regarding what constitutes the historical method. The overriding characteristic of it is that all sources should be approached critically or skeptically. Any witness, whether living or dead, may have made mistakes unconsciously or intended to mislead. Therefore, application of the historical method should act as a defense against any fanciful tale, willful distortion, and even honest error (Golder 2000). Sources located for historical research, whether primary or secondary, printed or electronic, and text or multimedia, must be evaluated before they are used as evidence to address a research question. Evaluation of sources in historical research typically comprises two kinds: external evaluation and internal evaluation, also respectively known as external criticism and internal criticism.

External criticism is concerned with the authenticity of a source. This evaluation deals with the document itself, not with its content or meaning. Frauds, hoaxes, and forgeries are common in historical sources, which can be particularly problematic for the naïve or novice researcher. Those frauds, hoaxes, and forgeries tend to be produced for major historical events or personages because this is where counterfeiters realize the greatest financial return for their effort. Researchers are thus advised to ask the following questions when assessing the authenticity of a source (Berg 2009, 303–7; Rüpke 2011, 292; Wood 1990):

- Who authored the source?
- What is the genuineness, authority, bias, interest, and intelligibility of the source?
- What motivated the production of the source? A bureaucratic process, public communication, or secret documentation?
- Who was the intended audience?
- What sources were privileged or ignored in the narrative?
- What evidence is offered or compiled?
- In what historical context was the source authored?
- How did it enter tradition or archives, and why has it been preserved?

Internal criticism, by contrast, deals with the credibility of a source, including the credibility of individual statements within a source. As noted above, sources for historical research may be readily separated into two categories: primary sources produced by eyewitnesses to an event and secondary sources authored by individuals who were not present when the events being described took place. Although we must keep in mind the fallibility of any witness, the account by one who was on the scene at the time of the event is generally more credible than that of someone who was not there; the testimony of a nonwitness is hearsay evidence in history as it is in law and must be regarded with suspicion (Wood 1990). Questions to be asked in internal criticism of sources include the following (Berg 2009, 307–9; Rüpke 2011, 292; Wood 1990):

- What was the author trying to say?
- What was the author's motive for making the statement or creating the source?
- What was the view of the event or phenomenon when the source was authored?
- How competent was the witness?
- What inferences were offered in the statement by the author?
- To what extent was the author in a position to provide first-hand information on the topic?
- Is the author's statement accurate?
- Was the statement of the author similar or contrary to another in the time period?
- How was the source used in later times?

The advent of the internet and other technologies serves as a double-edged sword for evaluating historical sources. On one hand, it becomes much easier for researchers to gain access to historical sources that are increasingly digitized and stored electronically. Researchers no longer need to physically travel to an archive or a similar site in order to obtain what they look for. On the other hand, frauds, hoaxes, and forgeries can be more readily made in historical sources with the internet and other technologies. Furthermore, fake news, misinformation (false or inaccurate information), and disinformation (deliberately misleading or biased information) are mushrooming on social media platforms and other similar venues on a scale and at a rate unprecedented in human history.

Evaluation of electronic sources in historical research (e.g., archived emails, blogs, tweets) also must go through external and internal criticism done in the analog world. Besides, additional criteria should be applied in assessing sources of electronic format. First, the site that hosts the source ought to be scrutinized. We normally value sources from top domains, such as .gov, .edu, and .org, while automatically checking sources from other domains like .com for their authenticity and credibility. Other aspects of the host site that should be inspected in the evaluation include whether the host site is authoritative and well regarded, whether it is qualified to host the source, and what relationship the author had with the host site (e.g., academic or paid service relationship). Second, we examine whether the source's author was an individual or an institution. An institutional author would carry more weight than an individual in terms of authenticity and credibility. We also check, for instance, if there is a link to the author's internet presence. Third, the web is at present the default platform for most, if not all, information and resources in electronic format, including electronic sources for historical research. Hyperlinks or links in short, a unique property of the web, enable people to make connections among different sources and sites. They should be used for evaluating electronic sources as well. For example, are there links to and from other sites? What are the linking and linked sources and sites? Are the links all live? Dead links could be an indicator that the site is not properly maintained, and sources on the linking site should be reviewed more closely. In sum, evaluation of electronic sources becomes a required component for historical research.

### *8.4.4 Conducting historical research*

Although researchers agree that the process of historical research, like projects using other research methods, is cyclical, iterative, and progressive, there is no consensus regarding how many different stages are in the process. Some authors (e.g., Mason, McKenney, and Copeland 1997; Witkowski and Jone 2006) believe that historical research commonly consists of seven steps, while others think that the historical research process contains five stages (e.g., Golder 2000) or two stages covering six steps (e.g., Smith and Lux 1993). The discrepancy in the number of steps in historical research is mainly due to the fact that those authors apply different criteria when labeling the steps in the entire process. In general, the process of historical research includes the following steps:

1) Recognizing a historical problem or identifying a need for certain historical knowledge;
2) Gathering as much pertinent information as possible about the problem or topic;
3) Forming hypotheses, if possible, that provide educated and tentative answers to the research questions derived from the research problem identified in Step (1);
4) Rigorously collecting evidence as well as assessing the authenticity and credibility of relevant sources;
5) Selecting, organizing, and analyzing the most relevant evidence collected to answer research questions and draw conclusions;
6) Reporting findings in a meaningful narrative.

Of the six steps in historical research, Steps (1) and (6) are quite representative in any research, especially in qualitative research. Chapter 2 of this book illustrates the general research process in detail, including how to identify a research problem and how to transform a research problem to research questions. What is unique about conducting research with the historical method lies in the middle four steps, namely Steps (2)–(5), and each of them is to be described and discussed below.

Using the historical method in research implies that data collection involves searching for relevant sources as well as gaining access to located sources. This is different from most other research methods covered in the present book in that the former must place its focus on not only how to search for relevant data sources but also how to gain access to them. Unlike in choosing the questionnaire method, for example, researchers only need to develop and administer the questionnaire to targeted participants to complete data collection. When using the historical method, gaining access to relevant sources already found via search is by no means a guaranteed outcome. Rather, serious efforts must be made in order to, for instance, obtain permission to visit an archive or interview a witness of interest for the purpose of data gathering.

Searching for sources about a research topic means that researchers should use a variety of search tools (e.g., library catalogs, databases, finding aids) and search techniques to locate what they want. Even though coverage on how to search is beyond the scope of this book, the researcher should get familiar with the major resources often used in the chosen topic area as well as fundamentals about searching. Section 2.2.1 provides an overview of search essentials and approaches before performing literature review. They are also to a great extent applicable to search for sources in historical research. Keep in mind that the search emphasis in this context ought to be placed on exhaustiveness or comprehensiveness to ensure most relevant sources for a given topic are located successfully.

Mason, McKenney, and Copeland (1997) recommend that we first search for public sources such as articles, books, annual reports, and trade association studies. Those sources are generally available from libraries' online catalogs, databases, government documents, special collections, and institutional repositories. Public libraries are accessible to anyone as long as one resides in the serving community, while academic libraries are open to students, faculty and staff of their parent institutions – colleges and universities. Access to private sources, in contrast, requires tactics specific to the source owner. For example, researchers wishing to read personal diaries would need to contact the diaries' owners regardless of if they are an archive, a special collection, or a diarist's descendants. Online sources at Internet Archive (https://archive.org) as well as online finding aids such as ArchiveGrid (https://www.oclc.org/research/areas/research-collections/archivegrid.html), a finding tool for over seven million archival material descriptions, should be of help to researchers in locating historical sources. Both Internet Archive and ArchiveGrid are freely accessible on the internet.

Subsequent to Step (2) in the historical research process, scholars should, if appropriate, form hypotheses that serve as educated and tentative answers to research questions. This constitutes Step (3) in the historical research process. As elucidated in Section 2.3.2, hypotheses are formulated only if researchers have an educated and tentative answer to a research question posed in a study based on existing theories, prior research as well as their own knowledge and experience. Historical scholars would do the same in their research. It is worth repeating that not every research question can have an educated and tentative answer before data collection and analysis. That is why the phrase "if appropriate" is placed in Step (3) of the historical research. Furthermore, hypotheses in historical research are qualitative in nature, which implies that they cannot later be tested quantitatively. However, the essentials and tips for hypothesis formulation discussed in Section 2.3.2 are all relevant to historical investigations. Always make sure that one and only one hypothesis can be made for one research question, no matter whether in historical research or in studies using other research methods.

The next step, Step (4), in historical research is to rigorously assess the authenticity and credibility of relevant sources before collecting evidence for the study in progress. As expounded in Section 8.4.3.3, researchers should perform both external and internal criticism in order to achieve the objective of source evaluation.

Step (5) in the historical research process includes selecting, organizing, and analyzing the most relevant evidence collected to answer research questions and draw conclusions. In historical research, evidence is gathered by scholars in their interpretation of the past through a careful sifting of many sources of various kinds: articles, diaries, memoirs, private correspondence, oral histories, government documents, and more (Alaszewski 2006, chap. 2, 8). The full context of those sources should also be taken into consideration when gathering evidence because each source must have been created in a certain context, regardless of whether it is a primary or secondary source. The context plays a vital role in facilitating researchers' interpretation of historical sources and evidence gathering.

According to Golder (2000), many historical researchers organize what they gather chronologically within each subject area (e.g., person, institution, country). This arrangement has two advantages: (1) Contradictory evidence can be viewed in the appropriate context, and (2) potential explanations are more easily recognized because causes precede effects. Proper synthesis of evidence, on the other hand, usually requires multiple readings of located sources. Researchers would use inferences in this step to complete information that is not specifically provided. When suitable quantitative data exist, researchers can conduct descriptive statistical analyses (e.g., frequency, average) to summarize their evidence.

Although the historical method seeks to minimize biases, the potential for biased analysis still exists. Previous studies have found, for example, that researchers may leap to conclusions based on limited data, be overly influenced by vivid descriptions, give more weight to elite participants' input, or other possible biased outcomes. Researchers using the historical method thus are advised of the following: gather as much pertinent data as possible, interpret data truthfully and impartially, and never be affected by research participants' status in a study no matter how influential they may be.

In addition to biased analysis, Barzun and Graff (1992, 145–48) identify four common historical fallacies: (1) Generalization – claiming more for one's argument than one's evidence warrants. For example, some people claimed that the smallpox vaccination in the 18th century was not safe because smallpox inoculations sometimes failed. As indicated in Section 8.4.2, one feature of historical research is that its findings cannot be generalized because no random sampling is possible when historical method is the selected research method. (2) Reduction – Attributing complex, multifaceted phenomena to single causes or factors. A historical event more often than not occurred due to multiple factors. When some considered the assassination of John. F. Kennedy, the 35th president of the United States, was a result of a conspiracy, the causes for this historical event were reduced to a conspiracy that failed to take into consideration an interrelated group of factors that caused the assassination. (3) Tautology – using different words for the same thing as though they are disparate. (4) Misplaced literalism – depicting historical events and activities facetiously and out of context. The last two historical fallacies are usually more about writing quality and hence not unique to historical research. All four fallacies can be avoided by following the recommended practices in research. When working with both published and unpublished sources in historical research, it is extremely important to follow the established practices in external and internal criticism of them because many of the sources, especially those unpublished, are hard to be evaluated for their authenticity and credibility.

Once all the collected evidence for the study is selected, organized, and analyzed, scholars are ready to answer their research questions and test their hypotheses, if any, formulated in previous steps. It must be pointed out that hypotheses for historical research can only be "tested" qualitatively by inference, reasoning, categorization, comparison, and other similar qualitative techniques. Unlike quantitative hypothesis testing, qualitative hypothesis testing has neither statistical test to perform nor significance level to serve as the basis for determining if a hypothesis is retained or rejected. Instead, researchers would rely on gathered evidence to test their qualitative hypotheses. Qualitative hypothesis testing is harder to perform. Extra caution thus should be exercised to ensure the validity and reliability of testing outcome. Research questions are answered similarly in historical research. Research question formulation and answering actually separates historical description from historical research. Without research question formulation and answering, the final product can only be regarded as historical description instead of historical research.

As indicated earlier, we do not plan to elaborate on Step (6) of the historical research process due to its similarity to that of the general research process depicted in Sections 2.6 and 2.7. However, we still would like to explain why in historical research we should report findings in a meaningful narrative. The keyword in the previous sentence is "narrative" because historical research rarely has the opportunity, if any, to present its findings in tubular or other visual format when compared to quantitative research. Historical researchers hence must depend on narratives to report their findings.

Compared to other research methods, the historical method is different in its implementation in that the data collection step cannot be distinctively separated from the rest of steps in the research process. As a result, all the steps have to be covered when illustrating how the historical method is utilized in research. Although all the steps in historical research are presented linearly above, it does not mean that their application in actual studies follows the same pattern. Rather, some steps may be taken simultaneously, while others might be done in an order different from the above illustration. For instance, Step (1) might be revisited after any step listed subsequently is taken (e.g., Step [2] Data Collection, Step [3] Hypothesis Formulation). Other steps besides Step (1) likewise may need to be redone in the research process. This reality confirms that the process of historical research, same as that of all other research, is iterative and recursive, and in most cases, to a greater extent.

### *8.4.5 Dos and don'ts in historical research*

Many of the recommended practices in historical research are already presented up to this point. However, several key points in using the historical method in research should be reemphasized.

First, be thorough and exhaustive in data collection, even though this may need extra time and efforts. Materials and data sources for historical research are generally scattered, hard to locate, and perhaps even harder to gain access. But those challenges should not become obstacles to reaching the goal of gathering as much relevant data as possible for a study. Avoid collecting only those sources that are conveniently available without trying one's best to obtain all materials that are necessary to carry out the research.

Second, do evaluate what is gathered because sources for historical research were usually created for purposes other than the current research project. In comparison with research using other methods, data for historical research cannot be gathered via instruments (e.g., questionnaire, interview guide) specifically developed for a study, nor are they generated during the research process. Instead, almost all the data sources of relevance to an investigation have been in existence before the commence of the present study, some of which could be there for decades or possibly for centuries. Yet there is commonly no opportunity for researchers to check with data creators for verification or clarification purpose, which is only possible in other kinds of research (e.g., experiment research, interview research). All these features of historical data sources determine that researchers must perform external and internal criticism of the collected data in order to judge their authenticity and credibility. Moreover, source evaluation becomes particularly important in the online environment because online sources (e.g., websites, social media posts) are notorious for their mixed quality and lack of quality control mechanism when compared to printed sources. Avoid taking online sources in historical research without questioning and rigorous assessment.

Third, be truthful in interpreting the data collected in historical research. Since historical research essentially relies on existing sources, as opposed to data solicited and collected using other techniques such as questionnaire and interview, how faithfully data are interpreted regulates the validity and reliability of historical research. Avoid generalization, reduction, and other possible mishaps in interpretation of historical sources.

Fourth, do take into consideration the contextual factors when making use of historical sources. Each source was created at a particular time and for a particular purpose. Thus, historical sources should never be analyzed and discussed out of context. For example, Vannevar Bush (1945) conceived and proposed the following two remarkable ideas close to the end of World War II: (1) the Memex machine – a conceptual prototype of the computer we know today, and (2) associative thinking – the theoretical predecessor of hypertext that has been implemented in the web architecture since the 1990s. Bush was able to envision both great inventions in the contemporary world, the computer and hypertext, at that time because he foresaw the need for effective information storage and retrieval after World War II was over. What Bush expressed in his seminal article "As We May Think" may be of little significance if World War II and all the R&D efforts during the war had not been considered in the interpretation of Bush's contribution in the time of the so-called information explosion.

Fifth, do present evidence accurately, consistently, and unambiguously in the final research report. Avoid generalization, reduction, tautology and misplaced literalism – the four common fallacies that exist in historical research. As indicated earlier, historical research is qualitative in nature, which makes the entire research more challenging because there are no established and explicit measures for judging its validity and reliability. However, difficulty in conducting and reporting historical research should not be used as an excuse for doing a substandard job. Historical researchers ought to instead take this challenge as motivation for carrying out quality studies.

### *8.4.6 Case study of historical research*

Mackintosh, Carolyn. 1997. "A Historical Study of Men in Nursing." *Journal of Advanced Nursing* 26 (2): 232–36. https://doi.org/10.1046/j.1365-2648.1997.1997026232.x

Mackintosh (1997) intends to retell the history of nursing with an emphasis on the frequently neglected place of men within it in the United Kingdom using a variety of historical sources that include primary and secondary sources. The study concludes that men have an equally valid historical role within nursing, and their contribution should be acknowledged in the nursing profession.

A total of 36 sources are referenced in this study: 16 books (mostly book chapters), 9 journal articles (including 3 editorials), 5 society documents or publications, 2 oral histories, 2 meeting minutes, 1 report, and 1 thesis. Some of those sources are obtained from hospital archives. The meeting minutes were kept for the Society of Registered Male Nurses then (July 1937 to January 1954) and for Royal College of Nursing. All those sources provide some indication of the historical place and value that men should have within the nursing profession.

Like many other historical researchers, Mackintosh (1997) organizes her research report chronologically in three sections:

1) Early history (before 1860) – This section supposedly describes the role of male nurses in poor law workhouses, the work of which at that time was considered low status. The author however writes: "... as an early workhouse visitor to Shoreditch Infirmary in 1861, Francis Cobbe, describes, recruits were far from ideal ..." (Mackintosh 1997, 233). The reference to 1861 in a time period before 1860 regarding male nurse history is perhaps an oversight of the author.
2) Reformed nursing (1860–1899) – Male nurses in this time period continued to fulfill roles outside the reformed or trained nursing sphere. A number of associations in the private nursing sector exclusively employed male nurses.

3) The 20th century (1900–1960s) – Before the outbreak of World War II, there were over 97,000 female registered nurses, while only 435 males belonged to the same category in the UK. The number of male nurses nevertheless got dramatically expanded later, "with a 542% increase in the number of male nurses registered between 1939 and 1947" mainly as a result of World War II (Mackintosh 1997, 235). Male nurses' contribution to the caring professional during and after the war had increased, but they were still "marginalized by their tendency to work in areas of nursing where limited historical research has taken place (e.g., asylums, military services and private associations)" (Mackintosh 1997, 235).

Mackintosh (1997) is apparently a shorter and simpler report of historical research in comparison with other case studies covered in this book. Additionally, the report contains no methodology section that details data collection and analysis – a feature that clearly distinguishes historical research from other kinds of research (e.g., experiment research, questionnaire research). Historical research reports also make many more quotes in their narratives when scholars attempt to demonstrate what was actually stated in the original sources. In contrast, the same usually does not hold true in other kinds of research reports. Exceptions nonetheless can be found in reporting interview, focus groups, and other similar qualitative research when scholars need to make frequent references to the data collected using techniques such as interview and focus groups.

Below are several additional comments on Mackintosh (1997). The author writes on page 233 that "[a]ttendants could therefore be described as '[t]he unemployed of other professions... if they possess physical strength and a tolerable reputation for sobriety, it is enough, and the latter quality is frequently disposed with'" (Dr Browne 1837, cited in Carpenter 1980, p. 134). Because the above quotation was cited in Carpenter (1980, 134) rather than directly from the source Browne authored, Mackintosh "cites from citations" (see Section 2.2.2), which should be avoided completely in scholarly writings, even though it can occasionally be seen in some publications. The reason for this recommendation is because the cited document is already interpreted and evaluated once by the citing author, and distortion is likely to occur in the process. If we rely on someone else's interpretation of the cited document without reading it by ourselves in research, our understanding of the original document is tinted with the citing author's viewpoint. When interpretation plays a crucial role in analyzing historical sources and gathering evidence from them, citing from citations in historical research would negatively affect the accuracy of its findings.

The author lists in her report that "[b]etween September 1921 and December 1938 a total of 97,028 general female nurses registered, compared with 435 males on their own supplementary register. The 1931 census indicated only 100 males who identified their occupation as nursing" (Mackintosh 1997, 234). Then the question is how many female nurses there were in the same year. Without a corresponding number of female nurses, how can the reader tell if 100 male nurses in 1931 is a very small number? Mackintosh provides both numbers in the first part of the above quote. Why cannot she do the same in the second part to facilitate comparison?

Lastly, Mackintosh (1997) does not describe how she selected and evaluated the sources for her research except listing the source types (e.g., hospital archives, meeting minutes, oral histories) in the abstract and introduction of her report. A brief section on methodology of this study should adequately address the issue and make this report more convincing as well as complete.

**Section Summary**

- The historical method refers to studying a research problem in history by critically collecting, assessing, analyzing, interpreting, and presenting evidence from both published and unpublished materials.
- Features of the historical method include no distinctive phase for data collection, iterative research process, qualitative in nature, and addressing research problems concerning the past.
- There are commonly six steps in historical research, although they are not necessarily executed linearly:

  1) Recognizing a historical problem or identifying a need for certain historical knowledge;
  2) Gathering as much pertinent information as possible about the problem or topic;
  3) Forming hypotheses, if possible, that provide educated and tentative answers to the research questions derived from the research problem identified in Step (1);
  4) Rigorously collecting evidence as well as assessing the authenticity and credibility of relevant sources;
  5) Selecting, organizing, and analyzing the most relevant evidence collected to answer research questions and draw conclusions;
  6) Reporting findings in a meaningful narrative.

- Sources for historical research must be gathered exhaustively and evaluated rigorously in addition to other dos and don'ts outlined.

## Note

1 Alaszewski (2006) is available to the author only in ebook format, which numbers pages in each chapter starting from 1. Thus, the chapter number must be listed along with the page number in order to differentiate the many same page numbers in the book.

## References


Alaszewski, Andy. 2006. "Using Diaries for Social Research." In *Using Qualitative Research*. London: Sage.

Allport, Gordon W. 1942. *The Use of Personal Documents in Psychological Science*. New York: Social Science Research Council.

Barzun, Jacques, and Henry F. Graff. 1992. *The Modern Researcher*. Boston: Houghton Mifflin.

Berg, Bruce L. 2009. *Qualitative Research Methods for the Social Sciences*. 7th ed. Boston: Allyn and Bacon.

Bloor, Michael, Jane Frankland, Michelle Thomas, and Kate Robson. 2001. *Focus Groups in Social Research*. Thousand Oaks: Sage.

Breen, Rosanna L. 2006. "A Practical Guide to Focus-Group Research." *Journal of Geography in Higher Education* 30 (3): 463–75. https://doi.org/10.1080/03098260600927575.

Browne, Brendan Ciaran. 2013. "Recording the Personal: The Benefits in Maintaining Research Diaries for Documenting the Emotional and Practical Challenges of Fieldwork in Unfamiliar Settings." *International Journal of Qualitative Methods* 12 (1): 420–35. https://doi.org/10.1177/160940691301200121.

Burton, Christopher, David Weller, and Michael Sharpe. 2007. "Are Electronic Diaries Useful for Symptoms Research? A Systematic Review." *Journal of Psychosomatic Research* 62 (5): 553–61. https://doi.org/10.1016/j.jpsychores.2006.12.022.

Bush, Vannevar. 1945. "As We May Think." *The Atlantic Monthly* 176 (1): 101–8. https://doi.org/10.1145/227181.227186.

Carr, Edward Hallett. 1961. *What Is History*? New York: Knopf.

Chu, Heting, and Qing Ke. 2017. "Research Methods: What's in the Name?" *Library & Information Science Research* 39 (4): 284–99. https://doi.org/10.1016/j.lisr.2017.11.001.

Colucci, Erminia. 2007. "'Focus Groups Can Be Fun': The Use of Activity-Oriented Questions in Focus Group Discussions." *Qualitative Health Research* 17 (10): 1422–33.

Corti, Louise. 1993. "Using Diaries in Social Research." *Social Research Update* (2). Accessed August 13, 2023. https://repository.essex.ac.uk/24606/.

Davie, Grace, and David Wyatt. 2011. "Document Analysis." In *The Routledge Handbook of Research Methods in the Study of Religion*, edited by Michael Stausberg and Steven Engler, 151–60. London, New York: Routledge.

Duhigg, Charles. 2012. "How Companies Learn Your Secrets." *The New York Times Magazine*, February 16. Accessed August 13, 2023. https://www.nytimes.com/2012/02/19/magazine/shopping-habits.html?pagewanted=1&r=1&hp.

Elliott, Richard, and Andrea Davies. 2006. "Using Oral History Methods in Consumer Research." In *Handbook of Qualitative Research Methods in Marketing*, edited by Russel W. Belk, 244–54. Northampton: Edward Elgar.

Erevelles, Sunil, Nobuyuki Fukawa, and Linda Swayne. 2016. "Big Data Consumer Analytics and the Transformation of Marketing." *Journal of Business Research* 69 (2): 897–904. http://dx.doi.org/10.1016/j.jbusres.2015.07.001.

Fry, Jenny. 2006. "Google's Privacy Responsibilities at Home and Abroad." *Journal of Librarianship and Information Science* 38 (3): 135–39. https://doi.org/10.1177/0961000606066571.

Golder, Peter N. 2000. "Historical Method in Marketing Research With New Evidence on Long-Term Market Share Stability." *Journal of Marketing Research* 37 (2): 156–72. https://doi.org/10.1509/jmkr.37.2.156.18732.

Gorman, G. E., and Peter Clayton. 2005. *Qualitative Research for the Information Professional*. 2nd ed. London: Facet Publishing.

Grigg, Susan. 1991. "Archival Practice and the Foundations of Historical Method." *The Journal of American History* 78 (1): 228–39. https://doi.org/10.2307/2078095.

Gullifer, Judith, and Graham A. Tyson. 2010. "Exploring University Students' Perceptions of Plagiarism: A Focus Group Study." *Studies in Higher Education* 35 (4): 463–81. https://doi.org/10.1080/03075070903096508.

Gullifer, Judith, and Graham A. Tyson. 2014. "Who Has Read the Policy on Plagiarism? Unpacking Students' Understanding of Plagiarism." *Studies in Higher Education* 39 (7): 1202–18. https://doi.org/10.1080/03075079.2013.777412.

Harvey, Laura. 2011. "Intimate Reflections: Private Diaries in Qualitative Research." *Qualitative Research* 11 (6): 664–82. https://doi.org/10.1177/1468794111415959.

Hill, Kashmil. 2012. "How Target Figured out a Teen Girl Was Pregnant Before Her Father Did." *Forbes*, February 16. Accessed August 13, 2023. https://www.forbes.com/sites/kashmirhill/2012/02/16/how-target-figured-out-a-teen-girl-was-pregnant-before-her-father-did/?sh=4b5fe0a86668.

Jamali, Hamid R., David Nicholas, and Paul Huntington. 2005. "The Use and Users of Scholarly E-Journals: A Review of Log Analysis Studies." *Aslib Proceedings: New Information Perspectives* 57 (6): 554–71. https://doi.org/10.1108/00012530510634271.

Jansen, Bernard J. 2006. "Search Log Analysis: What Is It, What's Been Done, How to Do It." *Library & Information Science Research* 28 (3): 407–32. https://doi.org/10.1016/j.lisr.2006.06.005.

Jansen, Bernard J., and Amanda Spink. 2006. "How Are We Searching the World Wide Web? A Comparison of Nine Search Engine Transaction Logs." *Information Processing and Management* 42 (1): 248–63. https://doi.org/10.1016/j.ipm.2004.10.007.

Jansen, Bernard J., Amanda Spink, and Tefko Saracevic. 2000. "Real Life, Real Users, and Real Needs: A Study and Analysis of User Queries on the Web." *Information Processing and Management* 36 (2): 207–27. https://doi.org/10.1016/S0306-4573(99)00056-4.

Jansen, Bernard J., Isak Taksa, and Amanda Spink. 2009. "Research and Methodological Foundations of Transaction Log Analysis." In *Handbook of Research on Web Log Analysis*, edited by Bernard J. Jansen, Amanda Spink and Isak Taksa, 1–16. Hershey: Idea Group Publishing.

Krueger, Richard A., and Mary Anne Casey. 2015. *Focus Groups: A Practical Guide for Applied Research*. 5th ed. London: Sage.

Kurth, Martin. 1993. "The Limits and Limitations of Transaction Log Analysis." *Library Hi Tech* 11 (2): 98–104. https://doi.org/10.1108/eb047888.

Logan, Judith, Kathryn Barrett, and Sabina Pagotto. 2019. "Dissatisfaction in Chat Reference Users: A Transcript Analysis Study." *College & Research Libraries* 80 (7): 925–44. https://doi.org/10.5860/crl.80.7.925.

Mackintosh, Carolyn. 1997. "A Historical Study of Men in Nursing." *Journal of Advanced Nursing* 26 (2): 232–36. https://doi.org/10.1046/j.1365-2648.1997.1997026232.x.

Madriz, Ester. 2000. "Focus Groups in Feminist Research." In *Handbook of Qualitative Research*, edited by Norman K. Denzin and Yvonna S. Lincoln, 2nd ed., 835–50. Thousand Oaks: Sage.

Mason, Richard O., James L. McKenney, and Duncan G. Copeland. (1997). "An Historical Method for MIS Research: Steps and Assumptions." *MIS Quarterly* 21 (3): 307–20. https://doi.org/10.2307/249499.

McCullagh, Declan. 2006. "AOL's Disturbing Glimpse into Users' Lives." *CNET News*, August 9. Accessed August 13, 2023. http://shawndra.pbworks.com/f/AOL's+disturbing+glimpse+into+users'+lives++.pdf.

Meroño-Peñuela, Albert, Ashkan Ashkpour, Marieke van Erp, Kees Mandemakers, Leen Breure, Andrea Scharnhorst, Stefan Schlobach, and Frank van Harmelen. 2015. "Semantic Technologies for Historical Research: A Survey." *Semantic Web* 6 (6): 539–64. https://doi.org/10.3233/sw-140158.

Merton, Robert K. 1987. "The Focussed Interview and Focus Groups: Continuities and Discontinuities." *The Public Opinion Quarterly* 51 (4): 550–66.

Merton, Robert K., and Patricia L. Kendall. 1946. "The Focused Interview." *American Journal of Sociology* 51: 541–57.

Meth, Paula. 2003. "Entries and Omissions: Using Solicited Diaries in Geographical Research." *Area* 35 (2): 195–205. https://doi.org/10.1111/1475-4762.00263.

Morgan, David L. 1989. *Focus Groups as Qualitative Research*. Newbury Park: Sage.

Morgan, David L. 1996. "Focus Groups." *Annual Review of Sociology* 22: 129–52. https://www.jstor.org/stable/2083427.

Morgan, David L., and Richard A. Krueger. 1998. *The Focus Group Guidebook*. Thousand Oaks: Sage.

Newbury, Darren. (2001). "Diaries and Fieldnotes in the Research Process." *Research Issues in Art, Design and Media* (1): 1–17.

Nicholas, David, and Paul Huntington. 2003. "Micro-Mining and Segmented Logfile Analysis: A Method for Enriching the Data Yield from Internet Log Files." *Journal of Information Science* 29 (5): 391–404.

Oard, Douglas W., and Jinmook Kim. 2001. Modeling Information Content Using Observable Behavior." *Proceedings of the Sixty-Fourth American Society for Information Science Annual Meeting* 38: 481–88.

Peters, Thomas A. 1993. "The History and Development of Transaction Log Analysis." *Library Hi Tech* 11 (2): 41–66.

Porra, Jaana, Rudy Hirshcheim, and Michael S. Parks. 2014. "The Historical Research Method and Information Systems Research." *Journal of the Association for Information Systems* 15 (9): 536–76. https://doi.org/10.17705/1jais.00373.

Rice, Ronald E., and Christine L. Borgman. 1983. "The Use of Computer-Monitored Data in Information Science and Communication Science." *Journal of the American Society for Information Science* 44 (4): 247–56. https://doi.org/10.1002/asi.4630340404.

Rüpke, Jörg. 2011. "History." In *The Routledge Handbook of Research Methods in the Study of Religion*, edited by Michael Stausberg and Steven Engler, 285–309. New York: Routledge.

Scheuren, Fritz. 2004. *What Is a Survey*. 2nd ed. Alexandria: American Statistical Association. Accessed August 13, 2023. http://people.umass.edu/biostat690c/pdf/ASA%20What%20is%20a%20survey%2068%20pp.pdf.

Shoaf, Eric C. 2003. "Using a Professional Moderator in Library Focus Group Research." *College & Research Libraries* 53 (2): 124–32. https://doi.org/10.5860/crl.64.2.124.

Smith, Ruth Ann, and David S. Lux. 1993. "Historical Method in Consumer Research: Developing Causal Explanations of Change." *Journal of Consumer Research* 19 (4): 595–610. https://doi.org/10.1086/209325.

Steward, David W., Prem N. Shamdasani, and Dennis W. Rook. 2009. "Group Depth Interviews: Focus Group Research." In *The Sage Handbook of Applied Social Research Methods*, edited by Leonard Bickman and Debra J. Rog, 2nd ed., 589–616. Thousand Oaks: Sage.

Sun, Xu, Sarah Sharples, and Stephann Makri. 2011. "A User-Centred Mobile Diary Study Approach to Understanding Serendipity in Information Research." *Information Research* 16 (3): paper 492. Accessed August 13, 2023. http://informationr.net/ir/16-3/paper492.html.

Toms, Elaine G., and Wendy Duff. 2002. "'I Spent 1 1/2 Hours Sifting Through One Large Box. ...': Diaries as Information Behaviour of the Archives User: Lessons Learned." *Journal of the American Society for Information Science and Technology* 53 (14): DOI: 10.1002/asi.10165.

Välimäki, Tarja, Katri Vehviläinen-Julkunen, and Anna-Maija Pietilä. 2007. "Diaries as Research Data in a Study on Family Caregivers of People With Alzheimer's Disease: Methodological Issues." *Journal of Advanced Nursing* 59 (1): 68–76. https://doi.org/10.1111/j.1365-2648.2007.04273.x.

Wildemuth, Barbara M., ed. 2017. *Applications of Social Research Methods to Questions in Information and Library Science*. 2nd ed. Denver: Libraries Unlimited.

Wilkinson, Sue. 2008. "Focus Groups." In *Qualitative Psychology: A Practical Guide to Research Methods*, edited by Jonathan J. Smith. 2nd ed. Los Angeles: Sage.

Witkowski, Terrence H., and D.G. Brian Jone. 2006. "6. Qualitative Historical Research in Marketing." In *Handbook of Qualitative Research Methods in Marketing*, edited by Russel W. Belk, 70–82. Northampton: Edward Elgar.

Wood, W. Raymond. 1990. "Ethnohistory and Historical Method." *Archaeological Method and Theory* 2: 81–109.

Yin, Robert K. 2016. *Qualitative Research from Start to Finish*. 2nd ed. New York: The Guilford Press.

Zimmerman, Don H., and D. Lawrence Wieder. 1977. "The Diary: Diary-Interview Method." *Urban Life* 5 (4): 479–98. https://doi.org/10.1177/089124167700500406.

# Part III

# Super-Methods, Meta-Methods, and Contingent Methods

# 9 Content Analysis

## A Super-Method

**Chapter Overview Questions**

- What is content analysis?
- What are the two major types of content analysis?
- What are the procedures for using content analysis as a data collection technique?
- What is the coding scheme and intercoder agreement?
- What are the dos and don'ts in research using content analysis?

### 9.1 Content analysis as a data collection technique

Content analysis is a method of textual examination used to describe and explain characteristics of messages embedded in texts (Nelson and Woods, Jr. 2011, 109). Content analysis in practice can be used as a technique for both data collection and data analysis. In the current chapter, content analysis is presented and discussed as a data collection technique. Namely, it is through the examination and analysis of the content in texts that the data collection for a study is accomplished, which differs from content analysis being utilized as a technique for data analysis. In the latter case, textual or qualitative data are already gathered via interview, observation, focus groups, and other similar techniques. Researchers would then apply various techniques (e.g., deduction, summarization, constant comparison) to analyze the content of collected data in order to address the research questions formulated in an investigation. This chapter focuses on content analysis as a data collection technique, while content analysis as a method for data analysis is to be covered in Chapter 13, Techniques for Data Analysis and Their Selection. There are nevertheless some overlapping areas (i.e., analytical techniques) in content analysis partly because content analysis must be performed regardless of whether it is utilized as a technique for data collection or for data analysis. The present book attempts to minimize such overlapping coverage of content analysis as much as possible.

This research method had its beginnings in the disciplines of journalism and mass communication, where it continues to be used for analyzing newspaper articles, advertisements, and other messages of various media (Krippendorff 2013, 11–12). The use of content analysis has been spreading to other disciplines (e.g., anthropology, psychology, sociology) in the past decades. The "content" in content analysis refers to words, sentences, paragraphs, passages, documents, pictures, symbols, concepts, ideas, themes, or any message that can be identified in the text. The text is anything written, visual, or

DOI: 10.4324/9781003021711-12

spoken that serves as a medium for communication. It includes books, newspaper or magazine articles, advertisements, speeches, official documents, videos, musical lyrics, photographs, emails, websites, and social media posts.

Content analysis can be done quantitatively and/or qualitatively. However, content analysis is mostly performed qualitatively when it is used for data collection purpose because researchers must analyze the content in texts qualitatively in order to gather data. For example, which idea is embedded in that paragraph? What concept is underlying that passage? It is rare that researchers would collect data through quantitative content analysis as Berelson (1971, 114–34) argued in his seminal book, although his definition of content analysis is drastically different from what content analysis is nowadays. If they do, that would be regarded as secondary analysis because such quantitative data (e.g., census data) are already gathered for other purposes. Discussion on secondary analysis and its comparison with content analysis will be provided at the end of this section. In short, content analysis in this chapter is presented as a data collection technique and via a qualitative lens.

Content analysis, as a research method, exhibits some salient features. First, content analysis is unobtrusive and thus eliminates any reactive effect commonly found in data collection from human participants (e.g., the Hawthorne effect in observation). Second, content analysis allows researchers to make use of large quantities of data that otherwise are hard to manage thanks to the explicit procedures and quality control checks developed for this data collection technique (to be discussed shortly). Third, content analysis can handle prose, images, and other unstructured texts in the data collection process. Fourth, many analytical techniques (e.g., comparison, induction, deduction, abstraction, summarization) can be used simultaneously to facilitate the task of data collection. Fifth, content analysis is context-sensitive and researchers therefore are required to process the context and textuality of the data that are significant, meaningful, and informative to others (Krippendorff 2013, 46–47). Sixth, content analysis can be used to collect data regardless of time and location, while many other research methods (e.g., experiment, observation) are only feasible when researchers and participants are at the same physical location simultaneously. Last but not least, content analysis incorporates analysis in the process of data collection because data collection must be done through analyzing the source data. We cannot separate data collection from data analysis when choosing content analysis as a method for data collection in the research process.

On the other hand, researchers have expressed some concerns with content analysis as a data collection technique. The most serious concern is that texts do not have purely objective qualities and they are not reader-independent. Such texts are produced by someone and likely having different meanings for someone else (Krippendorff 2013, 25). This implies that meanings tend to be subjectively brought to texts or texts are interpreted by content analysts based on their background, experience, research objective, and the like. Because of this feature, meanings in texts are not discovered but constructed through the act of interpretation, which could be different from what actually is contained in the text. Researchers infer answers to particular research questions from texts, but those inferences can only go so far. Additional limitations of content analysis as a data collection technique include the following points: First, it may be difficult to find representative samples of source texts. Unlike in questionnaire research, for example, investigators are able to design a sampling plan before the data collection process. By contrast, researchers are constrained by available source texts in content analysis and with little flexibility in choosing representative samples. Second, generalizing the results of one content analysis to another is difficult because replicability of any content analysis is close to impossible. Literally, no two content analyses can be conducted identically, even if researchers use a same coding scheme and follow a common set of steps in performing content analysis, as discussed in Section 9.3. Third, while content analysis is useful for describing message characteristics and trends, it does not allow researchers to draw cause-and-effect conclusions as with experiment and other research methods (Nelson and Woods, Jr. 2011, 116–17).

Content analysis is widely used in research as a data collection technique, even though it may not be explicitly so specified in reported studies. Often times, content analysis is implied when researchers use other research methods. For example, researchers need to analyze what they observe (i.e., content) in order to determine if the focus of observation should be changed or if the observation should continue. The interviewer likewise needs to analyze what the interviewee says before moving on to the next interview question. It is because of the ubiquitous presence of content analysis in most, if not all, research method applications that content analysis is labeled as a super-method of research in this book.

As indicated in the beginning of this chapter, content analysis is one form of textual analysis that shares some features with other forms of textual analysis, such as discourse analysis, secondary analysis, meta-analysis, systematic review, and scoping review. Table 9.1 outlines the similarities and differences between content analysis and other forms of textual analysis.

As shown in Table 9.1, content analysis also includes discourse analysis and secondary analysis since they all can be employed for achieving the objective of data collection by examining some kind of text data. It is the data source that essentially determines whether a research method is content analysis, discourse analysis, or secondary analysis. That is, a research method is called discourse analysis if the target source is written or spoken communication (e.g., dialogs, conversations, debates) and "text above the level of sentences" (Krippendorff 2013, 22). Conversation analysis is a spoken version of discourse analysis

*Table 9.1* Comparison of Content Analysis with Other Methods of Textual Analysis

| *Related Approach* | *Description* | *Similarity* | *Difference* |
|---|---|---|---|
| Discourse analysis | Analysis of discourse data (i.e., written or spoken communication such as debates and conversations) | Working with textual data | Analysis of language beyond the sentence in discourse data |
| Secondary analysis | Analysis of data collected previously for other purposes (e.g., census data) | | Data are previously collected for other purposes |
| Meta-analysis | Analysis of data, usually statistical in nature, from multiple related studies to summarize their results quantitatively | | Data are obtained from multiple studies and typically analyzed quantitatively |
| Systematic review | Review of research publications using systematic techniques to critically appraise them and synthesize findings qualitatively and/or quantitatively | | The focus is on critique and assessment |
| Scoping review | Usually done to get prepared and provide a scope for a systematic review | | Done as a preparation for a systematic review |

when the source data are conversations (Lehtinen 2011, 122). Hence, conversation analysis is by and large considered a synonym of discourse analysis. In comparison with content analysis, discourse analysis is performed with a focus on the language beyond the sentence level, examining and looking at patterns of the language used in the discourse.

A research method would be called secondary analysis if the target source involves data collected previously for other purposes or data already in existence (e.g., private diaries). Census data are a typical example of secondary data used in many studies, while they are always collected for a census initially. When the data are already in existence in the form of diaries, personal letters, or the like, secondary analysis functions more as the historical method (see Section 8.4 of this book). In either case, the primary emphasis of secondary analysis is placed on data analysis instead of data collection, while the focus of the present chapter is on content analysis as a technique for data collection.

Meta-analysis is for extracting data, usually statistical data, from multiple studies on a same topic, with each individual study reporting measurements that are expected to have some degree of error. The objective of meta-analysis is to use statistical approaches to derive a pooled estimate closest to the unknown common truth based on how this error is perceived. Meta-analysis also has the capacity to contrast results from different studies, identify sources of disagreement among those results or other interesting relationships that may come to light in the context of multiple studies.

Systematic reviews and scoping reviews belong to the category of content analysis in the broadest sense but basically make critique and assessment of what is gathered instead of for the purpose of data collection. Systematic reviews aim to critically appraise a set of related studies and mainly synthesize their findings qualitatively, as opposed to the afore-discussed meta-analyses that are mostly done quantitatively or statistically. Scoping reviews are one subtype of systematic reviews, as researchers normally use scoping reviews as a pilot or preliminary step, when needed, to identify the scope for a full-scale systematic review. In addition, both systematic and scoping reviews are reviews by nature. They are not intended as a data collection technique in the first place.

In sum, discourse analysis, secondary analysis, and meta-analysis are all not precisely a data collection technique as content analysis is, although each relates to textual analysis in one way or another. Systematic review and scoping review, strictly speaking, are simply one type of review rather than a research method.

## 9.2 Manifest and latent content analysis

Content analysis can be approached and categorized in different ways. For example, Krippendorff (2013, 355) posits that content analysis studies can be text-driven, problem-driven, or method-driven, while others (e.g., Berelson 1971; Neuendorf 2002) incline to label content analysis according to data analysis techniques: quantitative content analysis or qualitative content analysis. However, content analyses are most frequently categorized by content type, namely, manifest content analysis and latent content analysis.

Manifest content analysis means that the contents to be analyzed are visible, observable, and countable (e.g., personal names, geographical locations) in texts. Therefore, manifest content analysis is by and large based on examining the presence or absence of a term or terms. The contents in latent content analysis, by contrast, are hidden or underlying, conceptual, unobservable, and uncountable in texts (e.g., ideas, concepts, themes). Unlike manifest content analysis, latent content analysis must go beyond the presence or absence of a term. For instance, a document is essentially about discrimination, but the word "discrimination" is not present anywhere in the text. Researchers must perform analysis using various techniques, such as deduction, inference, and summarization, in order to reveal what the text actually covers. Latent content analysis thus is also known as semantic analysis, implying that the implicit meaning in the content of a text must be identified via analysis (Neuman 2004, 221).

Manifest and latent contents are analogous to "surface" and "deep" structures of language, respectively (Neuendorf 2002, 23). A distinction is often made between what is on the surface (i.e., manifest) and what is beneath the surface (i.e., latent) in content analysis. While manifest contents are easily identifiable, latent contents in texts require researchers to read between the lines, detect underlying meanings, hear silences, and discover hidden messages. All such activities in latent content analysis are harder for researchers to accomplish due to the subjective interpretations involved in the analytical process when compared to manifest content analysis. Further discussion on manifest content analysis and latent content analysis is to be done in Section 9.3.4.2.

## 9.3 Using content analysis in research

Although it is close to impossible to describe how content analysis, especially latent content analysis, is done in research partly due to the qualitative orientation in this chapter, efforts are made below to illustrate the major points in using content analysis as a data collection technique (Nelson and Woods, Jr. 2011; USGAO – United States General Accounting Office 1996).

### *9.3.1 Selecting source texts and determining units of analysis*

Selecting source texts relevant to the research problem under investigation is the first point that researchers need to consider when choosing content analysis as a method for gathering data. One of the features content analysis has is that texts, serving as data sources, are already in existence in such forms as books, newspapers, speeches, videos, photographs, and websites. Researchers can decide if they should use any sampling technique (see Section 2.5.2 of the present book) to collect data for their studies. For instance, Bie and Tang (2015) – the case study chosen for this chapter – purposively selected a sample of five leading Chinese newspapers as sources for their research. They then searched the term "autism" or its synonym "lonely disease" in Chinese to locate any newspaper reports that contained these two terms. Fifty percent of all retrieved news reports, selected via a systematic sampling method, became the source texts for content analysis in the study by Bie and Tang (2015).

Units of analysis is a concept commonly brought up after source texts for content analysis are obtained. This concept refers to the basic units of text that researchers would examine in content analysis. They could be entire texts (e.g., books, articles), text segments (e.g., words, sentences, paragraphs, passages), formatting units (e.g., rows, columns, pages), or thematic units such as chunks of text that reflect a single theme (Ryan and Bernard 2000, 780). The unit of analysis in Bie and Tang (2015) is individual news reports about autism.

### *9.3.2 Defining variables and generating content categories*

Any research, including content analysis studies, involves variables. Once the texts for content analysis are selected and units of analysis are determined, it is time to operationalize variables (see Section 2.3.1 of the current book) by developing categories or values that each variable would take. Again, let's take Bie and Tang (2015) as an example. The authors planned to examine six variables in their study: Issues discussed, role of the autistic individuals in the news report, sources cited in the news coverage of autism, causes of autism, solutions for handling autism, and types of discourse. Operational definitions are also given to each of the six variables listed above by specifying the values or categories each variable takes. For instance, "genetic/congenital", "unclear/complex", "individual behavior", "environmental causes", or "vaccine" constitute the five causes (i.e., categories) for the "causes of autism" variable. In other words, the "causes of autism" variable is measured by one of the afore-listed categories.

As discussed in Section 3.2.1 of the present book, categories for each variable must be collectively exhaustive and mutually exclusive – the two fundamental principles that researchers must observe in any categorization efforts. If the set of categories does not cover all possible values a variable can take, the variable may be miscategorized or not recorded at all due to the lack of a proper category. In Bie and Tang (2015), the authors enumerated five causes of autism. But more than 398 (i.e., 89.6%) of the 444 news reports they examined do not specify a cause. Hence, it would make more sense if one more category "no cause reported" is provided for the "causes of autism" variable to meet the criterion of collectively exhaustive in categorization. Likewise, if the created categories overlap or are not mutually exclusive, the variable in the content analysis may be improperly categorized or could be placed in two or more categories simultaneously. Bie and Tang (2015), for instance, established five categories for the "solutions for handling autism" variable regarding how to treat people with autism: "institutional interventions", "alternative interventions", "family-based interventions", "medical treatment", and "prevention". Are these five solutions mutually exclusive? No, because, for example, "institutional interventions" could involve "alternative interventions" and vice versa. Similarly, "family-based interventions" may include "medical treatment". The non-mutual-exclusiveness of the five categories is due to the fact that the authors applied more than one criterion in creating the categories. Specifically, Bie and Tang (2015) used the location criterion for forming the "institutional interventions" and "family-based interventions" categories, while "medical treatment" and "prevention" were created based on the purpose criterion (i.e., whether the solution is for treatment or prevention). Categories for a variable would not be mutually exclusive whenever more than one criterion is employed. This is an essential point that researchers must keep in mind in all categorization efforts.

### 9.3.3 The coding scheme

Researchers performing content analysis are known as coders, raters, or content analysts. This book would use "coders" to represent the other two synonyms of this title. The process of content analysis is often referred to as coding, which means to mark passages in the source texts with codes that represent the categories of variables. As shown in Figure 9.1, coding could be done with or without a coding scheme, a tool that helps coders to achieve consistency in coding between or among themselves (i.e., intercoder consistency). Generally speaking, a coding scheme is a must when more than one person does the coding and the texts to be coded are not a single entity (e.g., one book, one speech).

A coding scheme should at least comprise a list of categories (i.e., codes) as well as inclusion and exclusion criteria. Table 1.1 of the current book presents a simple coding scheme for research methods across disciplines. It also specifies the inclusion criteria for

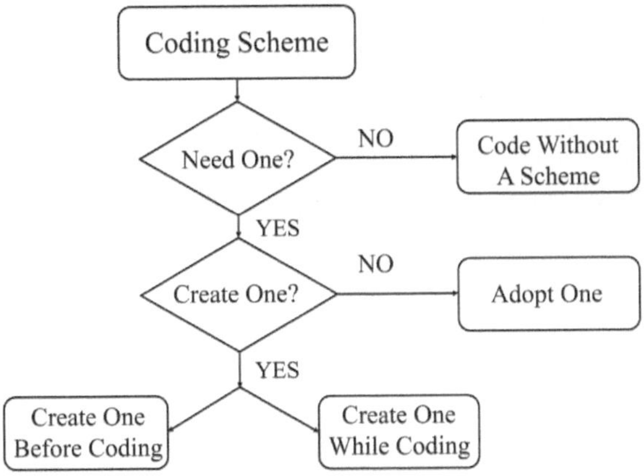


*Figure 9.1* Coding Scheme Options: A Decision Chart

some research methods like questionnaire and theoretical approach. A more comprehensive version of the coding scheme is called a codebook. According to MacQueen et al. (1998), a good codebook should include six basic components: the code, a brief definition, a full definition, guidelines for when to use the code, guidelines for when not to use the code, and examples taken from the text being analyzed. Researchers must adopt a codebook if the content analysis project is large, involving multiple coders and a great number of texts. Otherwise, a coding scheme should be adequate for the purpose of content analysis.

As indicated in Figure 9.1, a coding scheme can be either created from scratch or adopted from previous related research. If there is a coding scheme that other researchers have developed and utilized in early studies related to what you intend to investigate, you are encouraged to contact them and ask for permission to use the coding scheme in your own study. Those researchers usually would be happy to grant you the permission without any problem because sharing a coding scheme is an established practice in the scholarly world. An added advantage of adopting a coding scheme from a related study is that the coding scheme has already been validated in that research. However, the coding scheme developed by others often times may not be exactly what you want. If so, you can revise or adapt it to suit your research needs.

For the majority of novice researchers, it is desirable to learn how to develop a coding scheme. Thus, the present book is to illustrate the process of how to develop one. When researchers decide to create a coding scheme themselves, there are two common options available to them (see Figure 9.1). One is to develop a coding scheme before the full coding starts, while the other option is to create one while performing the coding. Creating a coding scheme before the full coding implies that researchers would develop one according to their research problem, associated research questions, as well as variables to be examined in the study. We may call this method the top-down approach to coding scheme creation. In contrast, a coding scheme can also be built while researchers perform coding. This approach can be considered the bottom-up approach to coding scheme development. Researchers adopting the bottom-up approach normally code a randomly selected set of texts to draft a coding scheme. The draft coding scheme is then altered and finalized by coding the remaining texts. In either approach, researchers need to revise the coding scheme multiple times until they are satisfied with it as a tool for coding.

As indicated earlier, researchers should explicitly specify all related inclusion and exclusion criteria besides providing a list of categories when developing a coding scheme. It is also recommended that decisions on those inclusion and exclusion criteria be recorded so that they can be followed consistently later. Categories in a coding scheme can be listed at one level (e.g., Table 1.1 of the current book) or at two levels (e.g., Bie and Tang 2015, 887–88 although the authors did not present so visually). Occasionally, a coding scheme may list categories of three levels or more. However, do note that more levels of categories in a coding scheme would increase the complexity in applying the categories in the coding process. Another point we must consider in creating a coding scheme is the number of categories at each level. Normally, we adhere to the guidelines George Miller (1956) set up regarding humans' short-term memory capacity by creating between five and nine categories at the second level or lower in the coding scheme. Avoid having more than nine categories at those levels of a coding scheme if possible because that would exceed the upper limit of our short-term memory capacity and consequently negatively affect the coding quality.

For the most part, coders can reasonably handle 30–40 codes (i.e., categories) at one time during the coding process (MacQueen et al. 1998). If the coding scheme contains more than 40 codes, the coding process needs to be done in stages. That is, separate the code scheme into three or four groups of related codes and have coders apply one set of codes to the entire source texts before moving onto the next set.

### *9.3.4 The coding process*

The coding process in content analysis involves highly intellectual activities, especially in latent content analysis. Deduction, summarization, constant comparison, and other analytic techniques should be applied integrally to identify and uncover concepts, ideas, themes, and other content to be gathered from texts. This process is easier said than done due to the complexity, subtleness, and vagueness of the language used in producing the text.

#### 9.3.4.1 *Coder training and intercoder agreement*

If a coding scheme is utilized in content analysis, proper training must be provided so that all coders are familiar with the coding scheme, from the overall framework to every component of it. It is also desirable to invite coders who have knowledge in the subject of the material being analyzed to perform the coding in order to ensure its accuracy and credibility. For example, in coding documents pertaining to Medicare claims, the coder's knowledge of medical terminology and practices would probably be useful (USGAO – United States General Accounting Office, 1996, 34).

During the training session, when coders work together, researchers ought to find out whether they agree on the coding. As part of the training, an independent coding test is normally offered to pilot reliability of the coding between or among coders (Neuendorf 2002, 51). This is called intercoder reliability, intercoder consistency, or intercoder agreement, meaning the degree to which different coders assign the same codes to segments of text. High inconsistency or disagreement can generate misleading data. From this point on, intercoder agreement is to be chosen to represent its other two synonyms, intercoder reliability and intercoder consistency, throughout the rest of this book.

The pilot of intercoder agreement is usually performed as follows: Two or more coders independently code a randomly selected sample of texts for a study using the adopted coding scheme. The phrase "a randomly selected sample of texts" should be underscored because it is rare that researchers would use the entire source texts for piloting intercoder agreement for cost and efficiency reasons. Only when a sample of texts representing the whole text sources is chosen can the pilot of intercoder agreement outcome be reliable. After all the coders complete their coding independently, they compare their coding results and calculate the intercoder agreement value to find out how many coding results are the same out of all that have been coded.

The simplest method for measuring intercoder agreement is percentage, which calculates the proportion of coding agreement between two or more independent coders. If two coders independently assign the same codes from a coding scheme to 80% of the texts they code, the intercoder agreement is 80% or 0.80. This simple and easy-to-understand method, however, has one major weakness, which leads to its failure to account for agreements that would occur solely by chance. Lombard, Snyder-Duch, and Bracken (2002) give the following example: "Two coders are given 100 units (news stories, words, etc.) to code as having or not having a given property. Without any instructions or training, without even knowing the property they are to identify, they will agree half of the time, and these random agreements will produce a percent agreement value of .50" (590).

In order to overcome the limitation of using percentage as an intercoder agreement measure, other methods have been developed that consider chance agreements in coding. Table 9.2 summarizes those major methods for evaluating intercoder agreement (Lombard, Snyder-Duch, and Bracken 2002, 591–93; Neuendorf 2002, 150–51).

Calculation of each measure listed in Table 9.2 is available from common statistics packages (e.g., SPSS, SAS), and no manual calculation is required nowadays. Generally speaking, Cohen's kappa (κ) is intended as an improvement over Scott's pi (π), which also explains why Cohen's kappa is often adopted as a measure for intercoder agreement beyond the percentage method. Fleiss' kappa (κ) and Krippendorff's alpha (α) are each created for their unique features in terms of number of coders and/or level of measurement. Nevertheless, Krippendorff's alpha (α) has little use as an intercoder agreement measure when content analysis is adopted as a data collection technique because coding in this context mainly involves nominal data, and there is little need to consider how to code data measured at ordinal, interval, and ratio levels. The complexity in calculating Krippendorff's alpha (α) also contributes to its low usage rate as an intercoder agreement measure.

Each intercoder agreement measure theoretically takes a value from 0 (no agreement) to 1 (perfect agreement) regardless of which calculation method is chosen. In practice, however, neither 0 nor 1 would be yielded as the final value of intercoder agreement because it is unlikely to achieve no agreement or perfect agreement in any study in the coding process. Instead, researchers would obtain a value between 0 and 1 such as 0.45 or 0.86. A question naturally arises regarding which value would be considered acceptable for measuring intercoder agreement. According to Neuendorf (2002, 143), an intercoder

*Table 9.2* Other Major Methods for Measuring Intercoder Agreement

| *Method* | *Level of Measurement* | *# of Coders* |
|---|---|---|
| Scott's pi (π) | Nominal | 2 |
| Cohen's kappa (κ) | Nominal | 2 |
| Fleiss' kappa (κ) | Nominal | 3 or more |
| Krippendorff's alpha (α) | From nominal to ratio | Any # of coders |

agreement value of 0.80 or greater would be acceptable in most situations, while a value of 0.90 or greater would be acceptable in all situations.

It is recommended that, in addition to the percentage measure, another calculation method (e.g., kappa) should be utilized to statistically assess the level of intercoder agreement in order to avoid chance agreements in the coding process. If coders receive an intercoder agreement value lower than 0.80 in the coding process, researchers who develop the coding scheme need to revise the coding scheme and possibly retrain the coders for enhancing coding agreement rate (MacQueen et al. 1998). Other measures for improving intercoder agreement include the test-retest technique when there is more than one coder and the split-half approach when there is only one coder. The test-retest technique implies that a same portion of data is deliberately coded twice with a time interval in between the two tests (e.g., two weeks of time) to erase the "memory effect" of the first test coding, and then coding results are compared with the aim to detect any possible inappropriate interpretation of the data. The test-retest technique is often used in developing a coding scheme as well. The split-half approach, on the other hand, enables the only coder to compute "intercoder" agreement by splitting the data into two halves, coding each half with a time interval in between, and then comparing the coding results. Although high intercoder agreement does not necessarily guarantee high quality of coding results, low intercoder agreement typically sets off an alarm about the coding quality.

Another point we should note about intercoder agreement in content analysis is that the pilot results actually serve two purposes. One is to determine if the coding scheme needs to be revised and/or if further training should be offered to the coders. The other purpose is that, if the pilot yields an acceptable intercoder agreement value, that value would be reported to demonstrate the reliability of the coding scheme as well as the coding. Virtually no study would measure intercoder agreement based on coding results of all the selected source texts. Take Bie and Tang (2015) as an example, the intercoder agreement value is calculated based on the 10% of their total data set that are coded by both authors of the study because the second author only coded that many news reports selected for the research.

#### *9.3.4.2 Coding for data collection purpose*

When content analysis is chosen as the data gathering technique, the essence of this method lies in coding, which powers the data collection process in research. Coding enables researchers to collect data from the source text through a process of analysis and interpretation while following explicit guidelines for determining the boundaries of the text associated with a particular code (i.e., category of a variable). When texts are coded according to research questions without a coding scheme, it is called preliminary coding, and such coding is general and global. In contrast, most of the coding is done by examining and assigning codes to the source text. This can be labeled as in-depth coding, which is specific and deals with individual variables explored in the study. In-depth coding may involve hierarchical codes if the coding scheme contains more than one level of codes.

Coders need to distinguish between manifest content analysis and latent content analysis if both types are involved during the coding process. Otherwise, coders can focus only on one of the two kinds. As indicated earlier, coding manifest contents is fairly straightforward because what is coded typically contains the codes researchers intend to use with or without a coding scheme. For example, the location names (e.g., New

York City, Paris) mentioned in news reports usually can be easily identified and readily extracted from the text. Performing latent content analysis, by comparison, is much harder because what needs to be coded is not explicitly presented, and coders must employ such analytical techniques as abstraction, synthesis, deduction, and induction in order to ascertain any codes, ideas, concepts, and the like underlying the text. For instance, if the source texts are some news reports covering Paris and New York City as popular tourist cities, coders are asked to code those news reports with the intention to find out what attracts tourists to those two cities. This type of coding falls under the category of latent content analysis.

Unlike other data collection techniques (e.g., questionnaire, interview), content analysis is not a standardized technique that, once developed, remains the same throughout a study. It instead must be tailored to suit the particular texts under examination and constructed especially for the research problem at hand (Mayring 2014, 39). Every analytical step and every decision in the coding process should be based on systematic and tested rules that consist of the coding scheme and other guidelines (e.g., how detail shall the coding be done). The coding scheme, as indicated earlier, constitutes the central instrument of analysis. It also contributes to the intersubjectivity of the procedure, helping to make it possible for other coders to reconstruct or repeat the analysis if needed. Although the rules governing the coding process are in general established beforehand, additional rules can be developed to handle cases that are not anticipated before the coding process commences. It must now have become clear that content analysis via coding is not a rigidly delineated technique, but a process in which new decisions regarding basic procedures and individual rules of analysis have to be made constantly (Mayring 2014, 42).

A systematic coding procedure should always specify which parts or segments of the text are to be coded and in what kind of order in the data collection process. If websites are chosen as source texts in a study, for example, rules ought to be spelled out regarding whether the full content of each page is to be coded or only the title, headings, and link names will be analyzed. For headings, shall headings at every level or only the top two levels of headings be coded? Websites normally contain many hyperlinks. Shall every link on the homepage be followed as far as possible or each link only be followed to its second level? Similar rules should also be established in coding websites by specifying whether headings, link names, and the rest of page content ought to be coded in a certain order or simultaneously. Moreover, the coding rules should be described specifically and detailed enough so that all coders are able to perform content analysis similarly by following the rules.

In the coding process, one other aspect of content analysis to which coders must pay attention is to interpret and code any text segments in context. That is, coders should relate their coding to what their study intends to explore. Using content analysis via coding as a data collection technique should still accomplish the same task: gathering what is needed to answer the research questions raised in the study. The contexts for coding hence also include the study's research questions and other elements (e.g., variables). No coding should be done in isolation or without reference to the study for which the coding is conducted. Specific types of coding (i.e., open coding, axial coding, and selective coding) are discussed in Section 13.2 because they are techniques for analyzing qualitative data.

Many authors (e.g., Nelson and Woods, Jr. 2011, 212–14; USGAO – United States General Accounting Office 1996, 15–46; Wildemuth 2017, 320–23) divide the content analysis process into multiple steps, ranging from selection of texts to review and revision of results. Although the number of steps in the process vary from author to author, coding always remains a key step in content analysis together with the usage of other analytical techniques (e.g., comparison, deduction, and induction) for achieving the data collection purpose. Furthermore, coding and other analytical techniques are not applied separately and sequentially in content analysis. Rather, all those techniques are often employed simultaneously and concurrently for gathering data from source texts in the research process. This is a unique feature of content analysis as a data collection technique. Coding, as one major technique in content analysis, requires the use of other analytical techniques in conjunction. But analytical techniques, such as comparison, deduction, and induction, are normally not covered individually in publications on research methods. The assumption underlying this practice is that those analytical techniques are already imparted in the education process about reasoning, critical thinking, and other similar subjects before the subject of research methods is introduced.

Coding can be performed with the aid of software for qualitative data analysis (e.g., Atlas.ti, Provalis Research's WordStat). However, the regular word processor can accomplish what such qualitative data analysis software can do (e.g., memoing, commenting) if supplemented by some basic counting ability (e.g., how many times a particular code is assigned to texts). The intellectual activities in coding (e.g., conceptual analysis of texts, determination of code assignment) still need to be done by coders rather than software because coding is a highly intellectual activity and only human coders have the intelligence to perform coding satisfactorily. Further discussion on the role of software for qualitative data analysis is to be made in Section 13.2.

## 9.4 Dos and don'ts in research using content analysis

As indicated earlier, content analysis is not a standardized technique for data collection like questionnaire and interview that, once developed, remains nearly the same in the research process. It has distinctive features and thus requires researchers to be extra cautious and observe the following dos and don'ts when choosing content analysis as a data collection technique.

First, define the variables to be examined in the study carefully to facilitate the establishment of categories for each variable. Avoid defining variables so ambiguously that it would defy reasonable categorization and interpretation. In addition, the categories developed for each variable must be collectively exhaustive and mutually exclusive. Otherwise, coders would run into difficulty (i.e., no category or more than one category available for coding) when attempting to apply categories in the coding process.

Second, budget sufficient time for content analysis because this data collection technique can be more time-consuming and less structured in application when compared to such research methods as questionnaire and metric-based techniques. Coders must be trained and given time to practice coding until their coding quality is satisfactory. The time needed for the full coding process depends on the amount of material to be coded, the number of variables, the number of coders, and the judgment required for coding decisions. Handling all those related factors would take time in content analysis via coding.

Third, complete a coding scheme development with proper testing and revisions if the decision is to create the coding scheme before coding (see Figure 9.1). Avoid only having a skeleton of a coding scheme and erroneously thinking that it can be expanded or completed later when coding is in progress. According to USGAO – United States

General Accounting Office (1996, 50), "[t]he time spent in being complete will be more than repaid by making the coders' task easier and faster and, especially, by ensuring coding of the highest quality". A complete and well-tested coding scheme is indispensable in content analysis.

Fourth, set up management procedures in the process of content analysis when more than one coder performs coding. A single person should be given overall responsibility in the coding process. The best choice is usually someone who has coding experience and will also perform some of the coding as a head coder. This person should develop detailed procedures for keeping track of source texts, assigning them to other coders, and maintaining a log of the process. The head coder should respond to queries from fellow coders, resolving ambiguities about categories in the coding scheme, and making other related decisions in the content analysis process.

Fifth, cross-check preliminary results from all coders to ensure a reasonable intercoder agreement rate. Avoid coding all the texts for one study and then realize that the intercoder agreement is lower than 80%. Percent agreement is a quick and forthright measure of intercoder agreement. Complement the percent agreement calculation with another measure (e.g., kappa) that statistically assesses the level of intercoder agreement and considers chance agreements in the coding process. Discussion among coders, revision of the coding scheme, and even retraining coders are some of the techniques that can be taken to improve intercoder agreement. When discussing results with other coders, do not blindly follow someone else's judgment or become frustrated. Instead, think calmly and logically. Walk away for a while if needed, and then return to the discussion.

Six, use various reading skills (e.g., scanning, skimming, intensive reading) to determine the content in the text as well as make use of the context a particular study provides in the form of research questions, variables, and text passages. If segments of the text do not seem to be adequate in the data collection process, read the entire text instead, especially when the source text is a monograph. In conducting content analysis of articles, reading the whole text might be a requirement rather than an option.

As always, it is hard to enumerate all the dos and don'ts for using content analysis as a data collection technique. On one hand, content analysis actually relies on the selection and usage of multiple analytical techniques (e.g., coding, comparison) simultaneously. On the other hand, this research method is widely employed and often gets utilized even if a study does not explicitly indicate that content analysis is adopted for data collection. In the examples given in Section 9.1 of the current book, both observation and interview need to make use of content analysis in their data collection processes as well. The same holds true for experiments (see Chapter 10) and some other data collection techniques covered in the present book. It is in this sense that content analysis is considered a super-method in research.

## 9.5 Case study of research using content analysis

As explained in Section 9.2, content analysis is commonly categorized into manifest and latent based on whether the content to be analyzed from the text is visible, observable, and countable. If the answer is yes, it is manifest content analysis. Otherwise, the content analysis is latent. The case study chosen for this chapter covers both manifest content analysis and latent content analysis. Therefore, only one case study (i.e., Bie and Tang 2015) instead of the commonly two, will be conducted for content analysis as a data collection technique. Below is its bibliographic information.

Bie, Bijie, and Lu Tang. 2015. "Representation of Autism in Leading Newspapers in China: A Content Analysis." *Health Communication* 30 (9): 884–93. https://doi.org/10.1080/10410236.2014.889063.

Bie and Tang (2015) report their study on how five leading newspapers in China covered autism from 2003 to 2012 through a content analysis. The authors intend to address the following five research questions (RQs) in their investigation:

RQ1: What are the main issues covered in relation to autism in Chinese newspapers?
RQ2: What are the major sources cited or quoted in Chinese newspapers' coverage of autism?
RQ3: How do Chinese newspapers frame the causes of autism?
RQ4: How do Chinese newspapers frame the solutions of autism?
RQ5: To what extent are human-interest discourse and scientific discourse used in Chinese newspapers' coverage of autism?

All the five Chinese leading newspapers chosen as data sources for the study are covered in Wisers, a widely recognized database of Chinese newspapers. The researchers used "autism" and "lonely disease", another Chinese expression of autism, to locate news stories on autism published between 2003 and 2012 in the chosen newspapers from the database. The authors, however, should have indicated whether the terms appear in the title or anywhere in the report in order for a news story to be selected for their study. Some explanation is also needed about why 2003–2012 was chosen as the time period for this study when key policies and services for autism were introduced between 2006 and 2012 (Bie and Tang 2015, 887). The initial search yielded 1,590 stories after eliminating duplicate ones, irrelevant entries such as advertisements, TV guides, and corrections to former news reports.

Bie and Tang (2015) used a systematic sampling method to select 50% of these news articles as the final data set (n = 795) for their study. The researchers cited Krippendorff (2013) in support of their choice of the systematic sampling technique, but that book is not a major source for the topic of sampling. Although Krippendorff (2013) devotes one chapter on sampling in his book, the inclusion of sampling theory and techniques in his book is mostly for providing a comprehensive coverage of content analysis. There are, however, a good number of well-regarded publications on the subject of sampling, and Babbie (2016, 182–220) is one of them. Moreover, two minor comments are due on this citation of Krippendorff (2013) by Bie and Tang (2015). One is that Krippendorff's book was published in 2013 rather than 2012, as the researchers listed. The other comment is that the authors should list page numbers whenever citing a book such as Krippendorff (2013) to enable readers to locate the cited content easily.

The units of analysis in this study are individual news stories about autism. The researchers developed a coding scheme, not "a codebook" as Bie and Tang themselves (2015, 887) indicated that a codebook should contain more information than what was provided in their report. According to MacQueen et al. (1998), a good codebook should include six basic components: the code, a brief definition, a full definition, guidelines for when to use the code, guidelines for when not to use the code, and examples taken from the text being analyzed. Therefore, what Bie and Tang (2015) present in their report at most fits the definition of a coding scheme.

The coding scheme created by Bie and Tang (2015) consists of the following six categories (i.e., variables) in their terminology, five of which were adapted from some previous sources. It is common for researchers to adapt another coding scheme for their own study. Nevertheless, it is atypical to see that Bie and Tang (2015) adapted values for each of the five variables from different research projects.

1) Issues discussed in the news story (i.e., topics)
2) Role of the autistic individuals in the news stories – However, this variable is not covered in any of the five research questions formulated for the study. It was later presented as part of the results for RQ1 and then used as a filter for results considered in RQs 2–5: "Only those articles with more than an incidental mention of autism (n = 444) were included in the data analysis conducted to answer RQs 2-5" (Bie and Tang 2015, 889). So the n is reduced from 795 to 444 in the middle of the content analysis. Such change in data size should have been explained in the Method section by stating that this coding category is intended to serve as a filter for the final data set.
3) Sources cited in the news coverage of autism – Multiple choices are allowed in this coding category.
4) Causes of autism
5) Solutions for handling autism
6) Types of discourse

As shown on the above list, the content analyses Bie and Tang (2015) performed include both manifest content analysis (e.g., sources cited in the news coverage of autism) and latent content analysis (e.g., issues discussed, types of discourse). The type of content analysis for the rest of the three variables (i.e., role of the autistic individuals in the news stories, causes of autism, and solutions for handling autism) remains to be identified, which is left to the reader as an exercise.

As discussed in Section 9.3.2, some of the values for the six variables in Bie and Tang (2015) are not collectively exhaustive and mutually exclusive. This is an important issue in content analysis when a coding scheme is adopted and should not be overlooked. Another comment on the coding scheme is its presentation. It would be clearer if all the variables and associated values in the coding scheme had been presented as a list or a table. Although Bie and Tang (2015) reported that two rounds of training were offered, no additional information was given about how they were done. A brief description about the training should have been provided. The intercoder agreement between the two coders was computed by using Cohen's kappa for all six variables and ranges from 0.90 to 1.00. The intercoder agreement rate is undoubtedly high. However, the authors should have pointed out that a coefficient value of 0.80 is acceptable to most and 0.90 to all according to established practices. Also note that the first author of this study coded all the news articles, while the second author only coded 10% of the total data set. This kind of coding practice is all right but not recommended.

Bie and Tang (2015) overall is a decent study using content analysis as the data collection technique. Its coding scheme and the presentation of it, however, need further improvement. In addition, it leaves much to desire for when only one of the two authors coded all the data set selected for the study.

**Chapter Summary**

- Content analysis is a method of textual examination used to describe and explain characteristics of messages embedded in texts. It can be used for both data collection and data analysis. In the current chapter, content analysis is presented and discussed as a data collection technique.
- Content analyses are most frequently categorized into two types: manifest content analysis and latent content analysis. Manifest content analysis means that the contents to be analyzed are visible, observable, and countable in texts. Latent content analysis, by contrast, deals with contents that are hidden, conceptual, unobservable, and uncountable in texts.
- Content analysis encompasses multiple steps that include selecting source texts, determining units of analysis, defining variables and generating their content categories, decision on the coding scheme and its development, coder training, and the actual coding along with the usage of other analytical techniques such as comparison and synthesis.
- A coding scheme is an instrument that comprises a list of codes or categories as well as inclusion and exclusion criteria. It is used by researchers in the coding process to ensure the quality of content analysis when the research project is large, involving multiple coders and a great amount of texts. Intercoder agreement refers to the degree to which different coders assign the same codes to segments of text in the coding process. It is usually calculated using the percentage measure plus other methods (e.g., kappa) that consider chance agreements in coding.
- Make sure to apply coding and other analytical methods while observing the dos and don'ts outlined in this chapter and beyond in research when using content analysis as a data collection technique.

## References

Babbie, Earl R. 2016. *The Practice of Social Research*. 14th ed. Boston: Cengage Learning.

Berelson, Bernard. 1971. *Content Analysis in Communication Research*. New York: Hafner.

Bie, Bijie, and Lu Tang. 2015. "Representation of Autism in Leading Newspapers in China: A Content Analysis." *Health Communication* 30 (9): 884–93. https://doi.org/10.1080/10410236.2014.889063.

Krippendorff, Klaus. 2013. *Content Analysis: An Introduction to Its Methodology*. 3rd ed. Los Angeles: Sage.

Lehtinen, Esa. 2011. "Conversation Analysis." In *The Routledge Handbook of Research Methods in the Study of Religion*, edited by Michael Stausberg and Steven Engler, 122–33. New York: Routledge.

Lombard, Matthew, Jennifer Snyder-Duch, and Cheryl Campanella Bracken. 2002. "Content Analysis in Mass Communication: Assessment and Reporting of Intercoder Reliability. *Human Communication Research 28* (4): 587–604. https://doi.org/10.1111/j.1468-2958.2002.tb00826.x.

MacQueen, Kathleen M., Eleanor McLellan, Kelly Kay, and Bobby Milstein. 1998. "Codebook Development for Team-Based Qualitative Research." *Cultural Anthropology Methods Journal* 10 (2): 31–36. https://doi.org/10.1177/1525822x980100020301.

Mayring, Philipp. 2014. *Qualitative Content Analysis: Theoretical Foundation, Basic Procedures and Software Solution*. Klagenfurt, Austria. Accessed November 12, 2021. https://www.ssoar.info/ssoar/handle/document/39517.

Miller, George A. 1956. "The Magical Number of Seven, Plus or Minus Two: Some Limits on Our Capacity for Processing Information." *Psychological Review* 63 (2): 81–97. https://doi.org/10.1037/h0043158.

Nelson, Chad, and Robert H. Woods, Jr. 2011. "Content Analysis." In *The Routledge Handbook of Research Methods in the Study of Religion*, edited by Michael Stausberg and Steven Engler, 109–21. New York: Routledge.

Neuendorf, Kimberly A. 2002. *The Content Analysis Guidebook*. Thousand Oaks: Sage.

Neuman, W. Lawrence. 2004. *Basics of Social Research: Qualitative and Quantitative Approaches*. New York: Allyn and Bacon.

Ryan, Gery W., and H. Russell Bernard. 2000. "Data Management and Analysis Methods." In *Handbook of Qualitative Research*, edited by Norman K. Denzin and Yvonna S. Lincoln, 769–802. 2nd ed. Thousand Oaks: Sage.

USGAO – United States General Accounting Office. 1996. *Content Analysis: A Methodology for Structuring and Analyzing Written Material* (Publication No. GAO PEMD 10.3.1). Accessed August 14, 2023. http://archive.gao.gov/f0102/157490.pdf.

Wildemuth, Barbara M., ed. 2017. *Applications of Social Research Methods to Questions in Information and Library Science*. 2nd ed. Denver: Libraries Unlimited.

# 10 Meta-Methods

## Experiments

**Chapter Overview Questions**

- What are experiments?
- What types of experiments are there? How are they categorized?
- What are the major steps a true experimental study should take?
- What are the factors that might jeopardize the internal and external validity of experimental research?
- What are the dos and don'ts in research using the experiment method?

### 10.1 What are experiments?

Experiments are a time-honored method for data collection by following a procedure to test or try out what is studied in either a laboratory or a real setting. Campbell and Stanley (1966, 1) defined experiments as "that portion of research in which variables are manipulated and their effects upon other variables observed" in their seminal book titled *Experimental and Quasi-Experimental Designs for Research*. Research using the experiment method for data collection is called experimental research.

According to Campbell and Stanley (1966), their book was originally published as a literature survey entry in *Handbook of Research on Teaching* in 1963. Later, its scope is expanded to the social sciences in general. While Donald T. Campbell (1916–1996) was a psychologist by education, Julian C. Stanley (1918–2005) was a high school mathematics and chemistry teacher initially and then did his doctorate in education at Harvard University. The educational background of those two prominent authors demonstrates that experimental research has been having a long tradition in psychology and education (Ross and Morrison 2004). When psychology emerged as an infant science during the 1900s, it modeled its research methods on the established paradigms of the medical, biological, and physical sciences, which for centuries relied on experimentation to derive principles and laws. Subsequent reliance on the experimental method was strengthened by behavioral approaches to psychology and education that predominated during the first half of the 20th century.

Consequently, experiments are not only common mostly in medical, biological, and physical sciences (e.g., the anti-AIDS AZT drug trials in the mid-1980s, the vaccine development in the COVID-19 pandemic in the beginning of 2020) but also are increasingly used in the social sciences for the following two reasons. First, the true experimental

DOI: 10.4324/9781003021711-13

design helps us understand the logic of all research designs; it is a model against which we can evaluate other research methods in terms of validity and reliability. Second, the true experimental design allows the investigator to draw causal inferences and observe, with relatively little difficulty, whether the independent variable caused changes in the dependent variable. In contrast, this causal relationship cannot easily be determined with other research methods such as interview and questionnaire (Frankfort-Nachmias, Nachmias, and DeWaard 2015, 84).

Sometimes the experiment method is reserved for situations in which both the independent variables and the environmental or concomitant variables – other sources of variation in the dependent variables – are controlled (Tague-Sutcliffe 1992). This means that experiments can be conducted to control both the independent variables and other associated variables in a laboratory setting. Experiments done in the real world are called field experiments while those conducted in a laboratory are referred to as laboratory experiments.

Because the problems each discipline attempts to solve are different, researchers in the natural sciences usually perform laboratory experiments in which the independent variable can be manipulated and controlled. By comparison, social scientists often carry out field experiments that occur in the regular course of social events or processes in which researchers are unlikely or impossible to control the independent variable in the studies even though it is still feasible for them to observe its effect on the dependent variable. Despite this major difference, this book aims to present the experiment method based on the fundamentals of experimental research while keeping a focus on its usage and application in the social sciences.

Another feature of experimental studies involves the application of multiple techniques for data collection. For instance, use questionnaire or interview for pretest and posttest, perform observation in the experiment process, or employ think-aloud protocols (see Section 12.1) for gathering cognitive data from participants. Because of this feature, experiment is considered a meta-method in research. Whenever experiment is adopted as a research method, it usually implies that more than one individual data collection technique is utilized in the study.

Experimentation has proven as an extremely useful research method for studying causal relationships. That is why it is preferred to other methods (e.g., questionnaire, interview) when the goal of the research is to produce reliable causal outcomes. That is, when we want to be reasonably confident that some event is caused by another event and not merely correlated with it, the best way to test our assumptions and beliefs is by experimentation (Haas and Kraft 1984). Experiments differ from other types of research methods precisely because in an experiment the researcher manipulates the "independent variable" (or treatment) under controlled conditions and then observes the behavior of the "dependent variable" (or outcome). Babbie (2016, 243) posits that researchers need relatively little time, money, and fewer subjects in experimental research because experiments are often limited in scope. It is often possible to replicate a given experiment several times using several different groups if necessary. This advantage outlined by Babbie (2016) nevertheless does not hold true for research in the social sciences because field experiments, unlike laboratory experiments, are hardly replicable. In other words, reliability in social science experiments remains an issue that needs to be addressed. The artificiality of experiments, on the other hand, can also lead to the validity issues Campbell and Stanley (1966) addressed throughout their seminal book.

## 10.2 Types of experiments

In addition to the laboratory and field experiments briefly mentioned above, experiments are typically categorized by how many requirements they each meet in their designs. The well-known classic or true experimental designs meet all the following requirements:

- Participants are randomly selected and assigned to the experimental and control groups.
- It permits the manipulation of the independent variable (i.e., treatment).
- It allows for pretest, posttest, and control group versus experiment group comparisons.

If an experiment does not meet all the above requirements, it is a quasi-experiment. If an experiment at most meets only one of the afore-listed requirements, it is a pre-experiment. Each of these three major types of experimental designs is to be explained and discussed below.

### *10.2.1 True experimental designs*

The classic or true experimental design is generally illustrated using the following notations (see Table 10.1). In the table, O stands for "observation" made after the pretest and posttest are done using data collection techniques such as interview or questionnaire. Observation in this context means checking and analyzing the pretest and posttest results instead of as a data collection technique. X represents an independent variable in the experiment which is often known as a "treatment" or an experimental stimulus. Examples of treatments include an advertisement to influence consumers' views on a product or the feedback Clariana and Lee (2001) provided in their experiment research – one of the two case studies chosen for the present chapter. Participants in both the experimental and control groups are randomly selected and assigned from a same population. The concepts of independent variables and sampling are introduced and presented, respectively, in Sections 2.3.1 and 2.5.2.

As shown in Table 10.1, the experimental group is exposed to or receives a treatment. The control group, in contrast, does not receive any treatment and serves as a comparison to the experimental group in order to see if it is the independent variable that actually causes any change in the dependent variable. Figure 10.1 visualizes the experimental design applied in research.

Figure 10.1 displays the main elements of an experimental design. One outstanding difference between the experimental group and control group is that the experimental group receives a treatment, shown by a solid line between the independent variable and dependent variable, while the control group does not, denoted by a dotted line between the independent variable and dependent variable. To form a base line for observing changes in the dependent variable before and after the treatment, a pretest (i.e., gathering data that measures the dependent variable) is taken with participants in both groups before the treatment is given. Likewise, a posttest (i.e., gathering data that measures the same dependent variable one more time) is performed with the same participants in both groups after the treatment is administered. Multiple techniques can be used for conducting the pretest and posttest.

*Table 10.1* The Experimental Design

| | *Pretest* | *Treatment* | *Posttest* |
|---|---|---|---|
| Experimental Group | O | X | O |
| Control Group | O | | O |

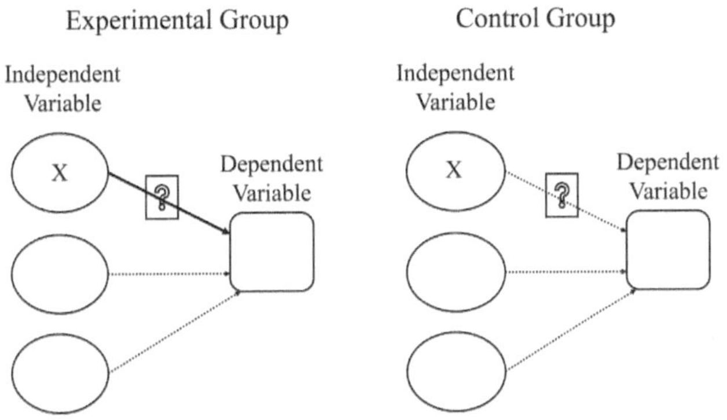


*Figure 10.1* Visualization of an Experimental Design[1]

The most commonly adopted techniques in experimental research in the social sciences include interview and questionnaire. When the pretest and posttest results are examined and compared, such activities are referred to as observation. As briefly noted above, observation in this context means checking and analyzing the pretest and posttest results rather than observation as a data collection technique described in Chapter 5.

The COVID-19 vaccine testing or experiment serves as a pertinent example for illustrating the true experimental design although this type of experiment is rarely performed in the social sciences. Nevertheless, this example is selected mainly due to its global impact and high methodological relevance. Drug or vaccine testing follows the true experimental design which consists of three phases. Of the three phases, Phase III of the COVID-19 vaccine testing best represents an experiment design by focusing on testing the efficacy and safety of the newly developed vaccine. Different COVID-19 vaccines have been developed by different companies in different countries since the outbreak of the pandemic in the beginning of 2020. The experiment extents of those vaccines are likewise different in terms of participant numbers. Some COVID-19 vaccine experiments had several hundred participants, whereas others might recruit over 10,000 participants. For the sake of simplicity and clarity, we choose 3,000 as the number of participants in a hypothetical Phase III COVID-19 vaccine trial to illustrate the true experimental design in six steps: selection, assignment, assessment (i.e., pretest), treatment, observation, and comparison (i.e., posttest) and evaluation.

1) Selection: Randomly select 3,000 people ages 18 and older from the general public to ensure that the 3,000 selected participants are representative of the entire population who would eventually receive the COVID-19 vaccine.
2) Assignment: Randomly assign 1,500 participants each to the experimental group and the control group, respectively. Make certain that the participants in both groups are identical in all aspects (e.g., age, gender, race) relating to this experiment.
3) Assessment (i.e., pretest): Perform a pretest by using a questionnaire or interview to collect data (e.g., demographics, health condition) from each of the participants selected for the COVID-19 vaccine trial. The most important piece of data that must be gathered in this pretest is that each participant is COVID-19 free.
4) Treatment: Administer the COVID-19 vaccine to all the participants in the experimental group while just giving placebos (i.e., harmless substance in place of the COVID-19 vaccine) to all the participants in the control group.

5) Observation: Ask the participants in both groups to live as usual for approximately ten weeks. During this time period, observe if and how many participants in either group contract COVID-19.
6) Comparison (i.e., posttest) and evaluation: Compare the number of confirmed COVID-19 cases in both groups, and then compute the vaccine's efficacy rate (i.e., percentage reduction in disease incidence = # of positive cases in the control group/ total # of positive cases in both groups). The World Health Organization (2021) specifies that a vaccine must have an efficacy rate of 50% and higher to get approved. If the vaccine being tested has an efficacy rate of 50% and higher, the experiment can be completed and results are reported. Taking the Phase III trial data for Pfizer's COVID-19 vaccine (Brothers 2020) as an example, there were 170 total confirmed COVID-19 cases with 162 of them occurring in the control group and only 8 in the experimental group among about 30,000 participants worldwide. The efficacy rate of Pfizer's vaccine is a little over 95% (162/170 = 95.29%), much higher than the minimum of 50% the WHO requires for approval. Hence, the experiment on Pfizer's vaccine could be concluded.

The above six steps depict how a true experiment is conducted in testing the COVID-19 vaccines in Phase III. It must be noted that Phase III of any vaccine testing usually takes six to nine months or longer to complete but was accelerated in the COVID-19 pandemic (John Hopkins Coronavirus Research Center 2022). In addition, other parameters should be examined as well in determining if the experiment can be completed. For instance, the entire Pfizer vaccine trial also reported an extremely low number (>2%) of adverse events, with none of them requiring immediate or critical medical attention. The true experimental design for testing drugs or vaccines is commonly known as randomized controlled trials (RCTs). Such RCT trials are double-blind, meaning neither the trial investigators nor the participants know who received the candidate drug or vaccine. The double-blindness can be applied in any true experiment.

Furthermore, the results of true experiments can be generalized because randomization is applied in selecting participants and assigning them to the control and experimental groups. Section 2.5.2.4 discusses the relationship between randomization in sample selection and generalization of research findings in a study. As long as sampling units in a study are randomly selected, its findings can be generalized to the population from which the sample is obtained. That explains why the COVID-19 vaccines can be given to people all over the world even though their clinical trials were conducted only with a small fraction of the 7.71 billion people on the earth (WolframAlpha 2023).

Variations of the true experimental design have been developed and extensively depicted in numerous publications (e.g., Campbell and Stanley 1966). For instance, the Solomon four-group design is devised to consider if the treatment might be sensitized by the pretest while the posttest-only control group design, as its name implies, does not include the pretest, arguing that "the most adequate all-purpose assurance of lack of initial biases" between the control and experimental groups can be achieved via randomization (Campbell and Stanley 1966, 25).

### *10.2.2 Quasi-experimental designs*

The prefix "quasi" means "resembling". Accordingly, quasi-experimental research is research that resembles experimental research but is not true experimental research (Price,

Jhangiani, and Chiang 2015). Researchers in the social sciences in general do not use the experiment method as widely as those in the natural sciences, primarily because the rigorous structure of true experimental designs often cannot be adapted to natural social settings (e.g., an online community, a city with over 8 million residents). Hence, social science researchers frequently use quasi-experimental designs that are weaker for drawing causal inferences but are more appropriate to the type of problems they examine. Quasi-experimental designs can also be implemented in a naturalistic setting instead of a laboratory.

In Campbell and Stanley (1966), ten quasi-experimental designs are presented and discussed. They include the time-series experiment, the equivalent time-samples design, the equivalent materials design, the nonequivalent control group design, counterbalanced design, the separate-sample pretest-posttest design, the separate-sample pretest-posttest control group design, the multiple time-series design, the recurrent institutional cycle design, and regression-discontinuity analysis. Quasi-experimental designs are adopted when some control over the experimental conditions can be exerted, yet full control is neither feasible nor desirable. While those quasi-experimental designs share some common features of experiments like treatment administration and observation, they may not have a control group (e.g., the time-series experiment), the control and experiment groups of participants are naturally formed rather than created by researchers (e.g., the nonequivalent control group design), or multiple treatments are applied to each of the participants but the order of treatments are rotated (e.g., the counterbalanced experimental design, also known as the rotation experiment or the Latin Square design). For example, a study attempts to evaluate the quality of four different search engines. Instead of asking each group to perform searches using one search engine, all four groups are invited to use all four different search engines ($S_n$) but the order of using the four search engines is randomly assigned (see Table 10.2) in order to eliminate concerns about the learning that might occur when participants conduct more searches during the course of the study. This rotation method is commonly utilized in quasi-experimental designs if multiple treatments are involved in the study.

As one or more requirements for the true experimental design are not met in quasi-experimental studies, results obtained from research using a quasi-experimental design have to be interpreted by keeping such limitations in mind. For instance, no finding of a study can be generalized if the quasi-experimental research did not randomly select and assign its participants in the experiment. In other words, quasi-experimental designs are appropriate and useful when we want to make causal inferences from research in which experimental control, especially randomization, is not feasible. But such causal inferences are weaker than those made in true experiments. In addition, we should be fully aware of the reduced internal as well as external validity in quasi-experimental research.

*Table 10.2* The Counterbalanced Experimental Design

| | $X_1$ | $X_2$ | $X_3$ | $X_4$ |
|---|---|---|---|---|
| Group A | $S_1O$ | $S_2O$ | $S_3O$ | $S_4O$ |
| Group B | $S_3O$ | $S_1O$ | $S_4O$ | $S_2O$ |
| Group C | $S_2O$ | $S_4O$ | $S_1O$ | $S_3O$ |
| Group D | $S_4O$ | $S_3O$ | $S_2O$ | $S_1O$ |

### 10.2.3 Pre-experimental designs

Compared with the quasi-experimental designs, pre-experimental designs exert even less control and may at most meet one of the three requirements listed in the beginning of Section 10.2 for a true experiment. Campbell and Stanley (1966, 6–13) expounded three pre-experimental designs in their influential book.

1) The one-shot case study – A single group is studied only once, subsequent to some treatment presumed to cause change with the following design notation:

   X O

2) The one-group pretest-posttest design – A single group is studied with pretest and post-test with the following design notation:

   $O_1$ X $O_2$

3) The static-group comparison – The group experiences a treatment X and then is compared with one group which has not, for the purpose of the establishing effect of X with the following design notation:

   X $O_1$
   $O_2$

Babbie (2016, 233) illustrates the three pre-experimental designs with an example of physical exercises and weight reduction. In order to make his graphical illustration clearer, Babbie (2016) uses individuals rather than groups. Each of the three pre-experimental designs is applied to explore whether exercise causes weight reduction.

1) The one-shot case study: In this case, we assume that a person exercises and is thin. As there is no before-and-after comparison nor comparison with another individual, we can only compare with our assumption based on the common knowledge that exercise leads to weight reduction. However, this pre-experimental design does not consider if the individual was thin before doing exercises or became thin due to other reasons (e.g., sickness or on a diet). Hence, the evidence between exercises and weight loss is very weak.
2) The one-group pretest-posttest design: Using this design, we can make the before-and-after comparison. That is, someone is overweight before exercises but loses weight after doing exercises. The comparison is made on the same individual. In this case, we can rule out that the individual was thin before doing exercises. But we still cannot exclude the likelihood that the person became thin for reasons other than exercises.
3) The static-group comparison: In this design, two individuals are involved, one exercises and is thin, while the other person does not exercise and stays overweight. The problem with this pre-experimental design remains that we do not know if the individual who exercises was thin before exercises. Neither can we determine if exercise was the only factor that caused weight loss even though there was another individual to compare with.

It is apparent that pre-experimental designs are weak in internal and external validity because neither experimental control nor randomization is performed. They also do not allow researchers to make causal inferences (Frankfort-Nachmias, Nachmias, and

DeWaard 2015, 117). As a result, pre-experimental designs are used only when researchers are unable to adopt any true experimental or quasi-experimental design.

As indicated earlier, selecting an experimental design for a study mostly depends on whether the following three requirements for a true experiment can be met:

1) Participants are randomly selected and assigned to the experimental and control groups.
2) It permits the manipulation of the independent variable.
3) It allows for pretest, posttest, and control group versus experiment group comparisons.

If all the three requirements are satisfied, we choose a true experiment design in research. If some of the three requirements are met, we select a quasi-experimental design for our research. If at most only one of the afore-listed three requirements is satisfied, a pre-experimental design is the viable option. Simply put, choose the true experiment design whenever possible, followed by the quasi-experimental option. Pre-experimental designs should always be our last choice in consideration of research validity and reliability. Figure 10.2 lists all true experiment, quasi-experiment, and pre-experiment designs.

Not every experimental design is equally utilized in research. Some designs are adopted more often than others. In the social sciences, researchers frequently turn to quasi-experiments and pre-experiments due to the kind of research problems they intend to explore in the field.

## 10.3 Using experiments in research

Although the requirements and implementation for true experiments, quasi-experiments, or pre-experiments are different, there are some essentials that researchers can observe when using the experimental method.

- One treatment that functions as the independent variable in the experiment. It must be produced or manipulated by the researcher. Multiple treatments are sometimes applied if a study requires so (e.g., the comparison of four search engines using the

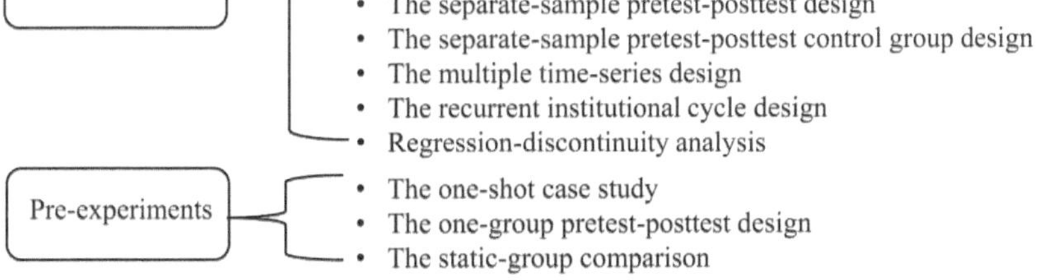


*Figure 10.2* True Experiments, Quasi-Experiments, and Pre-Experiments

counterbalanced experimental design depicted in Section 10.2.2). The electronic integration, for example, is the treatment administered by Venkatraman and Zaheer (1990) in their quasi-experiment research project – another case study chosen for this chapter.

- An outcome measure which assesses changes in the dependent variable after the treatment is administered. The measure should be explicit and quantifiable. Similar to treatment, multiple outcome measures are sometimes used in experimental research. In the same study by Venkatraman and Zaheer (1990), for instance, four performance measures are utilized: (1) total written premium, (2) commissions, (3) number of policies, and (4) number of new business policies.
- A comparison measure that assesses changes in the dependent variable of a study. Comparison measures are of two types:
    - Comparisons across groups – If there is more than one experimental group and if a different kind of treatment is given to each of them, then those groups may be compared to see whether the different treatments produced different outcomes. The research by Clariana and Lee (2001), one of the two case studies presented at the end of this chapter, compares outcomes obtained from five different groups. A true experiment can perform cross-group comparison likewise because it at least has one experimental group and one control group.
    - Comparisons across time – If the dependent variable is measured both before and after the treatment, then changes can be observed. All the experimental designs that perform pretests and posttests (e.g., the pretest-posttest control group design, the separate-sample pretest-posttest design) can make cross-time comparisons.
- Randomization, when it is feasible, constitutes another major component in experimental research. Randomization should be done whenever possible as it helps ensure the external validity of the experimental research.

In Section 10.2.1, the six steps in the true experimental design are outlined, which consists of selection, assignment, assessment (i.e., pretest), treatment, observation, and comparison (i.e., posttest) and evaluation. Not every experimental design listed in Figure 10.2 would go through all the six steps except the two true experiment designs: the pretest-posttest control group design and the Solomon four-group design. Generally speaking, taking fewer of the six steps or exercising less control in the experimental research process might jeopardize both the internal and external validity of experiment studies.

As discussed in Campbell and Stanley (1966, 5, 13–16), there are eight different kinds of extraneous variables, if not controlled in the experimental design, that might affect the internal validity of experimental research. Like in any research project, internal validity is the basic minimum without which an experiment is invalid and uninterpretable. For example, did the experimental treatment(s) in fact make a difference in this specific experimental instance? Factors jeopardizing the internal validity of various experimental designs include the following:

1) History – Events unrelated to an experiment occurred (e.g., a fire drill, a power outage) between the two measurements before and after the treatment was administered during an experiment may inadvertently affect the experiment's outcome. The longer the time lapse during an experiment, the higher the probability that events other than the treatment can influence the validity of experiment results.

2) Maturation – Biological, psychological, social, or political processes may lead to changes in the participants under study with the passage of time. The most straightforward example in explaining the maturation factor can be experiments with children as participants. If an experiment lasts a considerable amount of time, children participants will grow older and become more mature. This may affect the outcome of what is experimented. The hypothetical experiment study on evaluating four search engines can also serve as a less obvious example of maturation. That is, participants can become more familiar with or knowledgeable about search engines in general during the experiment when they evaluate one search engine after another. Thus, their evaluation of each search engine might be different because of the learning effect (i.e., more matured or knowing more about search engines as the experiment progresses) instead of the actual difference in the selected search engines.
3) Testing – The effects of taking one test may have on the scores of a subsequent test in the assessment and comparison steps of an experiment. Specifically, participants can learn from the pretest and then the changes in their posttest results may not be caused by the treatment.
4) Instrumentation – The measures in the data collection instrument used in one experiment should remain the same for repeated measurements under the same conditions. Such stability of measurement is also related to reliability, the other criterion for assessing research quality, because its absence can be a threat to the validity of all kinds of research (including experimental research). For example, student participants tend to overestimate hours spent on studying if the choices for time intervals used in a data collection instrument vary from one measurement (e.g., shorter intervals between 0.5 hour and 1 hour) to another (e.g., longer intervals between 0.5 hour and 2.5 hours). In that case, the experiment results obtained may not be caused by the treatment (e.g., positive feedback to participants who spent longer hours on study). Instead, it might be the different measurements in the data collection instrument that led to the longer time students perceived that they spent on study.
5) Statistical regression – Statistical regression (i.e., regression to the mean) can occur if participants in the experimental and control groups are knowingly selected on the basis of their extreme scores on the pretest for measuring dependent variables. For instance, in an experiment on using technology in classroom, if participants with higher scores on the pretest are assigned to the control group while participants with lower scores to the experimental group, the changes observed in the experimental group may not be a result of using technology in classroom. Rather, the obvious increase in posttest scores in the experimental group could be simply caused by statistical regression or by how the participants are assigned to the control and experimental groups.
6) Selection biases – This factor is similar to the afore-described statistical regression. The only distinction between the two is that statistical regression is done intentionally while selection biases might result from something beyond control or a mistake made unintentionally. Selection biases are likely to occur in experimental research in the social sciences because participants in such studies often cannot be randomly selected to minimize the biases in the participant selection process.
7) Experimental mortality – Experimental mortality refers to the loss of participants from the comparison groups and can happen in biomedical research (e.g., the AZT drug test for AIDS treatment; the COVID-19 vaccine trials). However, it is extremely rare to encounter experimental mortality in experimental research of the social sciences because research of this kind does not deal with life and death problems directly.

8) Selection-maturation interaction – This factor can affect those quasi-experimental designs with multiple groups such as the nonequivalent control group design (see Figure 10.2). Namely, participants in different groups of experimental research may change or mature at different rates due to selection biases.

In addition to the eight factors that might jeopardize the internal validity of experimental research, Campbell and Stanley (1966, 5–6, 16–22) also identified the following four factors that likely affect the external validity of experimental studies.

1) Reaction effect of testing – It refers to experimental group members' reaction to the testing as those in the control group do not receive any treatment and thus have no reaction. This reaction effect of testing is likely to affect the external validity of experimental research.
2) Interaction of selection biases and the treatment – Both the selection of participants and treatment can interact if not all the participants have an equal chance of being selected. This would eventually make the treatment function differently in the experiment and limit the generalizability of the experimental study.
3) Reactive effects of experimental arrangements – Experimental arrangements include the experiment setting, the way the treatment is given, or other similar influences. For example, when participants know that they will take part in an experiment and be observed, their behavior might change as a result. This kind of effects is known as the Hawthorne effect.
4) Multiple-treatment interference – When multiple treatments are applied in an experiment, the sensitizing effect can occur. This is similar to the hypothetical study of evaluating four search engines when discussing the counterbalanced experimental design (see Table 10.2). If the researcher does not purposely alter the order of each search engine tested in the experiment, participants would learn about the essentials of search engines via the ones presented to them first and hence would not evaluate all the four search engines equally.

The afore-listed four factors may sound elaborate, but the most crucial factor that determines if an experimental study has external validity is whether randomization is applied in participant selection and group assignment. The external validity of experimental research ultimately deals with the generalizability issue, namely, to what populations, settings, treatments, and measurements can the experiment findings be generalized? In other words, each aspect of the experiment must be considered before any attempt to generalize the findings of an experiment study.

Validity and reliability are the two gold standards for evaluating all research, including research using the experimental method. Section 2.8 is devoted to validity and reliability – the two essential criteria for evaluating research quality. Traditionally, publications on experiments always specifically cover factors affecting the internal and external validity of experimental research.

## 10.4 Dos and don'ts in experimental research

The dos and don'ts in experimental research are many partly due to the high degree of control exercised throughout an experiment. While some of the dos and don'ts are already embedded in the previous sections of this chapter, several points still need to be highlighted as follows.

Selecting an appropriate experimental design constitutes the first "do" in the long list of many after experiment is chosen as the research method. There are three major types of experiments: true experiments, quasi-experiments, and pre-experiments. As explained in Section 10.2, each type of experiments has its pros and cons. In addition, each of the three experiment categories can be further divided into individual experimental designs (see Figure 10.2). Thus, it is crucial for the researcher to choose one specific experimental design out of the multiple possibilities based on the research problem as well as the features of each individual experimental designs. As a rule of thumb, we would choose a true experimental design if all the requirements for an experiment (e.g., randomly selecting and assigning participants to the control and experimental groups) are satisfied. If a research problem we work on does not permit us to meet all the requirements for conducting experimental research, we can choose a quasi-experimental design for data collection. When a study methodologically can satisfy at most one of the requirements specified for experiments, a pre-experimental design is the only option left to use. As for choosing a specific experimental design within the broad experiment types, we in general rely on the characteristics of individual designs. For example, we would select the Solomon four-group design in the true experiment category if we intend to conduct a study to evaluate four products all at once. Never select an experimental design simply for the sake of familiarity and convenience.

Random selection and assignment of participants is a second major point to consider in true experimental research. This consideration is especially important when we anticipate to generalize our research findings. In some cases (e.g., testing a new medication), double-blind randomization is necessary in assigning participants to the experimental and control groups in order to eliminate any unintentional influence the researcher may have on the experiment process after knowing who is in which group. Although true experimental designs are rarely adopted by researchers in the social sciences, it is fundamental to select and assign participants randomly if such a design is chosen.

Third, experiments as one kind of research methods require the ability to systematically manipulate one or more treatments (i.e., independent variables) in the experiment process. Therefore, only those independent variables that can be manipulated may be considered treatments, while variables that cannot be experimentally manipulated (e.g., age, gender, race, ethnicity) should not serve as treatments. Because of this reason, we should never attempt to use such variables as treatments in experimental research.

Fourth, experiments involve a high degree of control over the variables of a research project. Researchers thus frequently favor laboratory environments in order to achieve proper control. The consequence of such decision, however, is that laboratory experiments are limited to what can be conducted in that environment. Researchers in the social sciences should pay particular attention to the experiment environment because the majority of problems social scientists aim to study using the experiment method cannot be done in a laboratory setting. Instead, field experiments are the only viable option.

Fifth, ethical issues tend to be at the forefront of experimental research. Those issues include physical and psychological impact on experiment participants although psychological or mental impact is the major concern for experimental research in the social sciences. While going through the IRB (Institutional Review Board) process (see Section 2.9) shall address some of the ethical issues in experimental research, caution still should be exercised to ensure that all ethical guidelines are faithfully followed. In experiments, participants might also feel excessive fatigue if the experiment tasks are too demanding. Likewise, participants might feel bored if one experiment session is too long. Another subtle ethical issue in experiments is that researchers should avoid introducing any bias (e.g., choosing participants who are already familiar with the treatment) to the experimental study.

In sum, we should strike a balance between the degree of control in the experiment and the degree of separation from the real world. Avoid emphasizing the controlling aspect of the experiment by sacrificing its reflection of the reality. At the same time, we should also keep in mind that "[a]n experiment almost always involves a caricature; one develops a simplified, artificial model of some natural process" (Batson 1977, 414).

## 10.5 Case studies of experimental research

There are two case studies for experimental research in this chapter. One of them uses a true experimental design (i.e., Clariana and Lee 2001), while the other (i.e., Venkatraman and Zaheer 1990) adopts a quasi-experimental design.

### *10.5.1 Case Study 1*

Clariana, Roy B., and Doris Lee. 2001. "The Effects of Recognition and Recall Study Tasks with Feedback in a Computer-Based Vocabulary Lesson". *Educational Technology Research and Development* 49 (3): 23–36. https://doi.org/10.1007/bf02504913.

#### *10.5.1.1 The overall research and associated concepts*

In education, students can learn and remember what has been taught by either recall or recognition. Recall is defined as bringing a fact, event, situation, and the like back into one's mind, whereas recognition refers to identifying a fact, event, situation, and the like from one's memory. In general, recognition tasks are easier than recall tasks for people to accomplish because the former usually provide hints and cues to facilitate recognition. Clariana and Lee (2001) designed a study that first asked students to learn the definitions of 35 common terms selected from a standard introductory textbook used in two introductory courses in the Instructional Systems program at a northeastern university in the United States. The computer-based (or programmed) vocabulary lesson consisted of either multiple-choice questions (i.e., recognition) or equivalent constructed-response questions (e.g., recall) with feedback.

The computer-based learning of the definitions for the 35 terms was assessed using a constructed-response posttest. The students in those two courses, 133 in total and randomly assigned to five groups, were then tested (i.e., posttested) on recalling or recognizing the definitions of the 35 terms by receiving one of the five experimental treatments (i.e., feedback) that are listed alphabetically below:

1) CR – Constructed-response questions with feedback
2) MTF – Multiple-choice questions with multiple-try feedback
3) MTF+OR – Multiple-choice questions with multiple-try feedback plus overt response (OR)
4) STF – Multiple-choice questions with single-try feedback
5) STF+OR – Multiple-choice questions with single-try feedback plus overt response (OR)

Specifically, each experimental treatment was provided via at least one of the following four different ways (Clariana and Lee 2001, 27):

1) AUC – Answer Until Correct response feedback which states try again until the student selects a correct response

2) Elaborative feedback which provides explanations, clues, hints, or guidance to the student
3) KCR – Knowledge of Correct Response feedback which provides the correct response to the student
4) KOR – Knowledge of Response feedback which states right or wrong to the student

Through this experimental research, Clariana and Lee (2001) attempted to examine the effects of recognition and recall study tasks with feedback on recall posttest performance.

The report on this experimental study is hard for readers to follow due to several reasons. First, Clariana and Lee (2001) did not explain the basic concepts (e.g., recall and recognition) involved in the study in its introduction. Yet, such concepts are fundamental in helping readers to understand this research. Second, many abbreviations (e.g., MTF) are freely used in the first several pages of the report without their corresponding full spellings and meanings until the Method section. This practice prevents readers from understanding the study, let alone appreciating it. Third, neither any research question nor hypothesis was clearly formulated throughout the research report even though the authors informally stated that "[i]t was anticipated that the constructed-response study tasks would be more instructionally effective than would the recognition study tasks" (Clariana and Lee 2001, 25). According to Section 2.3.2, a hypothesis is a tentative and educated answer to a given research question. We should always avoid stating a hypothesis without first clearly presenting a research question – the basis for making a hypothesis.

#### *10.5.1.2 The true experiment design*

Clariana and Lee (2001) adopted a posttest-only control group design with repeated measures for their experimental research. However, the authors did not offer any reason for choosing this true experiment design that Campbell and Stanley (1966) stated explicitly in their seminal book on experimental designs: "While the pretest is a concept deeply embedded in the thinking of research workers in education and psychology, it is not actually essential to true experimental designs. ... [I]n educational research, ... we must frequently experiment with methods for the initial introduction of entirely new subject matter, for which pretests in the ordinary sense are impossible, just as pretests on believed guilty or innocence would be inappropriate in a study of the effects of lawyers' briefs upon a jury" (Campbell and Stanley 1966, 25). The study by Clariana and Lee (2001) fits right in the case that students in the courses were introduced to the definitions of the 35 terms and it is unfeasible to pretest the students' knowledge on those definitions. However, randomization of the students participating in the study can suffice without the pretest. In other words, randomization can help achieve the purpose of pretesting and hence eliminate any initial biases between and among groups (i.e., familiarity with the definitions of the 35 terms) in the study (Campbell and Stanley 1966, 25).

The 133 participants in this investigation were graduate students, and everyone in the two introductory courses in the Instructional Systems program volunteered to take part in research. It is very typical for researchers to recruit participants from the courses they teach. The pros and cons of such a method for recruiting participants are beyond the scope of the current discussion. But it is appropriate to use students as participants in this study because students need to learn new terms and then are tested on what they have learned. All the participants were randomly assigned to one of the five experimental treatments below using a randomization computer program along with the number of participants in each group in parentheses.

1) STF – multiple-choice questions with single-try feedback (n = 23)
2) MTF – multiple-choice questions with multiple-try feedback (n = 26)
3) STF+OR – multiple-choice questions with single-try feedback plus the overt response of typing in the correct response after feedback (n = 29)
4) MTF+OR – multiple-choice questions with multiple-try feedback plus the overt response of typing in the correct response after feedback (n = 30)
5) CR – constructed-response questions with feedback (n = 25)

The authors did not specify which random sampling technique was adopted except explaining that "[t]he variations in sample sizes occurred as a factor of the random assignment by the computer program" (Clariana and Lee 2001, 27). However, we normally expect that researchers list the specific sampling technique (e.g., simple random sampling) used in their study as an assurance of the reliability of their study.

The 35 terms selected for this experimental study are listed in Table 1 of the research report and grouped in seven levels, with five items per level, based on average term difficulty which was determined as the proportion of individuals who defined that term correctly in the posttest. Term difficulty takes the value between 0 and 1 with 1 being the most difficult. Participants were posttested on recalling or recognizing the definition for each of the 35 terms that were given in random order to minimize any possible effect on this experiment's internal validity with regard to testing and instrumentation as depicted in Section 10.3.

The posttest-only control group design Clariana and Lee (2001) adopted in their experimental research can be illustrated as follows:

R $X_1$ $O_1$

R $X_2$ $O_2$

R $X_3$ $O_3$

R $X_4$ $O_4$

R $X_5$ $O_5$

In the afore-listed design, R stands for the randomized assignment of the 133 participants to each of the five groups. $X_1$–$X_5$ refer to the five experimental treatments each group received, while $O_1$–$O_5$ represent the observation Clariana and Lee made of each group in their experiment. Specifically, each participant was posttested on recalling or recognizing the definitions of the 35 terms in one 45-minute period, about 1.25 minutes for each question. Apparently, this posttest was rather intensive and it is unclear if the participants could perform as effectively in the beginning of the posttest as toward its end. Whenever we design an experimental study, we should be aware of the physical and mental capabilities of our participants because any activities during the experiment (e.g., intensive activities in a prolonged experiment time period) that exceed their limits physically and mentally would affect the internal validity of the experiment.

Clariana and Lee (2001) finally reported that the constructed-response study tasks were more instructionally effective than the recognition study tasks although the difference was not significant.

### *10.5.2 Case Study 2*

Venkatraman, N., and Akbar Zaheer. 1990. "Electronic Integration and Strategic Advantage: A Quasi-Experimental Study in the Insurance Industry." *Information Systems Research* 1 (4): 377–93. https://doi.org/10.1287/isre.1.4.377.

Venkatraman and Zaheer (1990) conducted this quasi-experimental study with the intention to find out if electronic integration had an impact on business insurance agents' performance. Electronic integration in this study is defined as "the integration of business processes of two or more independent organizations through the exploitation of the capabilities of computers and communication technologies" (Venkatraman and Zaheer 1990, 377–78). Insurance agents having access to a computerized system are considered "electronically interfaced". Insurance agents' performance is measured by four indicators: total written premium, commissions, number of policies, and number of new business policies (Venkatraman and Zaheer 1990, 387–88). The general assumption of this study is that the implementation of information technology such as electronic integration in business operations could provide potential for strategic advantage.

#### *10.5.2.1 Concepts and hypotheses of this study*

The property and casualty (P&C) market covers both personal and commercial lines. The study by Venkatraman and Zaheer (1990) includes only P&C insurance agents in the commercial lines because they "are less regulated and considerably more complex, and thus have only recently been subjected to computerized policy processing" (Venkatraman and Zaheer 1990, 381). Specifically, the authors "limit this study to the commercial lines for two reasons: (1) due to the significance of resources committed to computerized policy processing of complex commercial lines, it is an attractive setting for testing theoretical propositions; and (2) the recency of deployment of electronic interfacing allows a more systematic design for the collection of pre- and post-data described later" (Venkatraman and Zaheer 1990, 382). Such explanation is helpful for the reader to understand why the research participants – commercial P&C insurance agents – are chosen.

According to Venkatraman and Zaheer (1990), the study is designed to explore if electronic integration truly provides potential for strategic advantage or, simply put, increases insurance agents' performance. Instead of directly using total written premium, commissions, number of policies, and number of new business policies as performance measures or indicators, the authors introduced two additional concepts in between: effectiveness (i.e., total written premium and commissions) and efficiency (i.e., number of policies and number of new business policies) while attempting to operationalize strategic advantage or insurance agents' performance (Venkatraman and Zaheer 1990, 385). The extra layer of concepts may cause confusion to readers of this research report and should thus be avoided. It is not recommended at all to use fancy and additional terminology in research report writing without adding any practical value or clarity to the document.

The authors formulated hypotheses for their study but with no explicitly stated research questions. As indicated in Section 2.3.2, a hypothesis is a tentative and educated answer to a given research question. In other words, a hypothesis can be formulated only if the research question is already posed. When there is no research question, the foundation for formulating a hypothesis does not exist. This point holds true for all research projects without any exception, and we must keep that in mind whenever we formulate a hypothesis.

On the other hand, two more comments are due for the hypotheses Venkatraman and Zaheer (1990) presented in their report. First, a research hypothesis should always be formulated before its corresponding null hypothesis because a null hypothesis is simply a hypothesis that nullifies the corresponding research hypothesis. However, Venkatraman and Zaheer (1990) did exactly the opposite, presenting a null hypothesis before the two research (or alternative) hypotheses. Second, two research hypotheses were created, one as the "strong-form" and the other as the "weak-form" alternatives, for the study under examination. This violates the established practice of hypothesis formulation: one and only one hypothesis can be formulated for each research question in a study. Since Venkatraman and Zaheer (1990) did not explicitly specify any research question in their report, they were unable to create hypotheses in the way that they should have done for their study.

Up to this point, it becomes apparent that Venkatraman and Zaheer should have formulated four specific research questions, one for each of the four performance measures, in their research report: (1) total written premium, (2) commissions, (3) number of policies, and (4) number of new business policies. For instance, what impact does electronic integration in the P&C business insurance industry have on its agents' total written premium? The other three research questions can be posed accordingly. Because Venkatraman and Zaheer (1990) artificially introduced another layer of concepts (i.e., effectiveness and efficiency) for performance, they could not formulate research questions that are specific to each of the four performance measures actually utilized in the study. Yet, their study found that electronic integration in the P&C business insurance industry had a positive impact only on the number of new business policies (renamed "new business") while showing no positive effect on effectiveness (i.e., total written premium, commissions) and operating efficiency (i.e., number of policies). Even though the authors further broke down the concept of "efficiency" as "new business" and "operating efficiency" for their findings besides the creation of "strong-form" and "weak-form" hypotheses, they still could not handle the issues they unintentionally created when not following the established norms for hypothesis formulation (see Section 2.3.2). The above elaboration on this study's hypotheses demonstrates that it is obligatory for researchers to observe the norms for hypothesis formulation although this topic is not specifically or directly related to the quasi-experimental design this case study aims to discuss. Research basics and fundamentals are incorporated, discussed, or reviewed wherever appropriate throughout this book.

#### *10.5.2.2 The quasi-experimental design*

The quasi-experimental design developed for this study is fairly clean and follows the established practice. All the agents participated in this study worked in the P&C business insurance industry. About 70 agents who had access to or interfaced with the computerized system by one major insurance carrier (i.e., electronic integration) were recruited to be in the experimental group via an unspecified random sampling technique. Another group of about 70 agents who had no electronic integration at work were selected to form the control group based on three criteria (i.e., size, location, and geographic category) so that they matched with the agents in the experiment group. In this study, the size criterion is measured by insurance premium volume, whereas the location category includes metro, small city, suburban versus rural. The geographic criterion involves state or region. It is the selection of the control group without using a random sampling technique that defines the experimental design of this study as a quasi-experiment. The authors should have briefly mentioned this point in their research report.

The independent variable or stimulus for this quasi-experiment is the electronic integration used by the insurance agents, or interfaced agents in short, in the experimental group in the calendar year of 1987. Data on the four performance measures (i.e., total written premium, commissions, number of policies, and number of new business policies) were obtained from the internal records of the insurance carrier. For each interfaced agent, the performance data were collected for six months prior to the electronic integration (i.e., pretest) and for six months after (i.e., posttest) but excluding the month when the electronic integration was implemented. Strictly speaking, the data for this quasi-experimental study were not collected via the pretest and posttest we commonly do in experimental research with data collection techniques like interview and questionnaire because the data were already recorded and available from the internal records of the insurance carrier. This is a unique feature of this quasi-experimental study in terms of pretest and posttest.

The quasi-experimental design of this study is illustrated in Figure 1 on p. 387 in Venkatraman and Zaheer (1990) although that presentation is in fact a table, not a figure, because it contains texts and straight lines only without any graphics such as arrows and pointers. Furthermore, the notation for observation in experiment before and after the introduction of the treatment – electronic integration – is not the standard O as in most research reports of experimental studies. Another minor comment on that table is that the individual time values for "Performance Assessments at $t_i$ ..." should be listed as "i = 1, 2, 3, 4" rather than "i = 1, 4".

In addition, while Venkatraman and Zaheer (1990) did not specify which individual quasi-experiment design they used in their study out of the ten possible choices (see Figure 10.2), the equivalent time-samples design is their selected option according to their report. This quasi-experimental study was designed neatly despite its usage of unnecessary terms and improper hypothesis formulation.

**Chapter Summary**

- Experiment is a time-honored method for data collection by following a procedure to test or try out what is studied in either a laboratory or a real setting.
- All experiments can be broadly categorized as true experiments, quasi-experiments, or pre-experiments based on how many requirements each experiment can satisfy in its design. True experiments refer to those experiments that meet all the specified requirements. If an experiment does not satisfy all the requirements, it is a quasi-experiment. If an experiment at most meets only one of the stated requirements, it is a pre-experiment.
- A true experimental study usually involves six steps: (1) selection, (2) assignment, (3) assessment (i.e., pretest), (4) treatment, (5) observation, and (6) comparison (i.e., posttest) and evaluation.
- Eight factors jeopardizing the internal validity of experimental research are as follows: (1) history, (2) maturation, (3) testing, (4) instrumentation, (5) statistical regression, (6) selection biases, (7) experimental mortality, and (8) selection-maturation interaction. Four factors affecting the external validity of experimental research are as follows: (1) reaction effect of testing, (2) interaction of selection biases and the treatment, (3) reactive effects of experimental arrangements, and (4) multiple-treatment interference.

- Do be fully aware of the factors that might jeopardize the internal and external validity of experimental research while observing all other dos and don'ts in studies using the experiment method. Particular attention should also be paid to the experiment essentials presented in this chapter.

## Note

1 Figure 10.1 is adapted from Figure 2.5.1 in Barrett (2011).

## References

Babbie, Earl R. 2016. *The Practice of Social Research*. 14th ed. Boston: Cengage Learning.

Barrett, Justin L. 2011. "Experiment." In *The Routledge Handbook of Research Methods in the Study of Religion*, edited by Michael Stausberg and Steven Engler, 161–77. London, New York: Routledge.

Batson, C. Daniel. 1977. "Experimentation in Psychology of Religion: An Impossible Dream." *Journal for the Scientific Study of Religion* 16 (4): 413–18.

Brothers, Will. 2020. "A Timeline of COVID-19 Vaccine Development." Accessed March 5, 2022. https://www.biospace.com/article/a-timeline-of-covid-19-vaccine-development/.

Campbell, Donald T., and Julian C. Stanley. 1966. *Experimental and Quasi-Experimental Designs for Research*. Chicago: Rand McNally.

Clariana, Roy B., and Doris Lee. 2001. "The Effects of Recognition and Recall Study Tasks with Feedback in a Computer-Based Vocabulary Lesson". *Educational Technology Research and Development* 49 (3): 23–36. https://doi.org/10.1007/bf02504913.

Frankfort-Nachmias, Chava, David Nachmias, and Jack DeWaard. 2015. *Research Methods in the Social Sciences*. 8th ed. New York: Worth Publishers.

Haas, David F., and Donald H. Kraft. 1984. "Experimental and Quasi-Experimental Designs for Research in Information Science." *Information Processing and Management* 20 (1–2): 229–37. https://doi.org/10.1016/0306-4573(84)90053-0.

John Hopkins Coronavirus Research Center. 2022. "Vaccine Research & Development." Accessed August 20, 2023. https://coronavirus.jhu.edu/vaccines/timeline.

Price, Paul C., Rajiv Jhangiani, and I-Chant A. Chiang. 2015. "Quasi-Experimental Research." *Research Methods in Psychology*. 2nd ed. Washington: The Saylor Foundation. Accessed August 20, 2023. https://ecampusontario.pressbooks.pub/researchmethods/chapter/quasi-experimental-research/.

Ross, Steven M., and Cary R. Morrison. 2004. "Experimental Research Methods." In *Handbook of Research for Educational Communications and Technology*, edited by David H. Jonassen, 1021–43. Mahwah: Lawrence Erlbaum.

Tague-Sutcliffe, Jean. 1992. "The Pragmatics of Information Retrieval Experimentation, Revisited." *Information Processing & Management* 28 (4): 467–90. https://doi.org/10.1016/0306-4573(92)90005-K.

Venkatraman, N., and Akbar Zaheer. 1990. "Electronic Integration and Strategic Advantage: A Quasi-Experimental Study in the Insurance Industry." *Information Systems Research* 1 (4): 377–93. https://doi.org/10.1287/isre.1.4.377.

WolframAlpha. 2023. "World Population." Accessed August 20, 2023. https://www.wolframalpha.com/input?i=world+pupulation.

World Health Organization. 2021. "Vaccine Efficacy, Effectiveness and Protection." Accessed August 20, 2023. https://www.who.int/news-room/feature-stories/detail/vaccine-efficacy-effectiveness-and-protection.

# 11 Meta-Methods

## Ethnography/Field Study

**Chapter Overview Questions**

- What is ethnography or field study?
- What are the features of field research?
- What are the general steps in field research?
- What are the additional major considerations in field research?
- What are the major dos and don'ts in field research?

### 11.1 What is ethnography/field study?

Ethnos is a word originated from Greek, meaning nation. According to WordNet (2022) – an open-access lexical database of English – ethnos refers to people of the same race or nationality who share a distinctive culture. In the discipline of anthropology, ethnography initially focuses on providing scientific description of peoples and cultures about their activities, customs, habits, and characteristics. Peoples in this context could be members of a particular nation, community, or ethnic group although ethnographers previously committed to the study of traditional and predominantly non-Western societies. Nevertheless, present-day ethnography is no longer the sole province of anthropologists and expands its target subjects to communities of all kinds (Chambers 2000; Genzuk 2003). Schools, health care institutions, rural and urban developments, consumers and companies, online communities, and other human arenas all can be studied by ethnographic researchers. Ethnography aims to achieve a deep and holistic understanding of human activities, behaviors, and values within a broad sociocultural context (Zaharlick 1992).

Methodologically, ethnography implies some form of in-depth fieldwork invariably employing a combination of different data collection techniques such as participant observation, interviews, and content analysis. It is in this sense that some authors (e.g., Prasad 1997; Roberts 2009) argue that ethnography is a methodology instead of a single research method for data collection. That argument is meaningful although the present chapter title suggests that ethnography is a meta-method. As defined in Section 1.5, meta-methods are those research methods that involve multiple techniques for data collection in research. Ethnography and experiments are two examples of meta-methods.

Ethnographic researchers without exception adopt multiple or a variety of methods for data collection based on what they plan to investigate. Research methods commonly applied in ethnography include an amalgamation of the following: observation, interview,

DOI: 10.4324/9781003021711-14

focus group research diaries, and content analysis. Other data collection techniques (e.g., questionnaire, photo survey) are used less frequently in ethnography research. There are also certain procedures in conducting ethnography research, which will be introduced and described later in this chapter.

Jones and Smith (2017) indicate that ethnography is concerned with learning about people, as opposed to studying people, through fieldwork or immersion in native populations and cultures. The concept of "fieldwork" represents one key element or hallmark in ethnography (Schensul, Schensul, and LeCompte 1999, 70). Fieldwork is a form of inquiry in which the researcher is immersed personally in the ongoing social activities of some individuals or groups for the purposes of research (Wolcott 2005, 4). Researchers would definitely use "ethnography" for labeling fieldwork studies in anthropology while research in sociology and other disciplines that employs the fieldwork approach can also be called ethnographic studies. It is thus apparent that ethnography and fieldwork are synonymous. In order to avoid any confusion about these two terms and be more concise in expression, both ethnography and fieldwork will in general be referred to as field study or field research in the remainder of this book. If the subjects of a study are exclusively ethnical groups and the research is conducted from the perspective of anthropology, the research method involved will still be named ethnography.

Based on the explanation above, field study can be defined as a meta-method that is developed specifically for learning about or understanding the members of a community in their physical setting (i.e., field) through their actions, behaviors, exchanges, interactions, and other activities. Instead of limiting its subjects to a tribe or an ethnic group as in anthropology, the community in field study can be interpreted in the broadest sense. That is, a community can be a health care institution, a non-profit organization, a for-profit company, a group of consumers or social media users, or any other kind as long as the community members share a specific culture among themselves. Employees of one company, students of one school, patients receiving a same treatment are just a few examples of participants in a field study.

Speaking of participants, ethnographers usually name their subjects informants (people who provide information) or inhabitants (people who live in the community). As neither term seems appropriate in the vocabulary of most researchers doing field study today, all research subjects shall continue to be referred to as participants as done in the previous chapters of this book.

## 11.2 Features of field study

Compared to other research methods presented and discussed in this book, field study has the following distinctive features: use of multiple methods, immersion in the field and going native, research in context, evolving data collection plan, and research on a relatively small scale.

First, field research utilizes multiple techniques for data collection. That is why it is categorized and labeled as a meta-method in this book. Unlike questionnaire and most other research methods, field study cannot or should not be conducted with one single data collection technique (Athanases and Heath 1995). Among the possible method candidates, observation and interview constitute the two most frequently adopted methods for data collection. They are thus called essential ethnographic methods (Schensul, Schensul, and LeCompte 1999, xvii). Although both interview and observation are depicted as stand-alone methods in Chapters 4 and 5, respectively, further elaboration on each will be provided in Section 11.3.2.1 when we discuss how these two methods are normally applied in field research. Other methods that are also often used in individual field studies and covered in this book include content analysis (see Chapter 9), focus groups (see Section 8.1), and research diaries (see Section 8.3). In addition, methods other than those enumerated above (e.g., drawings, photo surveys) might get selected to form a methodology for conducting field research. In other words, researchers of field study select individual data collection techniques based on what they intend to investigate. Data collection techniques used in one field study hence could be very different from that in another field research project.

Second, field study researchers immerse in the field and go native. The field in field research refers to the place or setting where the target participants live, work, or stay. In traditional ethnography, researchers are expected to stay in the field for months or even years, while contemporary field studies may not require the researcher in the field for an extended period of time. Researchers instead can decide the time length of being in the field according to their research objectives. For instance, Athanases and Heath (1995) – one of the two case studies for this chapter – completed their major data collection in one school year. Bundgaard et al. (2011) – the other case study chosen for this chapter – in total performed observation for over 12 weeks. Regardless of the time length, being in the field is a must in any field study. Furthermore, it is the immersion in the field as well as going native that truly distinguishes field study from all other research methods. Bronislaw Malinowski (1984), one of the pioneers in field research, posited that the goal of field study "is to grasp the native's point of view, his relation to life, to realise his vision of his world" (25). Malinowski first published this seminal book in 1922 and the 1984 edition is just a reissue. For this reason, it is not surprising to note the extensive use of masculine pronouns in Malinowski (1984). Tedlock (2000), on the other hand, vividly illustrates ethnographers as "cross-dressers, outsiders wearing insiders' clothes while gradually acquiring the language and behaviours that go along with them" (455). All the above indicates that immersion in the field and going native are what researchers should do in any field study. We will elaborate on this feature of field study in Section 11.3, including how to go native and how to control the degree of immersion during research.

Third, data gathering for field study is done in everyday contexts of research participants instead of under artificial or experimental conditions created by the researcher. Unlike other research methods, field study can or should never be carried out outside the field where target participants live, work, or stay. For example, researchers must observe participants' behaviors, interaction, and activities onsite in the field if observation is one of the selected methods for data collection. Likewise, interviews are usually conducted in the field rather than in the researcher's office or another unfamiliar place. This feature of field study enables researchers to learn about their participants' behaviors, interaction, and activities in context. While all research is expected to be done in context, the word "context" is highlighted in all field research mainly because of its ready accessibility to the researcher. Take questionnaire research as an example, it is hard for the researcher to know the context when participants answer questions in a questionnaire. By contrast, researchers of field study can readily obtain contextual information in the field when gathering data with multiple methods such as observation and interview.

Fourth, the plan for data collection is evolving and needs to be revised or updated accordingly during the field research process. As Hammersley (2016, 2) points out, the approach to data collection in field study appears "unstructured" in the sense that it does not involve following through a detailed plan devised at the very beginning; nor are the categories used for recording and interpreting what participants say and do pre-prepared or fixed. Data collection in field research is by no means a one-time task. Instead, the process for gathering data is lengthy and might last years or even decades. The study by Athanases and Heath (1995) serves as a fitting example here because surveys and discussions plus periodical follow-up interviews were done two years after the initial data collection in order to gather some retrospective data. It is thus impossible and impractical to have a finalized plan for data collection before getting into the field. This, however, does not mean that the research is unsystematic and the researcher does not need to make preparation before and during the field study. Rather, this feature of field research creates a mechanism to ensure what to be collected is pertinent and current to the research in progress.

Fifth, the focus of field study is usually on a single community, group, organization, or other settings. In addition, field research commonly involves a small number of participants. A typical field study would not have participants in the hundreds and let alone thousands as, for instance, in some questionnaire research. This small-scale characteristic of field study is related to all the other features we have discussed so far (i.e., use of multiple methods, immersion in the field and going native, research in context, and evolving data collection plan). It is impractical to carry out field research on a large scale while attempting to meet all the basic requirements for using the meta-method of field study. This small-scale feature of field research is also reflected in the two case studies selected for the present chapter (see Section 11.5). Athanases and Heath (1995) conducted their study with two English classes from two different high schools, whereas Bundgaard et al. (2011) did observations in one endoscopic outpatient clinic and interviewed a total of 12 participants.

## 11.3 Using field study in research

Applying the meta-method of field study in research is by no means an easy and simple task. There are not only steps that researchers must follow but also several major points that researchers should consider throughout the research process. Both topics are to be deliberated below.

### *11.3.1 Steps in conducting field research*

Many scholars describe and discuss major steps in field research (e.g., Berg 2009, 190–225; Yin 2016, 116–36). The field study process in general comprises the following steps:

1) Choosing a field for study
2) Accessing the chosen field
3) Entering the field
4) Becoming invisible
5) Nurturing relationships in the field
6) Watching, listening, and learning
7) Disengaging and leaving the field

Variations in steps and focuses do exist in individual field study projects. However, some common practices have been established since the creation of the field study method to guide researchers in their investigations.

*11.3.1.1 Step (1) Choosing a field for study*

The first step in field research involves selecting a field for study. The field setting, once chosen, becomes the location or source for data collection. There are a great variety of possible field settings that researchers can choose from. Yin (2016, 118–19) categorizes them into four types as follows:

- Small groups of people sharing a common bond (e.g., classes, gangs, work groups)
- Residents in the same small geographic area (e.g., a tribe, a village, a town)
- Institutional scenes (e.g., health clinics, offices, schools)
- Unrelated groups of people who may share some common condition but do not interact with each other as a group (e.g., home bound people, patients with a similar medical problem)

The four kinds of field settings, however, are not mutually exclusive because, for instance, the participants can be students and teachers (the first field type) while they are also from schools (the third field type) as in Athanases and Heath (1995). The selection of a field setting is mainly determined by the research problem of a study and accessibility of the setting. As Bundgaard et al. (2011), for example, aim to find out what nurses should know and how they should get to know about their patients in endoscopic outpatient clinics in order to provide personalized services to them, it is only natural for them to choose an endoscopic outpatient clinic as their field setting. The accessibility of the field setting is another major factor that researchers must consider in the selection process because field study, unlike other research methods, features researchers' involvement or participation in the field. Although such involvement or immersion enables researchers to learn about the chosen field setting and target participants first-hand, it inevitably intrudes into or sometimes interrupts the regular activities at the site. Hence, the host site in general might be reluctant in opening itself to outsider researchers, which means that researchers must first assess the accessibility of a candidate field before seeking it. At this point, assessing the accessibility of a field is closely related to accessing the chosen field – Step (2) Accessing the Chosen Field.

*11.3.1.2 Step (2) Accessing the chosen field*

Before conducting any field study, researchers must gain entry to the chosen field. This step determines if the researcher can access the selected field and normally requires much time and efforts to complete this step. For instance, Chatman (1984, 428) illustrated vividly the "lengthy and involved" process she experienced in gaining entry to her chosen research site, a CETA (Comprehensive Employment and Training Act) program, after numerous phone calls and requests for interviewing with potential participants. The CETA program "was comprised of several job sites of employment, a variety of CETA activities, and a diverse array of employees: supervisors, secretaries, job counselors, CETA workers; and clients" (Chatman 1984, 427). Among the problems Chatman (1984) encountered in this step of her field study were delays and refusals of permission to be admitted to certain settings as well as the difficulty involved in obtaining the approval of assistance from various agency representatives. According to Chatman (1984), strategies that seemed most conductive to gaining entry in her research project included the following: (1) requesting interviews with directors in order to meet informally with them for explaining the nature of the research, and (2) being honest with them in response to such issues as confidentiality. Still, that process of gaining entry lasted several months and finally Chatman (1984) changed to a CETA program in another city before getting in. The process of accessing the field in Athanases and Heath (1995), one case study for this chapter, seems less complicated but by no means easier. Detailed description of that process of gaining entry is to be provided in Section 11.5.1.

In field studies, researchers often need to consider how to gain access to less or least accessible fields (e.g., drug traffickers, inmates, patients, victims of domestic violence). In those cases, we need to employ such specific techniques as using connections, building rapport (i.e., a two-way relationship), and offering incentives to gain entry to the selected field setting. Moreover, Hammersley and Atkinson (1989) have the following to say about gaining access: "In many ways, gaining access is a thoroughly practical issue ... it involves drawing on the interpersonal resources and strategies that we all tend to develop in dealing with everyday life. But the process of achieving access is not merely a practical matter. Not only does its achievement depend upon theoretical understanding, often disguised as 'native wit', but the discovery of obstacles to access, and perhaps of effective means to overcoming them, themselves provide insights into the social organisation of the setting" (54).

In addition, Yin (2016) points out that gaining access to a field setting is a process instead of an event (122). This suggests that the threat of losing access always exists throughout any field study if researchers take their field access for granted and exhibit any behavior that may appear as wearing out their welcome. Access can either be lost completely or limited as hosts may deliberately exclude researchers from certain activities. Neither scenario is within control of field researchers. Therefore, researchers must adhere to the dos and don'ts in field study to be discussed in Section 11.4 even after they have gained entry to their selected field initially.

#### *11.3.1.3 Step (3) Entering the field*

After gaining entry to the chosen field, researchers would enter the field to begin their study. There are two kinds of situations in this step. One is that the researcher is a natural member of the field (e.g., Karin Bundgaard in Bundgaard et al. 2011), and thus entering the field is not a step of field research that requires specific attention. The researcher does not need to make extra efforts in order to enter the field. In such a case, no introduction is necessary nor are any other formalities. The researcher in this scenario is ready to move onto Step (4) Becoming Invisible in the field study process. The other type of entering the field refers to the scenario where the researcher is an outsider and new to the selected field. Therefore, the researcher must first get familiar with the field before getting accepted and starting research. Between these two situations, the first type is apparently more desirable time-wise. But the possibility of having it in general is slim because few researchers would choose a field setting where they are already natural members. If that is the case, researchers would have to consider how they can approach the field without any biases and prejudice that might affect the objectivity of the field study.

Adler and Adler (1987, Ch. 1[1], 4) posit that field researchers should enter their settings, announce their intentions, and begin to interact with the people they encounter. Their initial activities are usually unfocused and merely involve hanging out. They should let members of the setting gradually get to know them as they hang around in this manner, so that those in the field realize the nonthreatening nature of their research intentions, overall interest, and sincerity. Researchers also seek to achieve their participants' confidence by presenting themselves as conscientious and likable individuals. Another piece of advice for researchers entering the field is to take part in what the setting members do while being decent and trustworthy. Harvey (2011, 231) recommends that honesty about being there as a researcher and humility about one's ignorance in the setting might be necessary in this step of field study and subsequently become accepted in the setting. This step is closely related to rapport building as well, a topic to be discussed in Step (5) Nurturing Relationships in the Field of the field study process: Nurturing Relationships in the Field.

In this step, field researchers are also expected to choose a role in the setting and set a stage for their study. There are a variety of roles researchers can adopt in the field (Adler and Adler 1987, Ch. 1, 4). The most widely known roles include the complete observer, observer-as-participant, participant-as-observer, and complete participant, ranging from the least to the most involved in the field and activities taken place there. Further discussion on the researcher's role in the field is to be conducted in Section 11.3.2 when specific issues in field research are considered. Closely associated with the role of researchers is whether researchers should be overt (i.e., openly admitting their role as researchers) or covert (i.e., not admitting the research dimensions of their participation) in the field. The consensus is that field researchers should take the overt role in the setting, with participants fully aware of their identities (Adler and Adler 1987, Ch. 1, 5) unless there is too high a risk for researchers to be overt (e.g., a field study of gangsters).

#### *11.3.1.4 Step (4) Becoming invisible*

In Step (3) Entering the Field, field researchers are recommended to be overt about the research dimension of their presence in the field in order to cope with ethical concerns in field study. However, researchers are also expected to become invisible after entering the field. Although members in the setting tend not to regard researchers as outsiders if they are in the field often and for a while, researchers still need to make certain that their presence and activities in the field are not intrusive, and they do not draw constant attention to themselves and to what they do. Rather, researchers should keep a low key and focus primarily on learning about what is going on in the field and on collecting data for their study.

Moreover, where to position oneself in the field is an issue that field researchers must consider as done in observation (see Section 5.3.3). Glazier (1985), for example, finds that the most ideal position for making observation in his study is behind the participant and slightly to the right. That describes only the case of observing one person. Observing a group of people or a field setting certainly requires a different strategy. As noted by many scholars (e.g., Berg 2009; Harvey 2011; Yin 2016), participant observation is one major data collection technique in the toolbox of field study. It is hence crucial for field researchers to keep a delicate balance between performing observation and becoming invisible. Participant observation, one of the essential ethnographic methods (Schensul, Schensul, and LeCompte 1999, xvii), is to be further elucidated in Section 11.3.2.

#### *11.3.1.5 Step (5) Nurturing field relationships*

This step in the field study process overlaps to some extent with the afore-described Step (3) Entering the Field because researchers should develop and nurture field relations with participants in both steps at the chosen setting in order to conduct field research as planned. Some tips are already offered in Step (3) Entering the Field regarding how to get accepted, gain trust, and the like. Still, nurturing field relationships requires more commitments and efforts from researchers in the setting.

First, researchers should portray their authentic self as that is the best approach to building field relationships. One feature of field study depicted in Section 11.2 is researchers' immersion in the setting, which means that researchers would stay and take part in the field for a prolonged period of time (e.g., several months or years). If researchers do not present their true self but "perform" a specific role, that cannot last long because they themselves would either become tired of pretending to be somebody else or participants in the field would be able to figure out that the researchers do not function and behave naturally. Research is not performance or game playing. Only when researchers wear their own authentic clothing, can they nurture relationship with participants in the field. In addition, Yin (2016) suggests that "[p]resenting yourself as doing a research study (not just 'doing some research') can be attractive because the identity connotes a serious and professional commitment rather than a casual curiosity into other people's lives" (125).

Second, researchers should display the right personal demeanor. According to Yin (2016), "[t]ypical demeanors include being respectful but not condescending, friendly but not ingratiating, and attentive to others but not pandering to them" (126). Similarly, Tedlock (2000) states that field researchers "are expected to maintain a polite distance from those studied and to cultivate rapport, not friendship; compassion, not sympathy; respect, not belief; understanding, not identification; admiration, not love" (457). Without doubt, maintaining such desirable demeanors is easier said than done. Researchers, however, should strive to show proper personal demeanors as they would in any public setting when trying to develop and nurture field relations. One other point to be kept in mind is that researchers are "observed and judged" too by those in the field as they themselves attempt to observe and interact with participants in the setting. In other words, researchers need to watch what they say and do in the field, including facial expressions and other body languages. Personal demeanor may not sound that important in steps of selecting and entering a field. Nevertheless, it is crucial in cultivating field relationships.

Third, researchers should properly handle the issue of reciprocity in field study. On one hand, researchers are expected to do favors to show their appreciation of help and services received from participants in the field. Reciprocity in field research does not have to be tangible awards. For example, Chatman (1984) often listened to her participants' concerns and then offered some advice or suggestions. Athanases and Heath (1995) in their ethnographic study of two high school English classes assisted the students with their essay writing and portfolio reviewing and helped the teachers by providing resources and teaching some lessons in areas of specialty. On the other hand, field researchers need to learn how to handle requests of doing favors from participants that cross the line (e.g., babysitting, small loans). In those cases, field researchers must decide for themselves if it is acceptable or comfortable for them to entertain such requests. Yin (2016, 127) suggests a few rules of thumb for dealing with them: (1) Avoid large favors at all. (2) Do small favors only on a rare occasion. (3) Maintain a principled enough position that no one would even think to ask for doing a favor bordering on anything illegal or harmful to others physically or psychologically. Nurturing good field relationships should never be interpreted as meeting all requests of doing favors from participants in the field without any principle.

Fourth, researchers should decide how to cope with unexpected questions and events while in the field. As field researchers, we are prepared to ask questions to gather data for our projects via interviews and other data collection techniques. Meanwhile participants too could ask us questions about our personal background, views, or any other possible topics. Researchers therefore need to be prepared for and think ahead of time about how to answer such questions by drawing a clear line between what can be answered and what cannot. Unexpected events in field study range from being invited to entertainment activities (e.g., football games, movies) to being asked to do illegal activities (e.g., using drugs) or undesirable activities (e.g., dating). Field researchers should use their best judgment in dealing with various unexpected events. For instance, going to a football game with participants should be alright if that is an opportunity for the researcher to foster their relationships. Researchers, however, should be vigilant and cautious if they are asked of doing anything undesirable or illegal (Yin 2016, 127–28). Unexpected events can also be various disasters and catastrophes occurred in the duration of a field study. The COVID-19 pandemic serves as a fitting example in this regard. Since early 2020, a lot of changes have taken place in all sectors of the society, including our daily lives. The plan for any field study undertaking during the COVID-19 pandemic must be altered accordingly to accommodate changes as a result of the pandemic. Because such events are unexpected, it is hard for researchers to anticipate. Nevertheless, adhering to the principles for doing field study should be the guiding light for researchers while paying attention to nurturing field relationships.

The purpose of establishing and maintaining pleasant relationships with participants in the field is to enable researchers to carry out their study as planned. This kind of relationships makes participants willing to trust researchers enough to answer questions and cooperate in other aspects of the study (Harvey 2011, 231).

#### *11.3.1.6 Step (6) Watching, listening, and learning*

All the steps we have discussed so far can be considered preparations in field research. It is in this step that researchers start to collect data by watching, listening, and learning. These three activities are by no means random or casual activities. Instead, researchers are expected to use all their five senses and adopt proper data collection techniques such as observation and interview for research data gathering. As indicated earlier, field study is a meta-method and always relies on multiple techniques to complete the task of data collection. More will be presented on the selection and usage of major data collection techniques for field research in Section 11.3.2.1.

Watching, listening, and learning of course should not be taken literally in field study. Rather, researchers need to follow the guidelines for applying each selected data collection technique while watching, listening, and learning in the field. Besides, Harvey (2011, 232) brings up the concept of epoché and suggests that field researchers should practice it while building rapport. Epoché is typically referred to as "suspension of judgment" and "withholding of assent" by being aware of research presuppositions and one's own ideas, thoughts, and beliefs as well as being vigilant against their untested influence on field research. Simply put, avoid bringing any assumptions and prior views into the process of field study, especially in the data collection stage. Like in all other kinds of studies, researchers also need to stay value free as opposed to value-laden in the data collection process and throughout the entire field study.

#### *11.3.1.7 Step (7) Disengaging and leaving the field*

After the data collection process is over, researchers would disengage and leave the field to begin their data analysis and report writing. This step, however, should not be taken lightly or carelessly. Researchers should not appear as if they would never return to the field or contact the participants there. Instead, Yin (2016) suggests that "[s]ome relationships are best left to linger rather than ending with a firmly established good-bye" (128).

It is also desirable to stay in touch with the participants in the setting because researchers may need to gather additional data, verify some points and the like after leaving the field. For example, Athanases and Heath (1995, 269), a case study for this chapter, gathered retrospective data two years after completing their major data collection from students at one school via survey and discussion. They also conducted periodic, informal follow-up interviews with both teachers in the third year of their study. Therefore, it is wise for researchers to keep the door open so that they can return to the field when necessary.

Even though the above seven steps of field research are presented sequentially, the actual process may not be as clearly cut. Researchers instead would proceed with their projects according to their research problem and other related factors before deciding when to move to the next step in the research process or if one of the previous steps ought to be revisited. Schensul, Schensul, and LeCompte (1999, 70–71) write that the process of field research is complicated and hence researchers must learn the following several things simultaneously: (1) How to function in a new setting where the language, rules of behavior, norms, beliefs, social relationships, dietary patterns, and other aspects of life in a community may be very different; (2) how to locate and build relationships with the people who have access to important information and other resources relevant to the study; (3) how to collect and record information in the setting in a way that is both unobtrusive and efficient; and (4) how to sort out what information is new, interesting, and likely to be useful for study purposes, especially in situations where everything the researcher encounters is new and different. As seen, the process of field research is exciting yet challenging. Field researchers thus need to be prepared as well as focused in each step of the research process.

### 11.3.2 Some considerations in conducting field research

In addition to the general steps in conducting field studies, there are several considerations researchers should take into when undertaking such investigations. Major considerations in field research include selection of multiple techniques for data collection, membership roles in the field, and note-taking in the field.

#### 11.3.2.1 Selection of multiple techniques for data collection

One outstanding feature of any field study is the need for choosing more than one research method for data collection. As noted earlier, observation and interview are the two essential techniques for data gathering, while other data collection techniques (e.g., content analysis, focus groups, questionnaire) may also be adopted based on the research objective of a field study. Although some field studies only report the use of the observation and interview techniques, it does not exclude the necessity of applying other research methods whenever appropriate. For instance, Athanases and Heath (1995) employed surveys as well as content analysis of student writing samples and other kinds of data (e.g., English language standards) listed in Table 3 of their report besides observation and interview.

Observation as one stand-alone data collection technique is extensively covered in Chapter 5. When observation is utilized in field study, participant observation (see Section 5.2.1) is the default type that field researchers would employ. Some scholars even equate participant observation with field study (e.g., Fox 1998; Harvey 2011), which does not seem appropriate because participant observation is only one of the principal data collection techniques used in field research and other research methods (e.g., interview, content analysis) also get utilized together with participant observation. As discussed in Section 5.2.1 and subsequent sections of Chapter 5, participant observation is usually further categorized into passive observation, moderate observation, active observation, and complete observation depending on the degree of the researcher's involvement in the field. These four categories of observation correspond closely to the three kinds of membership role (i.e., peripheral membership, active membership, and complete membership) to be discussed in Section 11.3.2.2.

In presenting participant observation as a data collection technique, Shah (2017) discusses four core aspects of participation observation: long-term engagement, understanding a group of people and their social processes, studying all aspects of social life, and befriending strangers. There are some risks and pitfalls in participant observation as well (Shah 2017; Wildemuth 2017, 222–23). According to Shah (2017), the first risk in participant observation is that some researchers may be unable to leave their shoreline due to the fear of the water they need to dive into. The second risk is that they may sink into the sea of other people's lives, unable to cope with the challenges of trying to swim back to the shore. The third risk is that there seems potentially no end to the inquiry. The fourth risk is that there is a tension between participation and observation. All these risks highlight the need to use other research methods for data collection to mitigate the limitations of participant observation. It is also likely that participants in the field may deliberately lead the researcher to incorrect conclusions. Researchers therefore should conduct longer and more active participation in the field to tell the truth from false conclusions. The setting or events in the setting may sometimes interfere with data collection (e.g., unrelated activities taking place near the observation site). In such cases, researchers should place themselves at the right places and times or perform multiple observations in order to minimize such interference. Moreover, the role of a researcher in participant observation is less clearly cut as that in interview or other research methods because of the participant role. This poses a major challenge in any participant observation: balancing between the role of a researcher and that of a participant. Researchers hence are advised to be open and clear about their researcher status when conducting participant observation in field study (see Section 11.3.1).

When participant observation is applied in field research, specific attention should be paid to the setting and other chosen data collection techniques. The field setting plays a major part in determining how participant observation is done because the same observation technique needs to be applied differently in different settings. For example, participant observation done in a clinic (e.g., Bundgaard et al. 2011) is obviously not the same as the one conducted in a high school (e.g., Athanases and Heath 1995). More is to be expounded about participant observation in field research in Section 11.5 when the two case studies selected for this chapter are examined.

Among the three types of interviews depicted in Chapter 4, semi-structured interviews (e.g., Bundgaard et al. 2011) and unstructured interviews (e.g., Athanases and Heath 1995) are commonly selected for gathering data in field research because of their flexibility in application. Sections 4.3.2 and 4.3.3 are, respectively, devoted to semi-structured interviews and unstructured interviews. Therefore, how to conduct these two kinds of interviews need not be repeated here as all the interviewing principles and practice discussed in Chapter 4 remain applicable in field research. Besides interviews, conversations, casual talks and other naturally occurring communications and exchanges (e.g., discussion) can also be made in field research for gathering additional data that otherwise cannot be captured.

As briefly mentioned earlier, additional techniques are often adopted in conjunction with observation and interview to collect data in field research. Of the research methods presented in this book, content analysis (see Chapter 9), focus groups (see Section 8.1), and research diaries (see Section 8.3) are also regularly chosen and applied in field study. Questionnaire (see Chapter 3) may occasionally be selected as a data collection technique as well (e.g., Athanases and Heath 1995). Among all the data collection techniques often used in field research, content analysis comes as the third top-chosen method and is only after the observation and interview techniques in terms of usage frequency. That is, observation and interview together form the methodological foundation in field study while content analysis helps researchers to gather additional data from a variety of written sources.

Sources for data collection in field research consist of many types and can be grouped into two broad categories: traditional and contemporary. Table 11.1 lists both traditional and contemporary data sources that are frequently utilized in field research.

As seen in Table 11.1, traditional data sources for content analysis in field research are generally writings or publications by individuals, while contemporary data sources are typically either generated by the public on a platform (e.g., a social media site) or created for internal use within an institution (e.g., memos, reports). Genzuk (2003) labeled the latter as site documents. Regardless of which type, they all can serve as data sources for content analysis and complement other data collection techniques such as observation and interview in field research.

The discussion above on selecting multiple data collection techniques for field research demonstrates the necessity as well as the importance of choosing the most appropriate individual research methods and developing a methodology for a particular field study. Unlike in other types of research projects (e.g., questionnaire research) where a single data collection technique may adequately meet the research need, multiple research methods should be selected in field study to create a methodology. For this very reason, some scholars (e.g., Prasad 1997) specifically emphasize that ethnography or field study is a methodology rather than a method because field study can only be done by using multiple, carefully-selected individual research methods (e.g., observation, interview, content analysis). "Carefully-selected" in this context is the keyword regarding research methods for field study. In no case can a field study be conducted with only participant observation or unstructured interview alone. After the methodology for one field study is developed, make certain to observe the established principles and practice of each individual research methods throughout the research process.

*Table 11.1* Traditional and Contemporary Data Sources for Field Study

| | |
|---|---|
| Traditional data sources | Biographies or life history<br>Memoirs<br>Narrative ethnographies (i.e., hybrid of life history and memoirs)<br>Field diaries, chronicles, travelogues<br>Novels, novellas, short stories, poems, and plays |
| Contemporary data sources | Community/organization documents (e.g., brochures, budgets, newsletters, reports, school records, teaching materials)<br>Emails, memos, meeting minutes, …<br>Social media posts, texts, …<br>Transaction logs of a system used by a target community<br>Websites, internet archives, … |

*11.3.2.2 Membership roles in field research*

Adler and Adler (1987) closely explore the membership roles in field research from the perspective of sociology and categorize them into three kinds: peripheral membership, active membership, and complete membership. These membership roles seem identical to the four kinds of participant observation discussed in Section 5.2.1 (i.e., passive observation, moderate observation, active observation, and complete observation) with the exception that peripheral membership encompasses two types of participant observation, namely, passive and moderate observations. This can be interpreted as that field researchers assuming the peripheral membership role perform passive and moderate observations. Every membership role consequently has an impact on each investigator's research in the field. Table 11.2 outlines and compares the three kinds of membership roles in four aspects: involvement, relationship with participants, role demand, and reciprocity.

The peripheral membership role in field study requires the least from researchers who assume such a role (i.e., peripheral-member-researchers or PMRs). They can be a researcher most of the time and get involved as a member in the field when needed. Because of their mostly detached involvement, PMRs' relationships with setting members are loose. Similarly, the demand for them to play the membership role in the field is minimum. When taking the peripheral membership role in the setting, PMRs contribute little to the field and are hence not expected to do reciprocity in the field.

In contrast to PMRs, researchers choosing the active membership role (i.e., active-member-researchers or AMRs) take part in the core activities in the setting. AMRs also do not have to rely exclusively on the bonds of friendship they establish with members to generate the rapport and confidence necessary to gather data. AMRs' relationships with setting members may therefore be activity oriented and businesslike. The role demands placed on AMRs are greater than those generally encountered by PMRs. They are supposed to behave like full-fledged members in many aspects in the field. All the above features determine that AMRs should reciprocate the support and cooperation they receive in the setting.

The complete membership role entails the greatest commitment on the part of the researcher. In this context, the researcher functions as the setting member and the research simultaneously in the field. Rather than experiencing mere participatory involvement, complete-member-researchers (CMRs) immerse themselves fully in the setting as "natives". CMRs are urged to embrace the native experience and let their dual member-and-researcher roles help each other by giving them equal balance and drawing on both. "Going native" (i.e., starting to behave or live like others in the field) might bring such problems as overinvolvement and overrapport (Adler and Adler 1987, Ch. 1, 7), which, however, is beyond the scope of this current chapter. CMRs are committed to the field so that they can have the opportunity to acquire "understanding in use" rather than

*Table 11.2* Comparison of Membership Roles in Field Research

| *Feature Dimension* | *Membership Role* | | |
|---|---|---|---|
| | *Peripheral* | *Active* | *Complete* |
| Involvement | Mostly detached | Involved | Immersed |
| Relationship with participants | Loose | Activity-based | Committed |
| Role demand | Minimum | Moderate | High |
| Reciprocity | Not expected | Expected | Diminished expectation |

"reconstituted understanding" (Adler and Adler 1987, Ch. 4, 11). The demand for CMRs is thus high when they function as natives in the field. Because CMRs acquire the complete membership role, the expectation for them to reciprocate via their outside or academic role is considerably diminished.

Among the three kinds of membership roles, it appears hard to distinguish CMRs from other members or participants in the setting. However, a close look at CMRs can reveal several differences between CMRs and other participants in the field: The researcher is both a participant and an investigator; the researcher is explicitly aware of what is happening in the setting, while other participants routinely employ "selective inattention"; the researcher tends to define the setting more broadly so that they can gather all the data needed for the project; the researcher is an insider as well as an outsider; the researcher takes notes and reflects on what is gathered, whereas other setting members hold no such responsibilities. These differences of the complete membership role undeniably highlight the challenges CMRs must meet in their field research.

Although the ultimate goal for field researchers to assume different membership roles in the field is the same – to gather data for their investigation, each of the three membership roles in field research does exhibit their distinctions in several dimensions (see Table 11.2). On the other hand, Adler and Adler (1987, Ch. 5, 3) point out at the end of their book that the membership role methodology may not be for everyone or fit for all field research because it requires more efforts, both academical and emotional, from the investigator. Therefore, researchers must decide for themselves whether they plan to assume a membership role in field study.

#### *11.3.2.3 Note-taking in the field*

What has been gathered by watching, listening, and learning using various data collection techniques in the field must be recorded, and such recorded accounts are known as field notes. Gorman and Clayton (2005) define field notes as "the recorded account of what a researcher observes, hears, experiences, and thinks when collecting data. They describe people, places, activities, interactions, and dialog. They also include the researcher's own ideas, reflections, and observations on what is occurring" (184–85).

In the past, note-taking was the primary vehicle for recording the collected data. Nowadays, various recording devices are readily available for capturing data collected in, for example, observations and interviews. Video recording is no longer a far-fetched technology that was only accessible to a few. Audio recording, by comparison, becomes even more accessible thanks to the ubiquitous smartphone technology. Despite the advancement of data recording technology, note-taking by hand remains an essential technique that researchers must master and utilize in order to capture data in the field mainly because the audio and video recording technology cannot be utilized in every minute when the researcher is in the setting. Likewise, researchers are unable to comment and reflect on what is recorded as they would in written field notes. In addition, many participants may feel uncomfortable when they are informed that what they say and do would be recorded even if they are assured of anonymity and privacy in any presentation of the research.

Note-taking is often done in the use of observation, interview, focus groups and other data collection techniques. As stated by Yin (2016): "The virtual simultaneity of doing fieldwork and taking notes, hour after hour and day after day, means that the notes and other recordkeeping will occur in a multitasking mode" (164). Ideally, field notes should be taken at the time when, for example, observation is performed although that may not be feasible all the time. If so, we should do note-taking as soon as possible or at the end of a day while the data are still fresh in our mind. In no occasion shall we delay note-taking till a later time because such postponement would affect the accuracy and details of what needs to be recorded. Only what gets recorded in field notes and other means can be considered data gathered. Otherwise, they may be lost forever.

The enterprise of note-taking is easier said than done. There are three language principles that need to be followed in field research: (1) The language identification principle means that field researchers need to "identify the language used for each fieldnote entry" to make sure that the field note taken "reflects the same differences in language usages as the actual field situation" (Spradley 1980, 66). The principle reaffirms the importance of using the language of the target participants in the setting for note-taking rather than using the language that researchers are familiar with. In fact, the language identification principle is not limited to note-taking. Researchers instead are supposed to use the field language throughout the entire research process. (2) The verbatim principle denotes that field notes should be kept word-by-word about what is said, what occurs, and what is done in the setting. Avoid taking notes by summarizing and rephrasing what is heard, watched, and observed. Spradley (1980, 67–68) further indicates that "both native terms and observer terms will find their way into the fieldnotes". But the native terms must be recorded verbatim so that what is recorded does not change the meaning of those native terms. (3) The concrete principle implies that field notes should be taken using concrete language. Note-taking in concrete language might be difficult because researchers have been trained in education to condense, summarize, abbreviate, and generalize when writing. However, taking field notes and writing scholarly articles are two different activities. In the former case, field researchers are expected to "expand, fill out, enlarge, and give as much specific detail as possible" (Spradley 1980, 68). If time does not permit writing in concrete language in the setting, we can jot down keywords or key phrases at that time and then expand and fill in details later when possible.

Yin (2016, 169–78) also offers many practical suggestions for taking field notes. For example, researchers need to decide whether they are most comfortable taking notes on standard-sized notepads, in a bound notebook, or with any other kinds of paper products. A laptop or tablet computer might be viable for taking notes. But keep in mind that it might be hard to view a computer screen when working outdoors in sunlight or daylight. Another suggestion for note-taking is to make a habit of writing the date and time of the note as well as briefly identifying the people and scene in the note. Additional note-taking tips include to write on one side of a page, leave wide margins, or write only on the left column so that the blank space can be used for making comments or for recording related comments made by any other participants in the setting. In addition to echoing the three language principles Spradley (1980, 66–69) proposes for taking field notes, Yin (2016, 170–74) strongly recommends the use of abbreviations and acronyms as well as quick drawings and sketches in note-taking to illustrate, for example, a field setting or seating positions of attendees at a meeting. Since smartphones are widely available nowadays, it is quite feasible to take a photo of a scene (with permission if there are participants) and later fill in specific descriptions about it. Researchers should always expand and review the field notes they take as soon as they can because such notes tend to be sketchy and contain a lot of "holes" that must be filled before being analyzed for addressing the research problem under investigation. To help reduce the amount of field notes to be taken, researchers are encouraged to obtain facts or duplicate materials on site as well.

Because field research projects typically last months or years, Yin (2016, 170–72) recommends that field notes should be "transcribed" so that researchers can add comments and reflections at the end of the day and in relation to field notes already taken for other days. In addition, researchers may want to enter their hand-written field notes into computers while expanding and enhancing them. Such a step might take more of the researcher's time but in the long run could be a time-saver as certain software is often used for analyzing text data like field notes. A brief coverage of software for analyzing qualitative data can be found in Section 13.2.4.

Virtually every field researcher uses field notes to record their observations, insights, and analysis (Peñaloza and Cayla 2006). However, what gets recorded in field notes is basically a pale reflection of reality (Gorman and Clayton 2005, 184) no matter how hard the researcher tries in note-taking. Still, field notes can facilitate the reconstruction of a believable account of what has been observed and experienced in a particular setting. Gorman and Clayton (2005, 189–90) provide several pieces of advice on taking field notes. For example, organize field notes into easily manipulated chunks based on a specific idea, event, and so on to assist data analysis later. Allocate plenty of uninterrupted time to record and refine the field notes. Do not worry about memory lapses when working on field notes but do put down what you can remember. Memory tends to return eventually with the right context and recall efforts in the process of reviewing and enhancing field notes.

In addition, researcher-kept diaries, depicted and discussed in Section 8.3.3.2, are often used in field research to complement field notes. As they are already presented in Section 8.3 together with participant-kept diaries, they are not to be repeated here.

## 11.4 Dos and don'ts in field study

In the previous three sections of this chapter, quite a few dos and don'ts in field research are already considered. There are, however, still several points that need to be emphasized and discussed as dos and don'ts in field study.

First, do develop a plan carefully before conducting any field research. Unlike other research methods, field study generally involves the selection and application of multiple data collection techniques as well as the commitment to a field setting for months, one year or even a longer time period. If a field study requires team work to take advantage of each member's expertise and skills, choose and train team members carefully before going to the field and manage the team sensibly once the field study commences. Occasionally, one field study might be conducted at multiple sites to meet its site requirement (Jones and Smith 2017). All such tasks require meticulous planning before starting a field study. Never undertake a field research project without appropriate planning and preparation.

Second, do gather a variety of data from different sources and perspectives. Different sources and perspectives in this context carry two layers of meaning. One is that various data sources (e.g., the setting, participants, site documents) should be utilized for data collection. The other layer of meaning is that data should be collected from different perspectives even when the same technique is chosen for gathering data. For instance, researchers should purposely invite participants of different backgrounds and opinions when interviewing them to ensure that the resulting interview data are gathered from different perspectives. Therefore, avoid using merely observation or interview for data collection. Instead, develop a methodology for any field study by integrating various data gathering techniques based on the research problem and other factors. For example, data gathered from observation and interview are further supplemented by focus groups and research diaries. Some field studies report the use of as many as half-dozen research methods for data collection. Similarly, never limit data collection to a homogeneous group of participants so that data can be gathered from different perspectives in return. When gathering a variety of data from different sources and perspectives, researchers can cross-validate and triangulate what they collect. This would enhance the quality of field research.

Third, strike a balance between being a researcher and being a member simultaneously in the field. As a field researcher, you are expected to keep your research objective in mind and behave accordingly throughout the research process. As a member of the field setting, you are required to establish a rapport with other members in the field and gain their trust before being able to carry out your study. On one hand, you cannot always keep a distance with other members and appear as if you are aloof and remote while in the field because behavior of this kind would hinder you from gathering any valuable data. On the other hand, avoid going to the oppositive direction by totally mingling with other members in the setting and forgetting about your role as the researcher. Without doubt, the dual-role of researcher-and-member is subtle and hard to play in field research. Yet keeping a balance between the two roles is a requirement rather than an option for any field researcher.

Fourth, be aware of and sensitive to the different steps of field research (Genzuk 2003, 5–6). Each step of field research has its own focus, and researchers must act and function accordingly. Avoid moving ahead of or behind the planned step in the field research process. For instance, building trust and rapport should be done in Step (3) Entering the Field. It would become awkward if the researcher attempts to do so in Step (4) Becoming Invisible. Because field research often lasts for months or even years, unexpected events or changes might occur during the research process. If that is the case, researchers should adjust their current study plan accordingly. Never attempt to stick to the original plan regardless because such a practice would only affect the field research negatively.

Fifth, do set and maintain ethical standards of conduct in field research. In Section 2.9, we discuss and elaborate on the topic of research ethics. While what is covered in that section remains relevant and applicable in field research, additional attention should be paid to issues unique to field study. Those issues include interaction with participants in the setting and decision on data exclusion (Yin 2016, 41–44). In field study, researchers without exception are expected to be in the field and interact with participants there. Such interactions can both enable researchers to conduct research if done properly and cause problems if ethical standards of conduct are not observed. It is hence imperative for researchers to know where the red line is and avoid crossing it all the time. In field research, investigators usually gather all they can while in the setting and decide later what to be included or excluded in the study report. At this point, researchers must make their decisions purely based on their research questions rather than anything else. Other acts (e.g., data falsification and plagiarism) that violate research integrity also have no place in field study, as in any research.

## 11.5 Case studies of field study

Two case studies are chosen to be examined in this chapter, one being an ethnography study on teaching and learning English by Athanases and Heath (1995) and the other reporting a fieldwork study conducted in an outpatient clinic by Bundgaard et al. (2011).

#### *11.5.1 Case Study 1*

Athanases. Steven Z., and Shirley Brice Heath. 1995. "Ethnography in the Study of the Teaching and Learning of English." *Research in the Teaching of English* 29 (3): 263–87. Accessed June 19, 2023. https://www.jstor.org/stable/40171429.

Athanases and Heath (1995) is not strictly a report of ethnographic research. Rather, it illustrates how to use ethnography to study teaching and learning English in high schools based on Athanases' dissertation research. In this sense, Athanases and Heath (1995) serves two purposes: (1) presenting the basics of ethnography research, and (2) reporting Athanases' dissertation research using the ethnography method. Although not explicitly specified in the report, Shirley Brice Heath, the second author of this case study, is likely Steven Athanases' dissertation advisor.

Because of the unique purposes Athanases and Heath (1995) set forth, the publication itself does not present the study's research questions and some other common components as the standard practice would. Instead, the authors only refer to some aspects of Athanases' dissertation research when they need to explain or discuss certain steps and elements in ethnography. As a result, we are unable to treat Athanases and Heath (1995) as a research report and apply the same criteria in examining it as we have done with the other case studies in this book. On the other hand, we understand that an ethnographic study is usually reported in several dozen or over 100 pages, and Athanases' dissertation should not be an exception. Such a length naturally cannot fit into the limited space a journal commonly allocates to an article. Since Athanases and Heath (1995) mainly focus on the ethnography method utilized in Athanases' dissertation rather than on its subject topic – teaching and learning English in high schools – it becomes possible for the authors to publish parts of the dissertation as a journal article. This article is then chosen as a case study for field research.

Athanases and Heath (1995) organize their article by the general steps of field study explained in Section 11.3.1. The authors first describe how the two research sites (i.e., two high schools) were selected (Athanases and Heath 1995, 269–71). Table 2 in Athanases and Heath (1995, 270) lists the criteria for choosing high schools, classes, and teachers although the selection process started with English teacher selections because once the teachers were chosen, the schools and classes were subsequently determined. The specific selection process is as follows: After sending letters of introduction, Athanases phone-interviewed 30 urban teachers who had initially been nominated by local organizations and sources relevant to high school teachers of English. Then, 10 of the 30 teachers were observed and interviewed face-to-face, followed by three weeks of preliminary information gathering and decision-making, and another three weeks of observation in four of the ten classes. Finally, two of the four classes were selected for the study (Athanases and Heath 1995, 270). The selection process indicates that Step (1) Choosing a Field for Study of field research – Choosing a Field for Study – involves multiple sub-steps rather than a straightforward procedure. Systematic planning and elaborative work are required to complete the task of site selection successfully.

Like Step (1) Choosing a Field for Study in field research, negotiating entry and building rapport, corresponding to Step (2) Accessing the Field and Step (3) Entering the Field, are also lengthy and require a great deal of advanced planning (Athanases and Heath 1995, 271). The authors detailed the gaining entry process (including how to obtain Institutional Review Board approval). However, they did not report specifically on how to build rapport in the two high schools although they did describe how to reciprocate the support received in the field. For example, Athanases helped students with essay writing, portfolio reviews, and editing of manuscripts accepted for publication in a school literary magazine. Assistance to teachers at the sites includes providing resources, teaching some lessons in areas of specialty, and team planning lessons. Reciprocity in field study cannot be overlooked as this is one of the important tactics to gain trust from and build rapport with participants in the setting.

Athanases and Heath (1995) particularly mentioned in their article the selection of informants or participants. This issue should be considered, especially in Step (5) Nurturing Relationships in the Field, in any field study. Field researchers are generally advised to choose participants who can provide a range of perspectives of the target group, community, or other settings. With that in mind, Athanases selected participants from various levels and locations within the schools, including administrators, department chairs, teachers, counselors, students, and parents. This shows that the kinds of participants selected by Athanases went beyond teachers and students even though his dissertation focused on the teaching and learning of English. School administrators, department chairs, other teachers, counselors, and student parents were also invited to take part in the study. This practice reflects the need for field researchers to take into consideration the different contexts and perspectives related to the setting when choosing participants.

The participants in Athanases' dissertation research were primarily the teachers and students from the two English classes of two different high schools in California. However, the authors did not specify that two different high schools were involved in the study until the 12th page of the article (i.e., p. 274). School administrators, department chairs, other teachers, counselors, parents of the students in the two English classes also involved in the study but to a much lesser degree. Athanases and Heath (1995) did not provide the number of participants in the study partially because it is not a typical research report as clarified above. The bulk of the study's data collection was performed over the course of one academic year. Additional data were gathered from students at one school, including survey and discussion responses regarding memories from their English classes and reactions to these classes from the distance of two years. Do note that asking students to recall what was done in their English classes two years ago seems to be a questionable practice because few people can accurately recall what happened two days ago, let alone two years ago. Moreover, the authors did not explain why only students in one of the two English classes were contacted for gathering such "retrospective data" (Athanases and Heath 1995, 269). Athanases also did periodic informal follow-up interviews with each of the teachers in the third year (Athanases and Heath 1995, 269) to gather more data for the study. Again, the accuracy of such data could not be guaranteed while the authors did not specify what "periodic" means in their report perhaps for the same reason of this nonstandard report of ethnography research.

As expected of any ethnographic study, Athanases used multiple data collection techniques in his dissertation research. Observations, interviews, surveys, and content analysis of site documents are among the data collection techniques chosen for the study although Athanases and Heath (1995) primarily reported about interviews, briefly mentioned observations, and merely touched on the rest of the data collection methods.

Interviews were conducted with teachers, students, parents, and other school personnel. At least three formal interviews were done with both English teachers (Athanases and Heath 1995, 273). The first interviews were conducted with each teacher separately and discourse-based (i.e., unstructured), in which each teacher discussed her purposes for literature study and discussion, the roles she hoped she and her students would play in discussions during the year, and what she had hoped students would take with them from the course. The initial interviews were taped, from which excerpts were taken as prompts for reflection in subsequent interviews. In the second formal interviews, each of the two teachers considered a set of questions while listening to the interview excerpts and signaled when she wanted to comment. In this way, the teacher's own words served as stimulus for reflection, as she considered her goals for the year and her work in achieving them. It would be helpful for the reader if the "set of questions" used in the second interviews was included in the article as an appendix. The final formal interview was carried out with both teachers together. The teachers reviewed a display of photographs from their own two classes and viewed short classroom video clips so that each teacher could see her own students on tape and get a feel for the characteristics of the other teacher's students. While the three formal interviews were meticulously designed and each was built on the previous interview plus certain prompts (e.g., photographs and video clips), the authors did not indicate when the three interviews were done during the time of one academic year. It would be nice if such a time line were given.

Teacher and student interviews generally occurred during lunch hours and teacher conference periods, and school and community data were gathered during free periods and after school. Even though the researcher is on site for conducting a field study, choosing an appropriate time for interviews and other in-person data collection is still central to minimizing the interruptions and inconvenience to participants. In this regard, Athanases did well and professionally.

Observations in this study were done during visits to the sites that generally happened at least twice weekly over the course of a school year, with the observation field notes recorded during each visit and supported with other data. This demonstrates that Athanases took field notes in a timely fashion, which is recommended for all field researchers. According to Athanases and Heath (1995), participant observations in this study were obtrusive and mostly passive (see Table 5.1) because it seemed that Athanases did not take part, for example, in class discussions.

Surveys were administered to the students in both classes along with one teacher survey. But the authors provided little information about survey contents and distribution approaches (e.g., self-administered). The same holds true regarding how data were collected from the site documents using content analysis. Table 3 in Athanases and Heath (1995, 275) enumerates all the data collection techniques used as well as the site documents for content analysis (e.g., district guidelines for 10th grade English, school newsletters) even though Table 3 was titled data on context. This table showcases the variety of techniques and sources used for data collection in ethnographic research. This represents what we should all strive to do when it comes to data collection in field research: adopting multiple techniques and sources for data gathering. In addition, Table 4 (Athanases and Heath 1995, 276) provides a sample data log from one high school site for one day, which gives us a glimpse of the kinds of data the researcher collected on one day. This table lists the raw data for that day, which include audiotape and videotape of full class discussion. It is, however, unclear why Athanases recorded both the audio and video parts of the class discussion. When the discussion was videotaped, why was the audio recording still needed because videotaping should also have recorded the audio parts of the class discussion?

Despite the lack of some details in reporting this ethnographic study, Athanases and Heath (1995) can serve as a decent example for researchers who intend to carry out an ethnography study.

### *11.5.2 Case Study 2*

Bundgaard, Karin, Karl Brian Nielsen, Charlotte Delmar, and Erik Elgaard Sørensen. 2011. "What to Know and How to Get to Know? A Fieldwork Study Outlining the Understanding of Knowing the Patient in Facilities for Short-term Stay." *Journal of Advanced Nursing* 68 (10): 2280–88. https://doi.org/10.1111/j.1365-2648.2011.05921.x.

Bundgaard et al. (2011) conducted a fieldwork study using participant observation and semi-structured interviews at a high-technology endoscopic outpatient clinic in Denmark. But the authors did not explain what high-technology is and why it was relevant to this study. The purpose of this study was to help nurses to know their patients before their scheduled procedures so that they can tailor nursing to the individual patients' needs and expectations despite the limited time and space available in this outpatient clinic.

Unlike Athanases and Heath (1995), Bundgaard et al. (2011) was presented as a full report of their fieldwork study. However, the authors completely skipped reporting on the first step of field study – Step (1) Choosing a Field for Study – and did not cover all the steps of field research in their presentation. Based on the authors' affiliation information as well as the part on Author Contributions at the end of the research report, Karin Bundgaard, the first author, was a Ph.D. student supervised by the three coauthors. She was fully responsible for this study while the three coauthors "made critical revisions to the article for important intellectual content" (Bundgaard et al. 2011, 2287). The field setting must be one of Karin Bundgaard's affiliations, which was easily accessible to her as a researcher. Having one's own group, community, or institution as a field setting for research is not common and with its pros and cons (see Step (3) Entering the Field in Section 11.3.1). While it is easy for the researcher to gain entry to the site in such a case, the researcher must avoid bringing any biases and prejudices to the field study that might affect the objectivity of the research.

Observations and interviews were chosen as data collection techniques for this field study (Bundgaard et al. 2011, 2282). Field observations were performed over 12 weeks of time and lasted approximately four hours each day in a time period from 2008 to 2010. No explanation was given about why the 12 weeks, about three months, of observations were done in approximately three years of time. Nor any specifics were offered regarding the distribution of the 12 weeks between 2008 and 2010. Could it be that observations were done one month each year or any other options? During the observation periods, the researcher stayed in the endoscopic clinic, trailed a nurse in her work, talked to patients in the resting and waiting area, and talked to nurses, physicians, and others in the clinic. But the report did not specify how many nurses in total Karin Bundgaard trailed. One nurse each day, each week, or any other option? That information should have been provided in the study report. The patients stayed at the clinic between approximately 20 minutes and 2 hours for having their gastroscopy. The participant observations were obtrusive with moderate (see Table 5.1) as the researcher did talk with patients, physicians, and others in the clinic but without providing any nursing services. It is commendable that the researcher also interacted with physicians and others besides the target patients and nurses. Field notes were made for each observed patient course in the endoscopic clinic and for each participant report although again the study did not mention when the field notes were taken.

Interviews were conducted with eight patients and four nurses. Patients scheduled to have gastroscopy at pre-fixed times on randomly selected days were asked if they would participate in an interview. The researcher contacted ten patients, nine accepted and one declined. One of the nine never came for his gastroscopy or the interview, which yielded eight interviews with the patients. Patient interviews were all carried out prior to gastroscopy. Several comments are due on the patient interviews. (1) It is good that the researcher randomly selected patients for interview. Because interviews usually are done with a small number of participants, random selection of them can help enhance their representativeness and subsequently the external validity of research findings. (2) Karin Bundgaard was at the clinic for data collection for over 12 weeks, each containing five work days. This implies that the researcher had a total of 60 days at the clinic but she only interviewed eight patients, averaging one patient interview every 1.5 weeks. Did the researcher ever think about interviewing more patients for gathering additional data? (3) Why were all the interviews done before the patients had gastroscopy rather than after the procedure? Bundgaard et al. (2011) reported that patients were anxious before having gastroscopy. For instance, the interview with Patient 6 had to be curtailed because of the obvious distress it caused. This should be a signal that the timing of patient interviews ought to be changed. Performing patient interviewing after the procedure when patients are calmer might enable them to provide more meaningful data for the research. Unlike the nurses involved in the study, patients were at the receiving end of nursing services. The patients would be able to say more about individualized services the nurses provided after the procedure than before. As can be seen, the timing of interview is always important in any research using this data collection technique. Researchers are recommended to select a time that is suitable for the researcher as well as for the participant. In addition, there is some inconsistency in reporting the patient interviewees' age. On page 2282 of Bundgaard et al. (2011), the authors wrote: "Interviewees were both men and women, over the age of 18 and fluent in Danish. ... The patients' age distribution ranged from 25–91 years". Why did the authors not use ages 25–91 directly and try to be more accurate in reporting when they know that over the age of 18 is not the same as ages 25–91?

Bundgaard interviewed four nurses who were working in the endoscopic clinic at the time of her field research. Participation was based on voluntariness and their nursing experience ranged from 4 to 21 years. The interviews were conducted after the researcher had trailed the individual nurse for one day. As indicated above, the researcher was in the clinic for 60 days to gather data for her research. Yet she only interviewed four nurses and did so right after trailing each of them for one day. Two questions arise in this context: (1) Why did the researcher only interview four nurses instead of a number more than four? Were there only four nurses at the clinic? (2) Why did the researcher conduct the interview right after merely trailing the nurse for one day during the three months of data collection? Since no answer was given to either of the questions in the report, the researcher seemed to have interviewed too few nurses if there were more than four nurses at the clinic and conducted each nurse interview too early in time. The researcher would have more relevant questions to ask if interviewing them after shadowing each for at least several days in that it appears unlikely that the researcher only had time to trail each nurse for one day.

In addition, no mention was made in the report about how long each of the patient or nurse interview lasted. That piece of information, however, should normally be reported to show the substantiality of the interviews. Bundgaard et al. (2011) stated that "[p]articipant observations and interviews were interrelated. Thus, the interviews helped focus participant observation which, in turn, informed the subsequent semi-structured interview guide" (2282). These statements themselves are appropriate. But when so much details were absent from the report about the observations and interviews, they sound more like empty talk. Moreover, Bundgaard et al. (2011) referred to "the semi-structured interview guide" in their statement above. Yet the interview guide was not ever mentioned again in the remainder of the study report, let alone providing it as an appendix according to the normal practice in reporting research using the interview technique. Hence, it is impossible for us to know what interview questions were asked of the interviewees. Likewise, it is hard to tell whether the interviews were indeed semi-structured.

One of the themes for this case study is to get to know patients' previous medical experience in a short amount of waiting time at the clinic before their scheduled gastroscopy. Such information, however, should be available from patients' medical records if they are returning patients and shall enable nurses to quickly learn about each patient. But Bundgaard et al. (2011) wrote as if the nurses did not consult their patients' medical records before receiving the patient and had to learn everything about each patient when interacting with them in person at the clinic. Perhaps this is a casual omission in the report. As discussed in previous sections of this chapter, data collection techniques other than observation and interview are commonly utilized in field research. This meta-method feature of field study nevertheless is missing from Bundgaard et al. (2011).

In sum, Bundgaard et al. (2011) applied observation and interview in their field research without adopting any other techniques (e.g., content analysis of patient records) for data collection. There are also too many specifics left to be desired for in the report regarding how observations and interviews were conducted in their investigation. Concise reporting is highly recommended in the scholarly world but a report with many omissions is not.

**Chapter Summary**

- Ethnography or field study can be called collectively as field study or field research. It represents a meta-method that is developed specifically for learning about or understanding the members of a community in their physical setting (i.e., field) through their actions, behaviors, exchanges, interactions, and other activities.
- Major features of field research include the use of multiple methods (i.e., meta-method), immersion in the field and going native, research in context, evolving data collection plan, and research on a relatively small scale.
- There are generally seven steps in field research: (1) choosing a field for study; (2) accessing the chosen field; (3) entering the field; (4) becoming invisible; (5) nurturing relationships in the field; (6) watching, listening, and learning; (7) disengaging and leaving the field.
- The following three main points require special considerations from field researchers: (1) selection of multiple techniques for data collection; (2) membership roles in field research; (3) note-taking in the field.
- The key dos and don'ts in field research comprise the following: Carefully select multiple data collection techniques to create an appropriate methodology for field study according to its research objectives and other related factors. Follow the principles and practice of each individual selected techniques when conducting field research. Also pay attention to all the dos and don'ts presented throughout this chapter.

## Note

1 Adler and Adler (1987) is available to the present writer only in electronic book format as individual chapters and without consecutive numbering. Thus, the chapter number must be added in order to identify a specific page number in the book.

## References

Adler, Patricia A., and Peter Adler. 1987. *Membership Roles in Field Research*. Thousand Oaks: Sage. https://dx.doi.org/10.4135/9781412984973.

Athanases, Steven Z., and Shirley Brice Heath. 1995. "Ethnography in the Study of the Teaching and Learning of English." *Research in the Teaching of English* 29 (3): 263–87. Accessed June 19, 2023. https://www.jstor.org/stable/40171429.

Berg, Bruce L. 2009. *Qualitative Research Methods for the Social Sciences*. 7th ed. Boston: Allyn and Bacon.

Bundgaard, Karin, Karl Brian Nielsen, Charlotte Delmar, and Erik Elgaard Sørensen. 2011. "What to Know and How to Get to Know? A Fieldwork Study Outlining the Understanding of Knowing the Patient in Facilities for Short-Term Stay." *Journal of Advanced Nursing* 68 (10): 2280–88. https://doi.org/10.1111/j.1365-2648.2011.05921.x.

Chambers, Erve. 2000. "Applied Ethnography." In *Handbook of Qualitative Research*, edited by Norman K. Denzin and Yvonna S. Lincoln, 851–69. 2nd ed. Thousand Oaks: Sage.

Chatman, Elfreda A. 1984. "Field Research: Methodological Themes." *Library & Information Science Research* 6 (4): 425–38.

Fox, Nick. 1998. *How to Use Observations in a Research Project*. Sheffield: Trent Focus Group. Accessed August 22, 2023. http://www.webcitation.org/5ZndSmDde.

Genzuk, Michael. 2003. *A Synthesis of Ethnographic Research*. Occasional Papers Series. (Eds.). Los Angeles: Center for Multilingual, Multicultural Research, Rossier School of Education, University of Southern California. Los Angeles.

Glazier, Jack. 1985. "Structured Observation: How It Works." *College & Research Libraries News* 46 (3):105–8. https://doi.org/10.5860/crln.46.3.105.

Gorman, Gary E., and Peter Clayton. 2005. *Qualitative Research for the Information Professional: A Practical Handbook*. 2nd ed. New York: Neal-Schuman.

Hammersley, Martyn. 2016. *Reading Ethnographic Research: A Critical Guide*. 2nd ed. New York: Routledge. https://doi.org/10.4324/9781315538457.

Hammersley, Martyn, and Paul Atkinson. 1989. *Ethnography: Principles in Practice*. London: Routledge.

Harvey, Graham. 2011. "Field Research: Participant Observation." In *The Routledge Handbook of Research Methods in the Study of Religion*, edited by Michael Stausberg and Steven Engler, 217–44. London, New York: Routledge.

Jones, Janice, and Joanna Smith. 2017. "Ethnography: Challenges and Opportunities." *Evidence-Based Nursing* 20 (4): 98–100. https://doi.org/10.1136/eb-2017-102786.

Malinowski, Bronislaw. 1984. *Argonauts of the Western Pacific: An Account of Native Enterprise and Adventure in the Archipelagoes of Melanesian New Guinea*. Prospect Heights: Waveland Press.

Peñaloza, Lisa, and Julien Cayla. 2006. "Writing Pictures/Taking Fieldnotes: Towards a More Visual and Material Ethnographic Consumer Research." In *Handbook of Qualitative Research Methods in Marketing*, edited by Russel W. Belk, 279–90. Northampton: Edward Elgar.

Prasad, Pushkala. 1997. "Systems of Meaning: Ethnography as a Methodology for the Study of Information Technologies." In *Information Systems and Qualitative Research*, edited by Allen S. Lee, Jonathon. Liebenau and Janice I. DeGross, 101–18. The International Federation for Information Processing Book Series. Boston: Springer.

Roberts, Taniya. 2009. "Understanding Ethnography." *British Journal of Midwifery* 17 (5): 291–94.

Schensul, Stephen L., Jean J. Schensul, and Margaret D. LeCompte. 1999. *Essential Ethnographic Methods: Observations, Interviews, and Questionnaires*. Ethnographer's Toolkit. Book 2. Thousand Oaks: Altamira Press.

Shah, Alpa. 2017. "Ethnography? Participant Observation, A Potentially Revolutionary Praxis. *Hau: Journal of Ethnography Theory* 7 (1): 45–59. https://doi.org/10.14318/hau7.1.008.

Spradley, James P. 1980. *Participant Observation*. New York: Holt, Rinehart and Winston.

Tedlock, Barbara. 2000. "Ethnography and Ethnographic Representation." In *Handbook of Qualitative Research*, edited by Norman K. Denzin and Yvonna S. Lincoln, 455–86. 2nd ed. Thousand Oaks: Sage.

Wildemuth, Barbara M., ed. 2017. *Applications of Social Research Methods to Questions in Information and Library Science*. 2nd ed. Denver: Libraries Unlimited.

Wolcott, Harry. 2005. *The Art of Fieldwork*. Walnut Creek: Altamira Press.

WordNet. 2022. "Ethnos." Accessed August 22, 2023. http://wordnetweb.princeton.edu/perl/webwn.

Yin, Robert K. 2016. *Qualitative Research from Start to Finish*. 2nd ed. New York: The Guilford Press.

Zaharlick, Amy. 1992. "Ethnography in Anthropology and Its Value for Education." *Theory into Practice* 31 (2): 116–25. https://doi.org/10.1080/00405849209543532.

# 12 Contingent Methods

As briefly mentioned in Section 1.5, contingent methods refer to individual research methods that have to be used together with another method in order to be functional. In other words, a contingent research method cannot be applied alone and its usage has to rely on the employment of another research method. For instance, the think-aloud protocol (TAP) method needs to involve the experiment method (e.g., performing a task, doing a test) in its implementation. No think-aloud can be performed if participants were not asked to take part in an experiment. Likewise, the Delphi method depends on the use of questionnaires for data-gathering purposes. Questionnaires must be developed before any Delphi study can be conducted. This chapter is devoted to present two contingent methods, TAPs and Delphi study, in two separate sections to help readers better understand each of them.

## 12.1 Think-aloud protocols

**Section Overview Questions**

- What are think-aloud protocols?
- What are the three different types of the think-aloud method?
- What are the chief features of think-aloud protocols as a research method?
- What are the general steps in conducting think-aloud research?
- What are the major dos and don'ts in think-aloud research?

### *12.1.1 What are think-aloud protocols?*

TAPs are a research method for collecting data about participants' cognitive and thinking processes by asking them to verbalize (i.e., thinking aloud) whatever crosses their mind as they perform a specified task (e.g., conducting a search, solving a problem). What the participants do, look at, feel, and think during the process can and should all be verbalized. Data collected via this method are called verbal reports (Ericsson and Simon 1980). TAP data should be gathered individually to ensure that each participant has the quiet environment to think aloud and the TAP process is not interfered by anything else.

The term "protocol" in TAP refers to a procedure for carrying out a scientific experiment as think-aloud must be done by performing a task in an experimental setting. It is in

DOI: 10.4324/9781003021711-15

this sense that TAP is considered a contingent method. Researchers may abbreviate think-aloud protocols simply as think-alouds.

The theoretical framework for TAPs is provided mainly in the works by Ericsson and Simon (1980, 1993), two noted figures in cognitive psychology. According to Ericsson and Simon (1993, 11), human cognition is information processing and information is kept in different memory stores with varying access and storage capabilities. Short-term memory (STM) is characterized by easy access and severely limited storage capacity, whereas long-term memory (LTM) is exemplified by more difficult access and larger storage capacity. Only information present in STM and being heeded by the participant can be directly accessed and reported in the TAP process.

Ever since the creation of the TAP method in cognitive psychology, it has been frequently utilized to explore participants' cognitive and thinking processes in domains and areas other than cognitive psychology. Those areas include clinical reasoning, information seeking, learning processes, translation research, and usability studies.

In addition to being a research method contingent on experiments, TAP is sometimes used with interviews (e.g., Fonteyn, Kuipers, and Grobe 1993), eye movements tracking (e.g., Cho et al. 2019), or transaction log analysis (e.g., Griffith, Hartley, and Willson 2002). Interviews and transaction log analysis are respectively covered in Chapter 4 and Section 8.2, while eye movements tracking is a data collection technique that is done through an eye-tracking application. It is not included in this book as a research method due to its specialized usage and its reliance on eye-tracking application for data gathering.

### *12.1.2 Think-aloud protocols: Types and features*

There are essentially two kinds of TAPs in practice. One is concurrent TAP in which participants think aloud while performing a specific task. The other kind is called retrospective TAP in which participants do not think aloud while performing a task. Rather, they recall what went through their mind soon after completing a task. In the retrospective TAP, researchers often use one of the following two techniques to collect think-aloud data:

1) The interview method – Ask participants some questions to help them recall their cognitive and thinking processes. Interviews are typically used together with the TAP method in this type of scenario.
2) The stimulated recall technique (e.g., Cotton and Gresty 2006) – Instruct participants to voice their thoughts retrospectively by using video recordings of their actions when completing the task as an "aide-mémoire", enabling participants to access and recall their cognitive and thought processes shortly after the task completion. Alhadreti and Mayhew (2018), a case study chosen for this section, adopted the stimulated recall technique for collecting retrospective think-aloud data.

The biggest advantage of retrospective TAP is that researchers do not interrupt or affect participants' task performance by asking them to think aloud. Instead, participants' think-alouds are gathered after task completion. But, how much such data can participants recall even with the help of the interview method or the afore-depicted stimulated recall technique remains an issue of debate and little consensus is ever reached. Retrospective reports are more likely than concurrent protocols to suffer from forgetting and from the increased opportunity for the participant to mix current knowledge with past knowledge, making reliable inferences from verbal data more difficult. In addition, retrospective TAP has also been criticized for the subsequent possibility of post-task rationalizations, implying that participants may rationalize their verbal reports after having time to think about the task they completed. For these reasons, Payne (1994) recommends that think-aloud data generally should be collected during the performance of the task. That is, concurrent TAP should be chosen in collecting think-aloud data whenever possible in order to avoid relying on participants' ability to recall think-aloud data.

Occasionally, researchers apply both the concurrent and retrospective TAP techniques in one study which is called the hybrid model. The case study chosen for the current section, Alhadreti and Mayhew (2018), represents an example of the hybrid method. The hybrid technique encompasses the pros and cons of concurrent and retrospective TAPs. On the other hand, the hybrid TAP might minimize the limitations of concurrent and retrospective TAPs to a certain extent if used wisely.

Among all the features of the TAP method, the most prominent one is that it enables researchers to collect data about participants' cognitive and thinking processes. Few other existing research methods are able to do the same. This feature makes TAP standing out whenever researchers intend to gather such data. As technologies for recording and capturing think-alouds have been improved since the beginning of the 21st century, the TAP method is becoming even more attractive. Researchers can do screen and video recording easily nowadays with proper software applications (e.g., Camtasia).

However, this feature of TAP also causes one of the major concerns about the method: whether the TAP is a valid measure of cognition and thinking. The answer from Ericsson and Simon (1980) is assertive and positive. They believe that "...verbal reports, elicited with care and interpreted with full understanding of the circumstances under which they were obtained, are a valuable and thoroughly reliable source of information about cognitive processes" (Ericsson and Simon 1980, 247). Furthermore, Durning and his colleagues (2013) used functional magnetic resonance imaging (fMRI) to investigate neuroimaging differences between thinking (e.g., answering multiple-choice questions in real time) and thinking aloud (e.g., on review of items) by examining the relationship between think-aloud data and brain activation patterns. Their findings add evidence to the notion that the TAP is a reasonable measure of thinking. Related to this afore-mentioned concern are two other features of the TAP method:

- Actual cognitive and thinking processes are faster than what one can verbalize.
- There is always the tacit and unconscious which cannot be verbalized.

These two features demonstrate that think-alouds can offer informative glimpses of the cognitive and thinking processes in progress but never constitute a complete account. In other words, only information present in STM and being heeded by the participant can be directly accessed and reported in the TAP process. In addition, participants are not used to think aloud while performing a task, which would further reduce the completeness of think-aloud data.

Ericsson and Simon (1980, 236) identify three different causes of incompleteness of verbal reports: (1) The information is not heeded, hence not stored in STM, hence not accessible for verbal reporting. (2) Not all the information available in STM at the time of the report is actually reported. (3) Not all of the information previously available in STM has been retained in LTM, or is retrievable from LTM. The first two kinds of causes are relevant to concurrent TAP, while the third cause relates to retrospective TAP. Appropriate training might help to some extent. On the other hand, Jääskeläinen (2010) suggests that think-aloud data can be complemented by short interviews, transaction logs, screen captures, and the like.

Many people (e.g., Cotton and Gresty 2006) believe that it is simply not realistic to expect participants to perform a task and simultaneously verbalize their thoughts in the process. This amounts to another salient feature of the TAP method. Although think-aloud is an activity that participants can do, performing a task while think-aloud would inevitably affect the quality of both. The request for thinking aloud might interfere with and alter participants' thought processes and may thus affect the ways in which they perform the experimental task – which can in turn affect the validity of the data obtained. In other words, the need to provide a verbal report, as a secondary task, may fundamentally alter the processes used in performing the primary task of interest (e.g., making a choice or solving a problem). Ericsson and Simon (1980) nevertheless argue that think-aloud would not affect participants' task performance unless verbalizations are required by researchers to cover information that participants otherwise would not pay attention to. For example, participants would not pay attention to the question about search engine selection when Google is the default search engine for almost all searchers. In this case, participants would not be able to verbalize their cognitive and thinking processes about search engine selection because that is not part of their search activities. Therefore, researchers should design their study by taking this feature of the TAP method into consideration.

Although the TAP method possesses other features such as lack of an established research paradigm (e.g., Bernardini 2001), the four characteristics portrayed above are outstanding when the TAP method is used for data collection purposes: (1) A unique technique for gathering data of participants' cognitive and thinking processes; (2) Actual cognitive and thinking processes are faster than what one can verbalize; (3) Participants are unable to verbalize the tacit and unconscious cognitive and thinking processes; and (4) Performing a task while think-aloud would inevitably affect the quality of both. All these features determine when and how the TAP method should be utilized.

### *12.1.3 Using think-aloud protocols in research*

Whenever researchers need to gather data about humans' cognitive and thinking process, TAPs should be the suitable method. Using the TAP method for data collection in research may appear easy (Wildemuth 2017, 198), the actual implementation nevertheless involves multiple steps and needs careful planning.

First, design an experiment for participants to perform so that they can think aloud. In Chapter 10, we discuss experiments as a data collection technique. However, the experimental task used in the TAP method is quite different from an experiment presented in Chapter 10 in that the experimental task of the TAP method focuses exclusively on the task itself without any consideration of the common components in an experiment such as random selection and assignment of participants to experimental and control groups. The task could be about clinical reasoning, information searching, learning processes, usability testing, or any other undertaking that the researcher aims to study. Because participants cannot verbalize what they do not need to heed (e.g., search engine selection when Google is the default search engine to use), researchers should make certain that the designed task must cover all the activities they wish to collect think-aloud data about. Take database searching as an example in this context. There are usually at least several databases, if not more, for a given search query when thousands of databases exist for various purposes. The searcher hence must choose one or more databases from an array of viable options according to a given search query. In this case, the task should include database selection as part of the search activities. By contrast, search engine selection generally is not an activity that searchers must perform in their search process because Google usually is the default option most of the time. Therefore, it is important to design a task for TAP research according to the research objective and never ask participants to think aloud what does not occur in the process of completing the task.

Second, set up an environment that is quiet and appropriate for thinking aloud so that participants can concentrate on the think-aloud process. When using the TAP method, it is a must to schedule think-aloud sessions individually, with one participant a time to avoid any possible interference from other participants (Fonteyn, Kuipers, and Grobe 1993). Although it would take longer time for researchers to complete data collection from participants individually, it is worth the effort in order to ensure the quality of think-aloud data gathered.

Third, offer training to participants before the data collection process if they have never done think-aloud before. Unlike completing a questionnaire or answering interview questions, few people have ever taken part in thinking aloud. This makes the training of participants absolutely necessary. The training normally includes some warm-up exercises along with clear instructions. In Ericsson and Simon (1993, 378), they asked participants to do some calculations (e.g., "What is the result of multiplying 24 x 36") with the following instructions to help participants get ready for thinking aloud:

> In this experiment we are interested in what you think about when you find answers to some questions that I am going to ask you to answer. In order to do this I am going to ask you to THINK ALOUD as you work on the problem given. What I mean by think aloud is that I want you to tell me EVERYTHING you are thinking from the time you first see the question until you give an answer. I would like you to talk aloud CONSTANTLY from the time I present each problem until you have given your final answer to the question. I don't want you to try to plan out what you say or try to explain to me what you are saying. Just act as if you are alone in the room speaking to yourself. It is most important that you keep talking. If you are silent for any long period of time I will ask you to talk.
>
> (Ericsson and Simon 1993, 378)

All the capitalized words are taken from Ericsson and Simon (1993) as printed. Even though the above instructions are for solving calculation problems, they are applicable for other kinds of data collection using the TAP method. The only phrase that might need to be changed is "the problem" if the task for participants to think aloud is something else (e.g., testing the usability of an information system). In such cases, corresponding terms would be used (e.g., the usability testing) instead of "the problem". During this step, researchers should remind participants to keep talking whenever they fall silent.

Fourth, record think-aloud data with an appropriate tool. Tools that can be utilized for capturing and recording think-alouds nowadays range from note-taking to software for video recording (e.g., Camtasia). The latter not only records think-alouds but also captures contents displayed on screen if the task is performed using a computer (e.g., searching). Audio and video recording applications have also been improved greatly compared to those used decades ago.

Fifth, transcribe the think-alouds recorded soon after the process is over in order to get ready for subsequent data analysis. Similar to interview data transcription, transcribing think-aloud data is a time-consuming process. Yet think-aloud data are not structured and organized as interview data because participants simply verbalize whatever goes through their mind while completing a given task. In addition, they are not supposed to organize or sort out what they say. Think-aloud data could also include words such as "uh", "um", and the like that carry little meaning about participants' cognitive and thinking processes. Therefore, it usually takes longer time to transcribe think-alouds when compared to transcribing interview data.

The afore-listed five steps constitute the data collection process using the concurrent TAP method. The application of RTA, as explained earlier, relies on two additional factors. One is participants' ability to recall their cognitive and thinking processes, while the other factor is the use of interview or other similar techniques (e.g., the stimulated recall technique) to help the participants to recall. In either concurrent or retrospective thinking aloud, there are a good number of dos and don'ts when using the TAP method for data collection. Those dos and don'ts are discussed in Section 12.1.4.

### *12.1.4 Dos and don'ts in think-aloud research*

As indicated above, the TAP technique may sound simple for researchers to use for data collection because once the experimental task is designed, it is the participants who will do all the think-aloud. However, that does not mean that researchers can just sit down and simply watch participants to think aloud. Instead, researchers still need to make sure that participants properly verbalize everything throughout the entire process while performing a given task. Below are the dos and don'ts in think-aloud research.

- Choose an appropriate kind of the TAP method as the concurrent, retrospective, or hybrid TAP technique each has its pros and cons in the data collection process. Although researchers (e.g., Payne 1994) recommend that the concurrent TAP should be chosen in collecting think-aloud data whenever possible to take advantage of this method, we still need to take into account in the selection process the participant population and the situation in which the study is conducted. For instance, if the researcher cannot be on site at the time of data collection, the retrospective TAP using either the interview method or stimulated recall technique might be a more viable option for collecting think-alouds.
- Try to minimize the cognitive load on the participants, allowing them to perform the task in a naturalistic way. Take Task 4 in Alhadreti and Mayhew (2018), the case study chosen for this section, as an example. This task asks participants to do the following: "You want to find the journal paper that has the title 'Building for the Future' written by Doyle Henry in 1963 to read before a coming seminar in an education project. Can you find it?" (Alhadreti and Mayhew 2018, 3). As this task specifically lists the journal article title, author's name, and publication year, participants can use the combination of these three pieces of information to locate the article easily and the cognitive load on the participants would not be large. If, for instance, only the author's name and publication year are given without the article title or merely a journal title is provided, it would place a heavy cognitive load on the participants because they have to sift through many search results to locate the journal article in the former case, while in the latter case it is not natural and possible to search for a particular journal article solely by a journal title.
- Offer practice/training opportunities if participants have never done think-aloud before. Avoid asking participants to directly take part in a study using the TAP method without any training or practice as think-aloud is not natural to most, if not all, participants.

- Minimize interactions with participants in the data collection process as recommended by Ericsson and Simon (1993, xiv). The experimental setting for thinking aloud should be arranged to make clear that social interaction between the researcher and participant is not intended, and the researcher should seat behind the participant and not be visible to the participant. Remind the participant to think aloud when s/he falls silent. The researcher tells participants to "keep talking" instead of making the social request such as "tell me what you are thinking". Ericsson and Simon (1993, 271) also suggest that the researcher might choose to play a "reminder tone" to keep the reminder short and completely contentless because reminders to keep talking often resulted in apologies, interrupting the task flow, and flustering the participant.
- Make use of applications such as Camtasia to capture what participants verbalize and view on the screen during the think-aloud process. Other methods such as interviews and transaction log analysis can also be used together with the TAP technique for collecting additional data to complement the think-aloud data.
- Don't interrupt or interfere while the think-aloud is in process so that the participant can focus on thinking aloud. Other don'ts in applying the TAP method in research include making the research environment comfortable for participants in terms of temperature and atmosphere. Never try to conduct think-aloud research with multiple participants in one setting because the interference among participants would drastically affect the validity of think-aloud data collected.

Needless to say, it is impossible as well as unnecessary to exhaust all the dos and don'ts in this section. However, research using the TAP method can be successfully conducted if we adhere to the general principles in performing research as well as specific guidelines for carrying out think-aloud studies.

### *12.1.5 Case study of research using think-aloud protocols*

Alhadreti, Obead, and Pam Mayhew. 2018. "Rethinking Thinking Aloud: A Comparison of Three Think-Aloud Protocols." *CHI 2018* Paper 44. April 21–26, 2018, Montréal, QC, Canada. 1–12. https://doi.org/10.1145/3173574.3173618.

Alhadreti and Mayhew (2018) is based on the first author's dissertation research, while the second author is the first author's dissertation advisor. This study compared three kinds of the TAP method: concurrent think-aloud (CTA), RTA, and a hybrid think-aloud (HB although HTA could be an abbreviation more consistent with CTA and RTA). The three TAP techniques were compared through an evaluation of a library website, which involved four points of comparison: task performance, participants' experiences, usability problems discovered, and the cost of employing the methods. The target of this study is the University of East London library (UEL-L) in the United Kingdom. The site was chosen because it possesses a certain number of potential usability problems and is geographically close to the authors' affiliated institution, the University of East Anglia (UEA), at the time of this research.

A total of 60 participants were invited to take part in the experiment using the TAP method, with 20 each for the CTA, RTA, and HB testing conditions. That assignment is atypical for TAP research because few studies would apply all three TAP techniques in a single project. All participants were students from the UEA. Although the authors did not specify if the students were undergraduates or graduates, they could be both based on the age information (i.e., age 18–39) provided in Table 1 of their report. The authors tried to make sure that the three groups of participants were "matched" on country of origin, gender, age, and internet use. But discipline, a factor that might have some major impact on the participants' ability to use the library's website, was not considered in the assignment of participants to the study's three groups. On the other hand, the participants were randomly assigned to the CTA, RTA, and HB groups using a random number generator. This randomization may not be that important in the present study because the 60 participants were not randomly selected from the UEA student population in the first place. The authors even did a Kruskal-Wallis H test to prove that the three groups of participants were identical or homogeneous. The authors went a bit too far in claiming that "the interval validity of the study is high" (Alhadreti and Mayhew 2018, 3). The internal validity of a study denotes if the study indeed investigates the research problem it intends to explore, which is not determined by the homogeneity of its participants at all.

The following seven tasks were designed in the study to collectively investigate the targeted website's main features and predicted problematic areas:

- Task 1 for evaluating the ease of navigating the site to find the name of a subject support.
- Task 2 for assessing the booking function for study rooms on the site.
- Tasks 3 and 4 for evaluating the site catalogue's "simple search".
- Tasks 5 and 6 for assessing the catalogue's "advanced search" and "sort results" functions.
- Task 7 for examining how participants worked with viewing search history on the site.

It would be better if the authors had listed the seven tasks in full as an appendix rather than as brief descriptions in the Method section. According to the task list, some were about the library catalogue but none about the databases available from the library. Yet Task 4, depicted by Alhadreti and Mayhew (2018, 3), apparently is a question about searching a journal article via databases. Thus, the authors appeared unclear about the difference between a library catalogue (for locating books) and databases (for locating non-books such as journal articles).

The authors depicted in detail the think-aloud method, practice, and instructions all in the "HB condition" subsection (Alhadreti and Mayhew 2018, 3–4). Two comments are due here: (1) All such information could have been offered in the Experiment Procedure section separately and before covering the subjections of "HB condition", "CTA condition", and "RTA condition" so that there would be no need to repeat the common information about the think-aloud method already presented (e.g., review the task instruction sheet) in any of the three subsections. (2) The authors described the experiment procedure for the "HB condition" first. It seems more appropriate if the parts for the "CTA condition" and "RTA condition" had been offered before the "HB condition" because HB is essentially the combination of the other two conditions.

Alhadreti and Mayhew (2018) made some good considerations to minimize the side effect of the TAP process in the experiment. For example, the participants were first asked if they were right- or left-handed for mouse configuration and were given a maximum of two minutes to familiarize themselves with the test laptop and regain their normal speed of interaction with the computer system. The screen-capturing software Camtasia was set up to collect both screen and think-aloud data during the experiment process. A neutral reminder "Please keep talking" was issued if the participant fell silent for 15 seconds.

The hybrid technique of TAP in this study nevertheless might cause some concern because the 20 participants in that group basically did think-aloud twice, once concurrently and once retrospectively. The concurrent TAP was performed as usual and recorded using Camtasia. By comparison, the retrospective TAP was done by watching the recordings of their own concurrent TAPs on muted video and giving retrospective reporting at the same time. This practice implies that the HB participants doing the retrospective TAP already went through the think-aloud process using the concurrent technique but were then instructed to voice their thoughts retrospectively while watching muted videos of their own activities. The learning and memory effects cannot be ignored in this type of retrospective TAP process and what was gathered perhaps consists of both concurrent and RTA data. In other words, the impact of the previous concurrent TAP on the subsequent retrospective TAP should not be overlooked and must be taken into consideration when analyzing data so collected.

One more point that needs to be brought up in this context is relating to how the think-aloud data collection was done in the study. Alhadreti and Mayhew (2018) did not specify if the 20 participants did the TAP experiment together in one location simultaneously or in different time periods. If the former, the validity of the study is in jeopardy as there must be interference among the participants doing the think-aloud together regardless of how large the research setting is.

Finally, Alhadreti and Mayhew (2018) reported that the concurrent method outperformed both the retrospective and the hybrid methods in facilitating successful usability testing. This finding echoes Payne's (1994) recommendation that concurrent technique should be chosen in collecting think-aloud data if all possible. According to Alhadreti and Mayhew (2018), the CTA method also requires much less time on the evaluator's part than the other two kinds of TAP techniques that needed almost twice as much testing and analysis time. Although it is rare for one study to employ three different kinds of the TAP method, Alhadreti and Mayhew (2018) provide an opportunity for readers to compare them side by side in one research report.

**Section Summary**

- Think-aloud protocols are a research method for collecting data about participants' cognitive and thinking processes by asking them to verbalize (i.e., thinking aloud) whatever crosses their mind as they perform a specified task such as conducting a search or solving a problem.
- The think-aloud method can be utilized in three different ways depending on how think-aloud data are collected. If participants think aloud while performing a specific task, that is concurrent think-aloud. If participants do not think aloud while performing a task but recall what went through their mind soon after completing the task with the help of video recordings and/or the interview method, that is called retrospective think-aloud. The third kind of think-aloud technique is known as the hybrid method which implies that both the concurrent and retrospective TAPs are applied in one study. Among the three kinds of think-aloud techniques, the concurrent one is the most often used, while the hybrid option is the least used. The RTA method falls somewhere in between the other two kinds of think-alouds.
- The TAP method has four main features: (1) A unique technique for gathering data of participants' cognitive and thinking processes; (2) Actual cognitive and thinking processes are faster than what one can verbalize; (3) Participants are unable to verbalize the tacit and unconscious cognitive and thinking processes; and (4) Performing a task while think-aloud would inevitably affect the quality of both. These features by and large determine when and how the TAP method should be utilized.

- The general steps in conducting think-aloud research consist of the following: First, design an experiment for participants to perform which provides the basis for thinking aloud. Second, set up an environment that is quiet and appropriate for participants to concentrate on the think-aloud process. Third, offer training to participants before the data collection process if they have never done think-aloud before. Fourth, record think-aloud data with an appropriate tool. Fifth, transcribe the recorded think-alouds soon after the process is over so as to get ready for the subsequent data analysis.
- Although there are many dos and don'ts in think-aloud research, the major ones include designing an experimental task that is natural to participants, providing adequate training to participants if think-aloud is new to them, and avoiding interaction with participants throughout the think-aloud process.

## 12.2 Delphi study

**Section Overview Questions**

- What is Delphi study?
- What are the four main features of the Delphi method?
- What are the major steps in the data collection process of Delphi research?
- What is the procedure for executing Delphi rounds?
- What are the dos and don'ts in using the Delphi method for data collection?

### *12.2.1 What is Delphi study?*

Delphi study is often known as the Delphi method/technique or Delphi survey (Belton et al. 2019) which was named after the Greek town of Delphi. The ancient Greeks believed that Apollo – son of Zeus and god of light, purity, the sun, and prophecy – killed the dragon Python in Delphi. The temple in Delphi then contained the famous oracle, Pythia, whom Apollo chose to speak through to predict the future. She would turn around in frenzy and utter strange sounds, which would then be used for prediction (Stone Fish and Busby 2005). The Delphi method is therefore closely related to prediction.

Delphi as a research method was first developed in the defense and military sector by the Air Force-sponsored Rand Corporation study called "Project Delphi" in the 1950s. "Project Delphi" was an attempt to forecast the probability of a particular event and focused on understanding the use of expert opinion (Dalkey and Helmer 1963). Additional Delphi studies were then published "on the superiority of group opinion over individual and on the justification of expert opinion in inexact sciences and its scientific use" (Landeta 2006, 468). The Delphi method proliferated in the 1960s and 1970s and continues to find applications in fields dealing with other complex problems that our society faces, such as education, health care, climate, and environment issues. The Delphi technique is also regularly used in psychology, sociology, political science, and other fields for one or more of the following purposes: (1) Obtaining consensus from a group of people on a topic when the consensus cannot otherwise be reached; (2) Forecasting future events; (3) Soliciting opinions from various people; (4) Assessing the relative importance of issues; (5) Developing a conceptual or theoretical framework. Of the afore-enumerated functions of the Delphi method, obtaining consensus and predicting the future are the most frequently intended purposes of Delphi research. Boulkedid et al. (2011) thus conclude that the Delphi procedure is valuable for achieving a consensus about issues where none existed previously.

While there are various definitions for Delphi study as a research method, Brancheau, Janz, and Wetherbe (1996) – the case study chosen for the Delphi method – provide a simple yet succinct explanation about it: "Essentially, Delphi is a series of linked questionnaires. Starting with a base line questionnaire, succeeding questionnaires summarize group responses to the preceding questionnaire and ask participants to re-evaluate their opinions based upon the new evidence" (226). A series of linked questionnaires or surveys mentioned above refers to the fact that after the first survey for one Delphi study is distributed, the surveys used in remaining rounds are updated on the basis of the previous one. All such surveys are hence linked or closely related. In no case are the subsequent surveys after the first one in a single Delphi study created anew. Simply put, Delphi study is a method for collecting data with a questionnaire or survey from a group of experts to address a research problem in order to reach consensus or make forecasts via several rounds of exchanges (Chu and Ke 2017). As explained in Chapter 3, questionnaires and surveys are often treated as synonyms in the research domain and these two terms will continue to be used interchangeably in the remainder of this book.

As the Delphi method relies on a series of linked surveys to be functional, it is natural for researchers to contrast the Delphi method with regular surveys. According to Lilja, Laakso, and Palomäki (2011), the Delphi method differs from surveys in two ways besides delivering a series of linked surveys in one study: (1) The participants are not picked randomly but are selected based on their knowledge and experience. (2) The number of participants in a Delphi study can be much smaller than what is thought to be sufficient to guarantee the external validity of a survey. Other researchers (e.g., Hsu and Sandford 2007) believe that surveys try to identify "what is", whereas the Delphi technique attempts to address "what could/should be". Furthermore, the Delphi method not only allows researchers to gather everyone's opinions about a topic but also enables participants to have the benefit of receiving feedback from other counterparts in subsequent rounds of data collection. This two-way communication in the Delphi process also distinguishes the Delphi method from regular surveys.

In addition, the Delphi method can overcome the following drawbacks of the traditional methods of pooling opinion (e.g., focus groups presented in Section 8.1): (1) the influence of dominant individuals, (2) irrelevant and biased communication, and (3) group pressure for conformity. Anonymity in the Delphi technique reduces the effect of dominant individuals; controlled feedback reduces irrelevant communication; and the use of statistical summaries of group responses reduces group pressure for conformity. These and other salient features of the Delphi method will be illustrated in detail in Section 12.2.2.

### *12.2.2 Features of the Delphi method*

The Delphi method possesses four basic features: anonymity, iteration, statistical summary, and aggregation of individual participants' responses, controlled feedback of responses to all participants (Belton et al. 2019; Hsu and Sandford 2007; Rowe and Wright 1999). That is to say, any study must have those four features in order to be regarded as Delphi research.

The anonymity feature of Delphi research denotes that all participants are encouraged to freely express their opinions while their identity is not revealed throughout the data collection process partly because participants do not meet face-to-face at any point of the study. Moreover, data collected from all participants in each Delphi round are analyzed in aggregation without unveiling each individual's responses to the linked surveys in the process of data gathering. Some participants in Delphi research may know each other as they are all supposedly experts or individuals with knowledge and experience in a particular domain. But they do not know who the other participants are in the Delphi process thanks to the anonymity feature. As a result, all participants can complete each round of the Delphi surveys according to researchers' instructions while having no concern about if others would be displeased or offended because of their responses.

The iteration characteristic of Delphi study involves two associated components: (1) Researchers distribute at least two rounds of linked surveys; (2) Participants are accordingly asked to complete at least two such surveys. Starting from the second round, participants are also expected to complete the survey in light of others' aggregated responses to the preceding survey. Most Delphi studies distribute three or more rounds of linked surveys in the entire data collection process.

The feature of statistical summary and aggregation of individual participants' responses in Delphi research serves two purposes. One is to ensure the anonymity of all Delphi participants. The other purpose is to lay a foundation for further analyses and interpretation of the data collected.

The controlled feedback feature of the Delphi method does not imply that participants can get each other's responses to the survey without any restriction. Rather, researchers eliminate all irrelevant information before sending the summarized and aggregated feedback to the participants through the next Delphi round. This controlled feedback feature informs the participants of others' perspectives as well as provides them with the opportunity to clarify or change their views.

The afore-depicted features of the Delphi method in combination greatly contribute to its wide usage in research projects. As a research method contingent on survey – a data collection technique that all scholars are required to learn and master before becoming a researcher, the remaining components of the Delphi data collection procedure appear closely related to and reliant on the survey instrument.

### *12.2.3 Using the Delphi method in research*

The basic data collection process for Delphi research starts with choosing participants and developing the initial survey. Both steps can be done simultaneously or one after the other. Starting from the second Delphi round, researchers need to aggregate all the responses via a statistical summary, sometimes along with the rationales given for the responses. Participants are then invited to submit a revised response after considering the statistically summarized and aggregated responses received or to resubmit their first response. This iteration and controlled feedback process continues over two or more rounds until a consistent pattern of responses (e.g., general consensus or notable dissensus) is reached. There are commonly several steps in using the Delphi method for data collection in research. Figure 12.1 visualizes the data collection process of Delphi research.

As in any research project, all the work (e.g., research question formulation, literature review) must be genuinely done before the data collection process for Delphi research commences. The description below for the data collection process of Delphi research

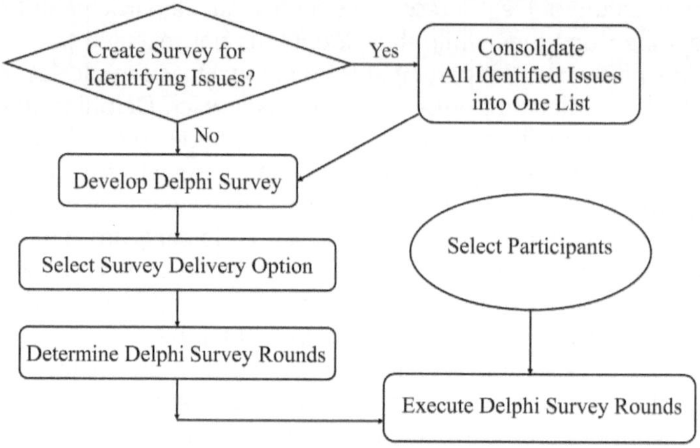


*Figure 12.1* Data Collection Process of Delphi Research

shall begin with survey development. If you, as the researcher, would like to recruit participants before survey development, that should be an acceptable practice too.

*12.2.3.1 Step (1) Survey development*

As mentioned earlier, survey is the main, if not the only, data collection instrument in Delphi studies. Some researchers may also choose other data collection techniques (e.g., interviews) to complement the Delphi method. For example, Brancheau et al. (1996) – the case study chosen for the Delphi method – used interviews as well in their Delphi research. Chapter 3 is devoted to questionnaire or survey as a data collection technique. Even though Delphi research depends on the use of surveys for data gathering, Delphi survey development specifically focuses on the creation of a survey for the first round and updates of the preceding survey for subsequent rounds.

There are two approaches to the initial Delphi survey development. One approach is to create the survey from scratch by first asking participants open-ended questions in order to generate issues for consideration in the Delphi research. For instance, Dickson et al. (1984) adopted this approach by developing their initial Delphi survey. They asked their participants to identify and briefly describe five to ten of what they considered to be the major information systems management issues over the next five to ten years. All the identified issues were then consolidated into one list which makes up the survey for the first Delphi round along with other survey components (e.g., a cover letter). Section 3.2 details questionnaire development, including questionnaire components, question types (e.g., open-ended vs. closed-ended questions), as well as dos and don'ts in the process.

The other approach to creating the initial Delphi survey is to take a list of issues from a previous, related study. A cover letter and closure will be added to complete the survey. Brancheau, Janz, and Wetherbe (1996) – the case study chosen for the Delphi method – chose this approach to developing their initial survey. Initial surveys so created can skip the procedure described in the first approach (e.g., asking participants open-ended questions to identify issues).

All the issues in a Delphi survey are presented as Likert scale statements or questions. Specifics about Likert scale statements or questions can be found under the heading

Opinion/Attitude Questions in Section 3.2.1. Researchers can choose from the three major types of Likert scale depicted there. One particular point that researchers should pay attention to at this moment is whether the data to be collected are intended to be measured at the ordinal level or interval level (see Step (2) Selection of Participants and Step (3) Survey Delivery Option Selection). Likert scale in nature collects ordinal data unless researchers purposely choose the interval level of measurement by explicitly asking respondents to treat the intervals between any two adjacent points on the scale as equal and by visually presenting the scale as such at the time of data collection. Delphi researchers should attempt to gather data measured at the interval level whenever possible so that they will have more techniques available for data analysis later. Brancheau, Janz, and Wetherbe (1996), for example, did visually present their scale with equal intervals in their Appendix A. However, they did not explicitly ask their participants to treat the intervals between any two adjacent points on the scale as equal.

When researchers have a list of issues and express them as scale statements or questions along with a cover letter and closure, the Delphi survey is ready to be distributed to participants for rating (e.g., Brancheau, Janz, and Wetherbe 1996) or ranking (e.g., Dickson et al. 1984). It is totally up to researchers to decide which activity, rating or ranking, the participants should do when designing the survey. In addition, many Delphi researchers may ask their participants to provide rationales for their ratings or rankings, which can help confirm the plausibility and strength of their responses. It is also common to invite participants to make comments and list any issues that are not already included in the survey.

Developing a Delphi survey essentially is similar to the creation of a regular questionnaire. Therefore, all the guidelines presented for survey development in Chapter 3 should be followed. Appendix A in Brancheau, Janz, and Wetherbe (1996) represents one example of a Round 3 Delphi survey. Figures 13.1, 13.2, and 13.3 in Stone Fish and Busby (2005) respectively show the excerpts of sample Delphi questionnaires (DQ) for generating issues (DQ1), for first round data collection (DQII), and for second round data collection (DQIII).

#### *12.2.3.2 Step (2) Selection of participants*

Research participants in Delphi studies are called experts because they have expertise (e.g., knowledge and experience) in the subject area under study. Delphi studies are also known as panel studies and briefly mentioned in Section 2.5. Delphi participants are thus also called panelists in some publications. For clarity reasons, however, only participants or experts will be used interchangeably in this chapter on Delphi research.

Unlike participant selection in survey research that commonly involves random sampling from a large population, experts for Delphi research are chosen according to a set of specified criteria. Dalkey (1969) indicates that participants' knowledge of the subject matter at hand is the most significant assurance of a high-quality outcome when the Delphi method is used. Delphi participants are hence chosen for their expertise rather than through a random procedure. This constitutes the very reason why Delphi research participants are considered experts.

The criteria for selecting experts for Delphi research vary from study to study. For instance, the criterion used for selecting experts in Brancheau, Janz, and Wetherbe (1996) is fairly straightforward: institutional and board members in the Society for Information Management (SIM). That is, any SIM institutional and board members can serve as experts in their study. On the other hand, some Delphi studies set up multiple criteria and anyone who meets some or all of them are qualified to serve as an expert in the research. The Delphi study by Zhang and Salaba (2009) stands for such an example. The two authors specify three criteria for their expert selection: (1) published on FRBR (Functional Requirements for Bibliographic Records) – a conceptual entity-relationship model for retrieval and access in online library catalogs and bibliographic databases, (2) served on FRBR review groups, or (3) directly involved in the FRBR model development. Anyone who met at least one of the three criteria enumerated above is regarded an FRBR expert.

Another consideration relating to expert selection in Delphi research is to decide whether the chosen participants should be homogeneous or heterogeneous. Nasa, Jain, and Juneja (2021) state that a diverse group of experts helps to achieve a broader perspective and generalization of consensus. By contrast, the homogenous group may be quicker to reach consensus. Belton et al. (2019) nevertheless recommend that it is generally desirable to use a heterogeneous group of experts rather than a homogenous one. The rationale behind this recommendation is that heterogeneity of participants encourages "more varied opinions and experience of the topic or issues in question, and are therefore more likely to represent the variety of perspectives that exist on a particular topic and achieve more accurate and plausible judgments" (Belton et al. 2019, 74). Other authors (e.g., Rayens and Hahn 2000) also echo that recommendation.

As for the number of experts one Delphi study should recruit, pertinent suggestions range from about one dozen to a couple of hundred. If the Delphi research is conducted nationwide, the number of participants could be larger than three digits. However, be aware that the volume of feedback for researchers to summarize and for participants to review after each round can quickly become unmanageable because of the iteration and controlled feedback features of the Delphi method. In practice, the number of participants in most Delphi studies normally falls in-between two dozen and a couple of hundred. Brancheau, Janz, and Wetherbe (1996), for instance, invited 217 experts to take part in the first round of their Delphi research. Investigators can determine an appropriate number of participants based on their research problem as well as available time and other resources. Avoid going to extremes by choosing either too few or too many experts in a Delphi study because the former would result in collecting non-representative data, whereas the latter might lead to an amount of collected data that easily overwhelms both experts and researchers.

#### *12.2.3.3 Step (3) Survey delivery option selection*

Survey delivery in Delphi research is similar to that in regular survey research discussed in Chapter 3. The major approaches to questionnaire administration include mail, telephone, email, survey site, and on-site or self-administrated (see Table 3.2). Each approach has its pros and cons, and Delphi researchers need to make their choice of survey delivery options based on their research project and available resources.

As Delphi surveys need to be distributed to selected experts at least twice and the survey site technology has become matured after over 20 years of development and improvement, the survey site approach seems to be a favored option that many researchers take to distribute their Delphi surveys. Common survey sites in the United States include SurveyMonkey (https://www.surveymonkey.com) – a subsidiary survey platform under Momentive as of 2021, QuestionPro (http://www.questionpro.com), Qualtrics (https://www.qualtrics.com), and Google Forms (https://www.google.com/forms/about/). Although the survey site approach appears to be a preferred option for survey administration, other approaches can still be chosen if the research environment so requires. For example, the email approach might be adopted if researchers do not have the budget to pay for the survey site service.

#### *12.2.3.4 Step (4) Delphi round determination and counting*

As discussed in Section 12.2.2, iteration is one major feature of Delphi research. Investigators thus need to determine the number of Delphi rounds before the data collection process begins. The number of Delphi rounds is what Brancheau, Janz, and Wetherbe (1996) refer to as "a series of linked questionnaire" (226). Seasoned Delphi researchers agree "that a point of diminishing returns is reached after a few rounds. Most commonly, three rounds proved sufficient to attain stability in the responses; further rounds tended to show little change and excessive repetition was unacceptable to the participants" (Linstone and Turoff 1975, 229). In other words, most Delphi studies choose three rounds of the survey question answering to complete the data collection process.

It must be pointed out that the step through which the Delphi survey is initially created by asking participants open-ended questions should not be counted as one round because that step is only for identifying or generating issues (see Figure 12.1) rather than for seeking participants' opinions and experience of the topic or issues in question. As mentioned earlier, Dickson et al. (1984) created a survey for their Delphi study by first asking their participants to identify and briefly describe five to ten of what they considered to be the major information systems management issues over the next five to ten years. They then consolidated those data into a combined list of issues before sending the list back to the SIM respondents for ranking and commenting. The consolidated list of issues along with other necessary survey components (e.g., a cover letter) ultimately makes up a Delphi survey. Only using a survey for gathering experts' opinions on the identified issues or questions can be considered one round of Delphi research while using a survey for identifying issues for Delphi survey development is not. Accordingly, the initial survey in Dickson et al. (1984) cannot be considered a Delphi round, while the first survey Brancheau, Janz, and Wetherbe (1996) distributed to their participants is counted as a Delphi round.

#### *12.2.3.5 Step (5) Execution of Delphi rounds*

The distribution of the Delphi survey in the first round or Round 1 is relatively simple. Researchers deliver the Delphi survey to participants the same way as they would in a regular survey study. Likewise, participants are asked to complete the survey as instructed. After receiving the completed Round 1 survey from participating experts, researchers need to do the following besides the normal survey processing (e.g., checking if a survey is complete and usable) before distributing the Round 2 survey:

1) Compute summary statistics that are appropriate for the responses to scale statements in the survey. As briefly reviewed in Step (1) Survey Development, the statistics researchers can calculate for summarizing experts' responses to the Round 1 survey should be based on the level of measurement chosen for data collection.

    a If the response data are measured at the interval level, average or the mean (in statistical term) and/or standard deviation are typically computed as summary statistics. Researchers should gather data using the interval scale in the Delphi survey whenever possible so that more techniques are available for analyzing such data later (see

Step (1) Survey Development). Brancheau, Janz, and Wetherbe (1996) – the case study for the Delphi method – followed this general recommendation and collected interval data via the survey.

b If the response data are measured at the ordinal level, the median and/or range are usually calculated as summary statistics. Although many Delphi studies report such summary statistics, gathering ordinal data in Delphi surveys is not recommended for the reason given in the previous paragraph.

2) If participants were asked to provide rationale for their ratings or rankings in the survey, the rationale data may be aggregated and shared with participants in the next Delphi round. The rationale data and the comments experts may make will also be kept for qualitative analysis after the entire data collection process is over. As for the additional issues experts suggested in the completed survey, they become candidates for being added to the list of issues rated or ranked in the previous round of the Delphi survey.
3) Aggregate and arrange all the issues in descending order by their corresponding summary statistics (e.g., mean for interval data; median for ordinal data). The issue rated or ranked the highest will be listed the first, and so on. If an issue receives no rating or low rating from the previous Delphi round, it is usually removed from the list in the next round to facilitate consensus reaching. Similarly, an issue will be added to the list if a good proportion of experts suggest it in the previous round. Thus, the list of issues included in the second round of Delphi survey is usually not the same as in the previous round.
4) Some Delphi studies also provide each participant's previous rating next to the summary statistic (e.g., the mean) to help one better learn about their own ratings in light of that by other participants (see Appendix A in Brancheau, Janz, and Wetherbe 1996). Such a design undoubtedly assists participants to view and assess their own ratings in the context of summarized group responses. This practice nonetheless adds additional work to researchers.

When all the above sub-steps are completed, the Round 2 survey is finalized and can be distributed to participants. Afterward, repeat the afore-listed sub-steps (1)–(4) in the next round of Delphi survey till researchers gather all the data they need for their study. Nasa, Jain, and Juneja (2021) suggest to set some closing criteria or criteria for ending Delphi rounds. Either percentage agreement based on the mean or median statistics computed from responses to the scale statements or the number of Delphi rounds can serve the purpose. Apparently, the latter is much more clear-cut and easier to be implemented. Also most, if not all, Delphi researchers agree that three rounds of the survey process should be sufficient for completing data collection in their studies as indicated earlier in Step (4) Delphi Rounds Determination and Counting.

What is an appropriate period of time between completing one Delphi round and moving to the next? Duffield (1993) reported that each round could take up to eight weeks to complete. That is probably due to the need to follow up non-respondents and the time needed to adequately summarize results and prepare feedback for the next round. The time length between two Delphi rounds can be reduced to four weeks (e.g., Zhang and Salaba 2009) if the Delphi rounds are done by electronic means such as the survey site option. The change in survey delivery method from the traditional postal mail to the present-day survey site saves the time needed for distributing the survey to experts and collecting the completed ones from them. If the time period between two Delphi rounds lasts too long (e.g., several months), experts' participation rate might be negatively affected. If the time interval is too short (e.g., one or two weeks), researchers would not have enough time to administer each round of Delphi survey and summarize experts' responses adequately.

As shown in Figure 12.1, the steps in the data collection process of Delphi research carry different weights. Step (3) Survey Delivery Option Selection and Step (4) Delphi Rounds Determination and Counting are less demanding, whereas the remaining steps require substantial efforts from Delphi researchers.

Even though response rate is not a primary concern in Delphi research as it is in survey studies, investigators still need to keep participant dropout rate as low as possible from one round to the next. Belton et al. (2019) propose several measures to ensure that experts complete all the Delphi rounds in research: (1) Use social rewards; (2) Use financial awards; (3) Use personal communication with experts. In addition, present academic awards (e.g., providing participants with a research report summary) when experts are from the academia as that could be another effective measure to minimize the participant dropout rate. Of all the applicable measures, designing, and conducting a quality Delphi study, including the creation of a quality survey, perhaps outweighs what is suggested above. Belton et al. (2019) also note that the greater the number of rounds, the greater the degree of dropout, which perhaps offers one more reason for Delphi researchers to choose a reasonable number of rounds.

Like any research method, the Delphi technique has its own weaknesses. Landeta (2006) summarizes its limitations as follows: (1) experts as its basic source of data when they unavoidably have biases; (2) use of consensus as a way to approach the truth; (3) restricted interaction involved in written and controlled feedback; (4) impunity conferred by the anonymity with respect to likely irresponsible actions on the part of the experts; (5) difficulty of checking the method's accuracy and reliability. The list of limitations goes on in the review of Delphi research by Landeta (2006). Nevertheless, the Delphi method stands out as an obvious choice when the problem at hand can benefit from collective, subjective judgments or decisions and when group dynamics do not allow for face-to-face communication due to time differences, geographic distance, personality conflicts, and other factors (e.g., the COVID-19 pandemic). Other research methods (e.g., interviews) can also be used together with the Delphi method to overcome some of its weaknesses Landeta (2006) described above.

### *12.2.4 Dos and don'ts in Delphi research*

When using the Delphi method for collecting research data, investigators are encouraged to pay attention to the following dos and don'ts in order to make certain that the Delphi study is conducted with good quality.

The first group of dos and don'ts in Delphi research concerns survey development. As explained earlier, creating a quality survey is instrumental in any Delphi study. Therefore, efforts must be made to ensure that the Delphi survey is properly developed by observing the dos and don'ts below.

- Do prepare the questionnaire for data collection based on research questions. This point may sound pedantic, yet it is quintessential for Delphi and all other researchers to keep in mind so that the study will not be astray from the very beginning.
- Don't impose your own opinion/judgment to Delphi participants in the survey. Since what is gathered from participants in Delphi research via several rounds of survey is nothing else but subjective judgments and opinions, it is important for investigators to

prepare all scale statements in the survey impartially so that they would not influence participants in one way or another. Section 3.2.2.2 details the don'ts in questionnaire development, including how to avoid asking leading questions in a survey.
- Do keep the number of scale statements at a reasonable level. In general, a Delphi survey can contain up to several dozen scale statements. Any number going beyond several dozen likely overwhelms the participants or might lead to incomplete or random responses from participants.
- Do formulate scale statements at the interval level. The reason for this position is already elaborated earlier: to facilitate better statistical summary and analysis of data in the Delphi research process.
- Do pretest the Delphi survey before Round 1 distribution if the survey is constructed from scratch. Refer to Section 3.2.2.1 regarding how to perform survey pretesting.

The second cluster of dos and don'ts in Delphi research relates to selection of experts. Because selected experts play a crucial role in Delphi research, additional considerations should be made to observe the following dos and don'ts in the process of choosing them.

- Do choose participants who have knowledge and experience in the research topic under study based on pre-established criteria. Delphi experts must have appropriate subject expertise before being selected to take part in the project.
- Do recruit participants who not only have expertise in the research topic but also are available and willing to participate in the research project because of the multiple Delphi rounds in the data collection process. Few other data collection techniques have this time commitment prerequisite of participants except ethnography or field study (see Chapter 11, Meta-Methods: Ethnography/Field Study).
- Do select heterogenous participants from the target population. The mere purpose of having heterogenous participants in Delphi research is to gather opinions and judgments from as diverse experts as possible. This heterogeneity helps ensure both the validity and reliability of Delphi research.
- Don't recruit participants only because they are available. Recruiting research participants is always a challenging task. The recruitment task in Delphi research is nevertheless even more challenging in that Delphi participants are experts who generally have a busy schedule. But choosing a convenience sample of participants is never an option in Delphi research.

The third set of dos and don'ts in Delphi research associates with executing the Delphi rounds. Of all the steps in Delphi research, executing the Delphi rounds is the hardest as there are many caveats that may result in a failure of the project. It is hence mandatory to carry out each Delphi round by adhering to the dos and don'ts below.

- Do encourage experts to offer explanations for their responses (i.e., rationales) as well as make comments and suggestions when completing the survey in each Delphi round. Although the ratings or rankings of scale statements are what researchers expect to gather, such data do not contain the richness and details about participants' responses. It is the participants' rationales and comments that may offer additional insights into their rating or ranking responses to the survey.
- Don't overlook the rationale data gathered from participants. Rationale data in Delphi research are explanations participants provide in support of their responses to the

survey. They are valuable to researchers for data analysis as well as to other participants to learn why a particular rating or ranking is given and then help them reconsider and/or revise their responses in the subsequent Delphi round. Because rationale data are qualitative, they need to be summarized qualitatively before sending the summary as feedback to participants in the next Delphi round.

- Don't summarize and present group responses inadequately. Three of the four outstanding features in Delphi research (i.e., iteration, statistical summary and aggregation of responses, controlled feedback) portrayed in Section 12.2.2 signify the importance of this point. Although summarizing and presenting group responses for the next Delphi round is a daunting and time-consuming task, feedback provision is an essential component of the Delphi procedure and there is no shortcut to accomplishing this task. Any corner cutting during the data collection process will eventually affect the quality of data collected in the next Delphi round and subsequently the quality of the Delphi project.
- Do select a time period of reasonable length between rounds. As explained earlier, avoid choosing a time period that is too short (e.g., one week) or too long (e.g., six months) from one round to the next. The ultimate purpose is to keep participants to stay in the data collection process while allowing researchers to have adequate time to summarize and present group responses for the next Delphi round.
- Don't purposely omit or misinterpret experts' responses to the survey. If such an act occurs, it violates research ethics and hence should be prohibited without any condition.

Again, it is impossible to exhaust all the dos and don'ts in using the Delphi method for research as in the case of any other research methods. What remains crucial is to stay concentrated on the general principles and established practices of applying this research method to Delphi studies.

### *12.2.5 Case study of Delphi research*

Brancheau, James C., Brian D. Janz, and James C. Wetherbe. 1996. "Key Issues in Information Systems Management: 1994–95 SIM Delphi Results." *MIS Quarterly* 20 (2): 225–42. https://doi.org/10.2307/249479.

Strictly speaking, Brancheau, Janz, and Wetherbe (1996) was a Delphi study to follow up several previous research projects under the auspices of the SIM. Those projects aimed to survey SIM members periodically since 1980 for key issues in the next several years. For this reason, Brancheau, Janz, and Wetherbe (1996) did not report any research questions specifically for their own study although that is rare for a scholarly publication. Nor did they include a section on literature review – another major component of a research publication – in their report. Searches were performed to see if more recent follow-up studies on the same topic were conducted after the one by Brancheau, Janz, and Wetherbe (1996), and the answer is no. Even though a case study published more recently is desirable, the essence in conducting Delphi research remains the same and is usually independent of the passage of time.

Brancheau, Janz, and Wetherbe (1996) summarized the preceding studies, including their own research under discussion, in Table 1 of their report. They indicated that they followed a similar Delphi approach Dickson and his colleagues (1984) developed but mainly in terms of Delphi rounds. Dickson et al. (1984) began with an open-ended survey asking SIM institutional members to identify and briefly describe five to ten of what they considered to be the major information systems management issues over the next five to ten years. They then consolidated these data into a combined list of issues along with other survey components (e.g., a cover letter) to form a Delphi survey before sending it to the SIM respondents for ranking and commenting. This iteration of ranking and commenting was continued for two additional Delphi rounds, three rounds in total, while the initial step of creating the survey is for generating issues.

Brancheau, Janz, and Wetherbe (1996) started their Delphi study with the top 21 IS management issues derived from Niederman, Brancheau, and Wetherbe (1991) – another prior Delphi research project the SIM sponsored. This is considered a common and acceptable practice when researchers adopt or adapt a survey instrument already used in a similar study for their own investigation so that the instrument has been validated and no pretest of it is necessary. Table 12.1 summarizes the three-round Delphi study by Brancheau, Janz, and Wetherbe (1996) in terms of survey distribution time, number of participants, number of IS management issues included in each round of survey, number of usable responses received, and response rate for each round.

As seen in Table 12.1, Brancheau, Janz, and Wetherbe (1996) conducted a three-round Delphi study, which is a frequently chosen and recommended number of Delphi rounds by many research projects. The three rounds of surveys were distributed three months apart, giving the study's busy participants enough time to respond. If the same study were replicated today, however, the time interval between each round perhaps can be greatly shortened because survey sites (e.g., SurveyMonkey) make the survey process much easier and shorter. For instance, Zhang and Salaba (2009) administered their Delphi survey for each round online and one month apart. Around two dozen issues or scale statements were included in each Delphi round of Brancheau, Janz, and Wetherbe (1996), close to the general and acceptable practice regarding the number of scale statements in the survey. In addition, the authors provided the number of usable responses and response rate for all three Delphi rounds in their article, which is highly desirable but not always reported in many Delphi studies. The survey instrument for Round 3 is available in Appendix A of Brancheau, Janz, and Wetherbe (1996), conforming to the suggested guideline for any study that utilizes a data collection instrument.

The 217 targeted participants of this study were all SIM institutional and board members. In this sense, they are homogeneous. On the other hand, 95% of them were from the United States and 5% from Canada. They were also from different industries (e.g., commercial or non-profit sectors) and held different positions (e.g., chief information officer, senior vice president), which implies some degree of heterogeneity in the participants. As reported by Brancheau, Janz, and Wetherbe (1996), 217 were the targeted population of their study. The actual number of participants was 78, 87, and 83 respectively in Rounds 1–3. Those numbers of participants are within the reasonable limit, especially when the study was beyond nationwide and 5% of the participants were from Canada.

Brancheau, Janz, and Wetherbe (1996) stated that the issues included in their survey were presented in random order to avoid any possible unintended biases. For instance,

*Table 12.1* Summary of Key Features of the Three-round Delphi Study

| *Round* | *Time* | *# of Experts* | *# of Issues* | *# of Usable Responses* | *Response Rate (%)* |
|---|---|---|---|---|---|
| 1 | 1994.3 | 217 | 21 | 78 | 36 |
| 2 | 1994.6 | 217 | 23 | 87 | 40 |
| 3 | 1994.9 | 108 | 20 | 83 | 76 |

respondents may incline to consider the issues listed early in order more carefully than the ones presented later. Additionally, the authors did the randomization so meticulously that they created four different randomized versions of the survey in their study. Such a randomization practice is a nice feature of the current case study.

Brancheau, Janz, and Wetherbe (1996) asked their participants to rate each of the issues listed in every Delphi round on a 10-point scale, where 10 indicated their highest priority issue(s) and 1 the lowest. Although a 10-point scale is often used for rating purpose, such a large number of choices perhaps make it impractical for participants to do their ratings precisely because the number 10 exceeds the maximum capacity of people's STM: five to nine chunks (Miller 1956). Both Sections 3.2.2.2 and 9.3.3 discuss this guideline Miller (1956) established regarding humans' STM capacity. As a result, participants may arbitrarily choose a number from the 10-point scale instead of carefully considering a best choice from 1 to 10. In other words, a 9-point or even a 7-point scale would be more in agreement with humans' STM capacity and thus might yield more accurate ratings of the scale statements.

The three-round Delphi process in Brancheau, Janz, and Wetherbe (1996) followed the typical Delphi procedure. In Round 1, "participants were encouraged to comment on the list of issues and associated rationale as well as write in issues they felt may have emerged" (Brancheau, Janz, and Wetherbe 1996, 227) since the study by Niederman, Brancheau, and Wetherbe (1991). In Round 2, feedback summarizing the results of the first round was sent to the participants along with their individual ratings as a base line for comparison. Issues were listed in rank order of importance from highest to lowest mean rating. Five new issues were added to the list, while three issues with markedly lower ratings were dropped (Brancheau, Janz, and Wetherbe 1996, 227). In Round 3, summary results from the second round were sent to the participants and the three lowest rated issues were dropped from the list of 23 issues, yielding a 20-issue list in the final round of the Delphi study. Adding or deleting survey items (e.g., issues in this case study) based on ratings is a common approach to achieving consensus in a Delphi study. It is also highly commendable that Brancheau, Janz, and Wetherbe (1996) asked for comments and rationales from participants as well as provided participants' individual ratings as a base line for comparison starting from Round 2 of their study.

Appendix C in Brancheau, Janz, and Wetherbe (1996) summarizes results by Delphi round which includes rank, average rating score, and standard deviation. Generally speaking, average and standard deviation are more indicative summary statistics than medium range or interquartile range that numerous other Delphi studies often report. We will further elaborate on the rationale behind the above statement in Chapter 13, Techniques for Data Analysis and Their Selection, from the statistical perspective. Furthermore, Delphi researchers must have interval data to calculating means and more advanced statistics (e.g., standard deviation) because scale data by default are ordinal and cannot be used for computing means and standard deviation (see also Section 3.2.1). This is a statistical fact that we should be aware of when planning to use scales to collect interval data. Specifically, if the researcher wishes to use the Likert scale to collect interval data for computing means and other more advanced statistics, a statement indicating that "all the intervals of two adjacent points are treated as equal" must be included in the survey itself or the scale is visually presented as such at the time of data collection. Such a statement though is not included in the survey instrument for Round 3 of this Delphi study, the visual presentation of the equal intervals between two adjacent points on the scale is prominently illustrated in Appendix A of Brancheau, Janz, and Wetherbe (1996).

A four-page summary of results was mailed to all the 217 target participants of the Delphi study in February 1995 to facilitate a follow-up interview. Seven of the SIM officers and board members who had participated in the Delphi study made themselves available for the telephone interview. In addition to asking for their general reaction to the results, the researchers asked which forces were driving the top issues, what made the issues so troublesome, and whether they thought the framework had the right balance between management and technology issues. The interviews seemed properly done and are a fine complement to the Delphi method in that interviews can solicit in-depth data from participants that would not be able to be gathered via the Delphi survey. Therefore, it is generally recommended to use interviews or other research methods in conjunction with the Delphi method when appropriate and feasible. In sum, Brancheau, Janz, and Wetherbe (1996) can serve as a decent example of Delphi research methodologically.

**Section Summary**

- Delphi study is a method for collecting data with a questionnaire or survey from a group of experts to address a research problem in order to reach consensus or make forecasts via several rounds of exchanges.
- The Delphi method possesses four main features: anonymity, iteration, statistical summary and aggregation of individual participants' responses, and controlled feedback of responses to all participants.
- The major steps in the data collection process of Delphi research include survey development, selection of participants, survey delivery option selection, Delphi round determination and counting, and execution of Delphi rounds. These steps are also visualized in Figure 12.1.
- The procedure for executing the second or later Delphi round involves the following sub-steps: (1) Compute summary statistics of group responses to scale statements in the survey; (2) Summarize participants' rationale data qualitatively to be included in the next Delphi round, and gather all issues participants suggested in the previous round; (3) Aggregate and arrange all the issues in descending order by their corresponding summary statistics, and update the list by adding issues participants suggested in the previous round and remove any issues that receive no or low ratings; (4) Provide each participant's previous rating next to the group summary statistics; (5) Distribute the survey for the next Delphi round.
- The dos and don'ts in using the Delphi method for data collection primarily concern survey development, participant selection, and Delphi round execution. Refer to Section 12.2.4 for details in addition to following the general guidelines for Delphi research.

## References

Alhadreti, Obead, and Pam Mayhew. 2018. "Rethinking Thinking Aloud: A Comparison of Three Think-Aloud Protocols." *CHI 2018* Paper 44. April 21–26, 2018, Montréal, QC, Canada. 1–12. https://doi.org/10.1145/3173574.3173618.

Belton, Ian, Alice MacDonald, George Wright, and Iain Hamlin. 2019. "Improving the Practical Application of the Delphi Method in Group-Based Judgment: A Six-Step Prescription for a

Well-Founded and Defensible Process." *Technological Forecasting & Social Change* 147: 72–82. https://doi.org/10.1016/j.techfore.2019.07.002.

Bernardini, Silvia. 2001. "Think-Aloud Protocols in Translation Research: Achievements, Limits, Future Prospects." *Target* 13 (2): 241–63. https://doi.org/10.1075/target.13.2.03ber.

Boulkedid, Rym, Hendy Abdoul, Marine Loustau, Olivier Sibony, and Corinne Alberti. 2011. "Using and Reporting the Delphi Method for Selecting Healthcare Quality Indicators: A Systematic Review." *PLoS ONE* 6 (6): e20476. https://doi.org/10.1371/journal.pone.0020476.

Brancheau, James C., Brian D. Janz, and James C. Wetherbe. 1996. "Key Issues in Information Systems Management: 1994–95 SIM Delphi Results." *MIS Quarterly* 20 (2): 225–42. https://doi.org/10.2307/249479.

Cho, Hwayoung, Dakota Powell, Adrienne Pichon, Lisa M. Kuhns, Robert Garofalo, and Rebecca Schnall. 2019. "Eye-Tracking Retrospective Think-Aloud as a Novel Approach for a Usability Evaluation." *International Journal of Medical Informatics* 129: 366–373. https://doi.org/10.1016/j.ijmedinf.2019.07.010.

Chu, Heting, and Qing Ke. 2017. "Research Methods: What's in the Name?" *Library & Information Science Research* 39 (4): 284–99. https://doi.org/10.1016/j.lisr.2017.11.001.

Cotton, Deborah, and Karen Gresty. 2006. "Reflecting on the Think-Aloud Method for Evaluating e-Learning." *British Journal of Educational Technology* 37 (1): 45–54. https://doi.org/10.1111/j.1467-8535.2005.00521.x

Dalkey, Norman. 1969. "An Experimental Study of Group Opinion: The Delphi Method." *Futures* 1 (5): 408–26. https://doi.org/10.1016/S0016-3287(69)80025-X

Dalkey, Norman, and Olaf Helmer. 1963. "An Experimental Application of the Delphi Method to the Use of Experts." *Management Science* 9 (3): 458–67. https://doi.org/10.1287/mnsc.9.3.458

Dickson, Gary W., Robert L. Leitheiser, James C. Wetherbe, and Mal Nechis. 1984. "Key Information Systems Issues for the 1980's." *MIS Quarterly* 8 (3): 135–59. https://doi.org/10.2307/248662

Duffield, Christine. 1993. The Delphi Technique: A Comparison of Results Obtained Using Two Expert Panels. *International Journal of Nursing Studies* 30 (3): 227–37. https://doi.org/10.1016/0020-7489(93)90033-Q

Durning, Steven J., Anthony R. Artino Jr., Thomas J. Beckman, John Graner, Cees Van Der Vleuten, Eric Holmboe, and Lambert Schuwirth. 2013. "Does the Think-Aloud Protocol Reflect Thinking? Exploring Functional Neuroimaging Differences With Thinking (Answering Multiple Choice Questions) Versus Thinking Aloud." *Medical Teacher* 35 (9): 720–26. https://doi.org/10.3109/0142159X.2013.801938.

Ericsson, K. Anders, and Herbert A. Simon. 1980. "Verbal Reports as Data." *Psychological Review* 87 (3): 215–50.

Ericsson, K. Anders, and Herbert A. Simon. 1993. *Protocol Analysis: Verbal Reports as Data*. 2nd ed. Cambridge, MA: MIT Press.

Fonteyn, Marsha E., Benjamin Kuipers, and Susan J. Grobe. 1993. "A Description of Think Aloud Method and Protocol Analysis." *Qualitative Health Research* 3 (4): 430–41. https://doi.org/10.1177/104973239300300403.

Griffith, Jullian R., R.J. Hartley, and Jonathan P. Willson. 2002. "An Improved Method of Studying User-System Interaction by Combining Transaction Log Analysis and Protocol Analysis." *Information Research* 7 (4). Accessed September 5, 2022. http://InformationR.net/ir/7-4/paper139.html.

Hsu, Chia-Chien, and Brian A. Sandford. 2007. "The Delphi Technique: Making Sense of Consensus." *Practical Assessment, Research, and Evaluation* 12: Article 10. https://doi.org/10.7275/pdz9-th90.

Jääskeläinen, Riitta. 2010. "Think-Aloud Protocol." In *Handbook of Translation Studies*, Vol. 1, edited by Yves Gambier, and Luc van Doorslaer, 371–73. Philadelphia: John Benjamins Publishing Co.

Landeta, Jon. 2006. "Current Validity of the Delphi Method in Social Sciences." *Technological Forecasting & Social Change* 73 (5): 467–82. https://doi.org/10.1016/j.techfore.2005.09.002.

Lilja, Kari K., Kimmo Laakso, and Jari Palomäki. 2011. "Using the Delphi Method." *2011 Proceedings of PICMET'* (Portland International Center for Management of Engineering and Technology) *11: Technology Management in the Energy-Smart World*, 1004–13.

Linstone, Harold A., and Murray Turoff. 1975. *The Delphi Method: Techniques and Applications*. London: Addison-Wesley.

Miller, George A. 1956. "The Magical Number Seven Plus or Minus Two: Some Limits on Our Capacity for Processing Information." *Psychological Review* 63 (2): 81–97. https://doi.org/10.1037/h0043158

Nasa, Prashant, Ravi Jain, and Deven Juneja. 2021. "Delphi Methodology in Healthcare Research: How to Decide Its Appropriateness." *World Journal of Methodology* 11 (4): 116–29. https://doi.org/10.5662/wjm.v11.i4.116

Niederman, Fred, James C. Brancheau, and James C. Wetherbe. 1991. "Information Systems Management Issues for the 1990s." *MIS Quarterly* 15 (4): 475–500. https://doi.org/10.2307/249452

Payne, John W. 1994. "Thinking Aloud: Insights into Information Processing." *Psychological Science* 5 (5): 241, 245–48. https://doi.org/10.1111/j.1467-9280.1994.tb00620.x

Rayens, Mary Kay, and Ellen J. Hahn. 2000. "Building Consensus Using the Policy Delphi Method." *Policy, Politics, & Nursing Practice* 1 (4): 308–15. https://doi.org/10.1177/152715440000100409

Rowe, Gene, and George Wright. 1999. "The Delphi Technique as a Forecasting Tool: Issues and Analysis." *International Journal of Forecasting* 15 (4): 377–81. https://doi.org/10.1016/S0169-2070(99)00018-7

Stone Fish, Linda, and Dean M. Busby. 2005. "The Delphi Technique." In *Research Methods in Family Therapy*, edited by Douglas H. Sprenkle and Fred P. Piercy, 238–53. 2nd ed. New York: Guilford.

Wildemuth, Barbara M., ed. 2017. *Applications of Social Research Methods to Questions in Information and Library Science*. 2nd ed. Denver: Libraries Unlimited.

Zhang, Yin, and Athena Salaba. 2009. "What Is Next for Functional Requirements for Bibliographic Records? A Delphi Study." *Library Quarterly* 79 (2): 233–55. https://doi.org/10.1086/597078

# Part IV

# Research Design and Method Selection

## A Further Look

# 13 Techniques for Data Analysis and Their Selection

**Chapter Overview Questions**

- What are the two major kinds of data analysis techniques?
- What are the functions for descriptive and inferential statistical data analyses?
- What are the features of qualitative data analysis?
- What are the two main techniques for analyzing qualitative data?
- What should be considered in selecting data analysis techniques in research?
- What is the integrated approach to data analysis?

From Chapters 3 to 12, more than one dozen research methods are presented and discussed for collecting various kinds of data (e.g., questionnaire data, interview data) in studies. Those collected data can be broadly categorized into two kinds: quantitative and qualitative. Table 13.1 summarizes the quantitative and qualitative data that each research method covered in this book can be used to gather.

Table 13.1 confirms that any research method can be used for collecting both quantitative and qualitative data – a point brought up in Chapter 1. However, each method may focus more on one of the two data types. It is therefore inappropriate to label a research method, named after data collection technique, as entirely quantitative or qualitative. What we should do instead is to analyze the quantitative data gathered for one study quantitatively and likewise analyze qualitative data qualitatively. In other words, the quantitative and qualitative categorization only applies to data analysis but cannot to data collection. Accordingly, there is no research method or data collection technique that can be called quantitative and qualitative.

This chapter comprises two main sections: one on techniques for analyzing quantitative data and the other on techniques for analyzing qualitative data. After those two kinds of data analysis techniques are presented, an elaboration on their selection in research will be provided.

## 13.1 Techniques for analyzing quantitative data

Analysis of quantitative data is also known as statistical analysis. This type of analysis deals with numerical data and can be subdivided into descriptive statistics and inferential statistics. Statistics often constitute a prominent portion of books on research methods or are the only subject of such books. The present monograph, however, attempts to cover

DOI: 10.4324/9781003021711-17

*Table 13.1* Research Methods Versus Data Type (Quantitative or Qualitative)

| *Data Collection Technique* | *Data Collected* | |
|---|---|---|
| | *Quantitative* | *Qualitative* |
| Bibliometrics and other metric-based methods | Citation frequencies, download frequencies, ... | Reasons for citation, ... |
| Content analysis (including discourse analysis, and secondary analysis) | Some frequency data | Text data of various kinds |
| Delphi study | Rating/ranking data | Rating/ranking rationales, comments, suggested issues |
| Ethnography/field study | Depending on respective techniques used in ethnography | Depending on respective techniques used in ethnography |
| Experiment | Frequencies and other experiment measures | Observation data, answers to open-ended questions in pretest and posttest |
| Focus groups | Demographic and summary data | Focus group discussion data |
| Historical method | Summary frequencies, ... | Text data of various kinds |
| Interview | Demographic and summary data | Responses to interview questions |
| Observation | Summary frequencies, ... | Observation notes |
| Questionnaire | Frequency and other quantitative data | Answers to open-ended questions |
| Research diary/journal | Some quantitative data, depending on what is kept in the journal | Text data kept in diaries/journals |
| Theoretical approach (e.g., conceptual analysis, modeling, and theory building) | Rarely | Text data of various kinds |
| Think-aloud protocols | Limited numerical data | Text think-alouds |
| Transaction log analysis | Frequency and other numerical data | Text data from the logs |

this subject in the space of one chapter section and with an emphasis on the essentials of quantitative data analysis rather than a comprehensive coverage of statistics for two reasons: (1) Numerous excellent books on statistics are readily available from various domains and fields. (2) The orientation and plan of this book focus on how to select proper quantitative data analysis techniques for a given research project.

### *13.1.1* ***Descriptive statistical data analysis***

Descriptive statistics enable researchers to describe, summarize, and gain an overview of what is collected before deciding what advanced statistical procedures can be utilized to further analyze the data on hand. Frequency and percentage are among the first descriptive statistical measures researchers try to obtain in their initial analysis.

Frequency (f) is a measurement often first selected in summarizing data. It refers to the number of cases in one data category (e.g., number of male or female participants in an experiment study). We attain a frequency by counting (manually for a small data set) or computing (using software for a large data set) the number of cases in the data. The second row in Table 13.2 shows the frequency of female and male participants in a hypothetical experiment study.

*Table 13.2* Female and Male Participants in a Hypothetical Experiment Study

| | *Female* | *Male* | *Total* |
|---|---|---|---|
| Frequency (f) | 14 | 8 | 22 |
| Percentage (%) | 63.6 | 36.4 | 100 |

Frequencies alone are not good for comparison purposes. Researchers hence usually report percentages as shown in the last row of Table 13.2 in addition to frequencies. Percentage is calculated by dividing the frequency for one data category (e.g., female participants) with the total frequency for all the categories (e.g., female and male participants). For example, percentage of female participants in the hypothetical experiment study: 14/22 = 0.636. We then multiply it with 100% to convert it to 63.6%. We can do the same to calculate the percentage of male in the hypothetical experiment study: 8/22 = 0.364 which can be converted to 36.4%. There are thus about twice as many female participants as male participants in the hypothetical experiment study. Frequency and percentage, though very basic, can say a lot about data. Most researchers report them in the beginning of their result presentation. Frequency also establishes the basis for other statistics, including central tendency and dispersion measures to be discussed shortly after.

#### *13.1.1.1 Measuring central tendency*

Central tendency is a statistical jargon and indicates how typical (central, similar) the individual cases in the data set being analyzed are. The mode, median, and mean (aka average) are three central tendency measures often seen in the literature.

##### 13.1.1.1.1 MODE

The mode is the value that appears most frequently in a distribution or a list of data, especially when they are visualized in a graph or figure. For example, female is the mode of gender data among the participants in the hypothetic experiment study (see Table 13.2). If we record the number of times ten teens visit social media sites daily as follows: 54, 200, 34, 60, 80, 80, 45, 80, 20, and 35. The number of 80 would be the mode because it appears three times in the distribution, while the rest occurs only once each.

We discussed levels of measurement in Section 2.5.1. The mode can be used for data measured at all levels, including the nominal level. It is relatively easy to determine the mode as no calculation is involved in the process. The mode only considers the value that appears most frequently and disregards the rest in the data set. For this reason, mode is not a regularly used central tendency measure in research reports.

##### 13.1.1.1.2 MEDIAN

The median is a positional measure that divides a distribution into two equal parts. Assuming we have a data set of 9, 11, 24, 12, 16, 6, 18, and 21, we want to identify its median. The steps are as follows:

- Put all the values in the data set in order: 6, 9, 11, 12, 16, 18, 21, 24. The order can be either ascending or descending, although an ascending order seems more natural to most of us.

- Count the number of cases in the distribution
    - If the outcome is an even number as in 6, 9, 11, **12**, **16**, 18, 21, 24, add the middle two values and divide the sum by 2 to obtain the median: (12 + 16)/2 = 14.
    - If the outcome is an odd number as in 6, 9, 11, 12, **16**, 18, 21, 24, 30, the middle value is the median: 16.

The median only considers the middle value(s) of a distribution in the calculation. The median is suitable for data measured at the ordinal level or higher. In other words, we cannot calculate median for nominal or category data such as gender, race, or religion. Often the median is reported for housing price, salary, and more. Based on how the median is identified, we know that half of the cases in the data set have a value lower than the median and the other half have a higher value.

#### 13.1.1.1.3 MEAN

The mean or average is the sum of all values in a data set divided by the total number of cases involved. Assume we have a list of data: 9, 11, 6, 12, 18, and 16, each representing the number of publications six individuals read. What is the average number of publications read among the six individuals? Below are the steps for calculating the mean:

- Add all the values (i.e., total value) in the data set: 9 + 11 + 6 + 12 + 18 + 16 = 72.
- Find out the number of cases in the data set: 6.
- Divide the total value 72 by the total number of cases in the data set 6: 72/6 = 12 (the mean).

The mean calculated as above is called the arithmetic mean. Other kinds of means are beyond the scope of this book. The mean is suitable for data measured at interval or ratio level but not applicable to nominal and ordinal data. For example, you know something is not right if you see that the mean is calculated with ordinal data such as the ordinal rating data collected in some Delphi research (see Step (1) in Section 12.2.3.1). The mean also serves as a foundation for calculating other statistical measures.

The calculation of the mean takes account of all the values in a data set but is sensitive (i.e., prone to change) to extreme values. Note "extreme" implies really big difference. For instance, the average household income in your community will be increased significantly if Mark Zuckerberg, co-founder of the social media site Facebook and its parent company Meta Platforms, had moved there.

In summary, the mode indicates the point in a distribution with the highest density. The median is the midpoint of a distribution. The mean is an average of all the values in a distribution. Researchers choose central tendency measures according to their study objective (e.g., interested in average or middle value?), nature of their data (e.g., containing extreme values?), and levels of measurement. Table 13.3 recapitulates what central tendency measures can be obtained for data measured at a given level. Generally speaking, the mean is more appropriate than the median, and the median is more suitable than the mode for measuring central tendency if the data are measured at the proper level (i.e., nominal, ordinal, interval, or ratio). On the other hand, avoid using a less indicative and representative central tendency measure if the data are measured at a higher level. If one has interval data but only analyzes them using the mode, for example, all the ranking and

*Table 13.3* Levels of Measurement Versus Central Tendency Measures

| *Central Tendency Measure* | *Level of Measurement* | | |
|---|---|---|---|
| | *Nominal* | *Ordinal* | *Interval/Ratio* |
| Mode | √ | Losing | |
| Median | | √ | Data |
| Mean | | | √ |

equal distance information in the data is not considered when using mode as a central tendency measure.

The procedures for manually calculating central tendency measures are depicted above to help better understand them and also because they are essential yet not complicated in calculation. These measures as well as all other statistics can certainly be calculated using spreadsheet and other statistical applications.

### *13.1.1.2 Measuring dispersion*

As shown in above discussion, measures of central tendency only identify the most representative or typical value of a data set. However, a complete description of the data set also requires measuring its extent of dispersion, variation, or atypical value. Measure of qualitative variation, range, and standard deviation represent three conventional statistics for determining dispersion.

#### 13.1.1.2.1 MEASURE OF QUALITATIVE VARIATION

The measure of qualitative variation is a dispersion measure for nominal data. When researchers gather data measured at nominal level and still wish to analyze them with a desirable degree of rigor and sophistication, they need measures appropriate to such data. Indexes of heterogeneity or qualitative variation can be used to achieve the purpose of measuring statistical dispersion of nominal data.

Wilcox (1967) first proposed the so-called indexes of qualitative variation by presenting a series of indexes for assessing the deviation of each value in a distribution from the mode, mean, et cetera. Such indexes, regardless of their computation dissimilarities, reflect the number of differences among the categories of the distribution and are based on the number of categories and their respective frequency. "In general, the larger the number of categories in a data set, the greater the overall differences among them and the greater the degree of variation. Likewise, the smaller the number of categories and their difference, the smaller the variation within the distribution" (Frankfort-Nachmias, Nachmias, and DeWaard 2015, 318).

The measure of qualitative variation is in essence based on the ratio of the total number of differences in the distribution to the maximum number of possible differences within the same distribution. The ratio thus always takes a value between 0 and 1 which facilitates comparison among different indexes of qualitative variation for the same distribution. Variation is 0 if and only if all cases in the data set belong to a single category. Variation is 1 if and only if cases in the data set are evenly divided across all categories (Wilcox 1973). In addition, Wilcox (1973) performed an evaluation of the six different indexes of qualitative variation in terms of formula simplicity, computational ease, information utilization, multivariate extension, sampling distribution, distributional characteristics, and stability after he applied those indexes of qualitative variation with the 1968 US presidential election data. However, no single index of qualitative variation outperforms other counterparts in his evaluation. Wilcox (1973) hence concludes that the "evaluation is inevitably somewhat arbitrary because it is based on an estimation of the relative importance of the seven desiderata. There simply does not exist a more general criterion that we can invoke to support such a judgment. The effort is nevertheless made in the belief that it might stimulate others to form their own judgments and perhaps develop less subjective rationales for them" (340). Although there are a wide variety of indexes of qualitative variation and they are valuable for measuring dispersion in nominal data, they have been relatively little studied in the statistics literature (Wikipedia Contributors 2022).

A pertinent example for measuring qualitative variation can be found when we consider the racial composition of China and the United States. The racial composition of China is quite uniform, implying that most, if not all, are of the same race and the number of racial differences in its population is therefore small. By comparison, the United States is known as a melting pot of different races, which denotes a large racial variation as its population is made of many different races. In other words, the qualitative variation of races in the United States is much larger than that of China. The index value of qualitative variation for China should be close to 0, whereas that for the United States must be a value well above 0 although it is not 1 either. This is because all the racial categories (e.g., Caucasian, African American, Latin American, Asian) in the United States are not evenly divided in number. In short, the measure of qualitative variation provides a method for calculating dispersion for data measured at the nominal level.

#### 13.1.1.2.2 RANGE AND INTERQUARTILE RANGE

The range is the distance between the highest and lowest values of a data set or distribution at the ordinal level. We commonly call it difference. Its calculation is pretty straightforward. We just subtract the lowest value (i.e., minimum) in the data set from the highest value (i.e., maximum) to obtain the difference. According to the list of countries/territories ranked by GDP (gross domestic product) per capita in 2021 (Wikipedia Contributors 2023), for instance, Luxembourg had the highest value of 133,590 USD, while Burundi had the lowest value of 222 USD. The difference between the two GDP per capita values is 133,368 USD. That is, the range in the distribution of GDP per capita by country/territory in 2021 is 133,368 USD. In addition to the range, maximum and minimum values themselves in a distribution can also serve as measures of dispersion. That is why these two values are often reported in research publications too.

The range can provide a quick dispersion measurement of the data, especially when measures of central tendency produce a distorted picture of the reality. For example, a billionaire moving to a middle-class community would "inflate" the average household income there. However, the range is a crude measure of dispersion because it only takes account of the two values at both ends of a distribution, while all other values in between are completely ignored. The range hence is sensitive to any changes to these two end values.

The interquartile range (IQR) is a measure of statistical dispersion based on dividing a data set into quartiles. Quartiles divide an ordered data set into four equal parts. The values that divide these parts are known as the first quartile, second quartile, and third quartile (Q1, Q2, Q3). The IQR is equal to the difference between the upper and lower quartiles: IQR = Q3 − Q1. The IQR is an alternative to the range in that it measures the spread of the middle half of the distribution and is less affected by the possibly extreme values at both ends of a distribution (Rowntree 1981, 52–53). That is the reason why some Delphi researchers suggest that the IQR be used when analyzing and summarizing participants' responses to the scale statements measured at the ordinal level (see Section 12.2). Even though the IQR is less affected by the possibly extreme values of a distribution, this statistic again only takes into consideration two interquartile values of a distribution. Both the range and IQR are appropriate for data measured at the ordinal level or higher.

###### 13.1.1.2.3 VARIANCE AND STANDARD DEVIATION

Both the variance and standard deviation (SD) are important measures of dispersion. We, however, will not discuss both measures in detail because the mathematical calculations involved go beyond the scope of this book. Instead, we will merely indicate that SD is derived from variance and used more often in quantitative data analysis. Now, let us use some hypothetical data (see Table 13.4) to explain the meaning of SD.

Simply put, the SD is the average difference from the mean in a data set. As displayed in Table 13.4, the mean life satisfaction values for the four countries are very similar. This implies that people in these four nations have similar level of life satisfaction. But the SD score for Italy (i.e., 3.2) shows that people there have a wide range of life satisfaction when compared with the other three countries. Some Italian people are very satisfied about their lives while others not. The SD reflects the large variation in the data set.

In contrast to the range, the SD is more stable because it takes account of all the values in the data set. It is also heavily used in inferential statistics. Furthermore, the SD should always be considered in relation to the mean rather than in isolation. For example, an SD of 3.2 with the mean of 6.6 is large (see Italy in Table 13.4). If the mean value were 66, an SD of 3.2 would signal a small variation or dispersion in the data. In other words, most of the data values are around the mean and not widely dispersed.

In summary, the selection of a dispersion measure in research depends largely on study objective and levels of measurement of the data collected. Table 13.5 reiterates what dispersion measures can be obtained for data measured at a given level.

If nominal data are gathered, the measure of qualitative variation is the only viable choice. If collected data are measured at the ordinal level, then the range or IQR can be computed. Use the SD if interval or ratio data are gathered in research. Similar to the selection of central tendency measures, we should avoid choosing a dispersion measure that does not make full use of the data measured at the proper level (i.e., nominal, ordinal, interval, or ratio) even though the calculation can be done. If we calculate the range for data measured at the interval level, for instance, the equal distance information in the interval data is not taken into consideration. As a result, such information is lost in analysis and thus should be precluded by all means. In data collection we should gather

*Table 13.4* Life Satisfaction in Four Western Nations (Hypothetical Data)

| *England (N=903)* | | *Germany (N=950)* | | *Italy (N=998)* | | *United States (N=998)* | |
|---|---|---|---|---|---|---|---|
| *Mean* | *SD* | *Mean* | *SD* | *Mean* | *SD* | *Mean* | *SD* |
| 6.7 | 1.0 | 6.7 | 1.2 | 6.6 | 3.2 | 6.5 | 1.3 |

*Table 13.5* Levels of Measurement Versus Dispersion Measures

| *Dispersion Measure* | *Level of Measurement* | | |
|---|---|---|---|
| | *Nominal* | *Ordinal* | *Interval/Ratio* |
| Measure of qualitative variation | √ | Losing | |
| Range and interquartile range | | √ | Data |
| Standard deviation | | | √ |

data measured at the highest possible level as initially discussed in Section 2.5.1.5. For example, Delphi researchers are encouraged to prepare interval scale statements instead of ordinal ones for their survey instrument. Likewise, in data analysis we should analyze what is gathered to the highest possible level of measurement in order to utilize all the information embedded in the data collected to the fullest.

In conclusion, measures of central tendency exhibit the most typical or average characteristics of the data set, while dispersion measures indicate how many members of the distribution deviate from the central tendency measure and the extent of the deviation. A small deviation denotes that most cases of a data set are clustered around the measure of central tendency. In contrast, a large deviation demonstrates that the measure of central tendency is a poor representation of the distribution. If researchers want to present a complete picture of their data, they should use both central tendency and dispersion measures of descriptive statistics. Presenting only one type of measures could be misleading in some cases. Descriptive statistical data analysis, as mentioned earlier, signifies the beginning of statistical data analysis as well as lays the foundation for subsequent inferential statistical data analysis. Furthermore, descriptive statistical data analysis is often made to provide an overview of some parameters in the qualitative data collected (e.g., number of categories) to help readers better comprehend the ensuing qualitative data analysis results.

### *13.1.2 Inferential statistical data analysis*

Inferential statistical data analyses or just inferential statistics denote analyses of quantitative data collected from a sample that represents a population for research purposes. Refer to Section 2.5.2 if you would like to review sample, population, sampling, and other related concepts. The notation convention for sample and population is depicted in Table 13.6. As seen, properties of a population are called parameters, whereas that of a sample are labeled as statistics.

Table 13.6 lists the two prominent notations for the mean and SD. Other less common notations are named accordingly. Without going into the technical details, inferential

*Table 13.6* Notation Convention of Sample and Population

| | *Property* | *Mean* | *Standard Deviation* |
|---|---|---|---|
| Population (Greek letters) | Parameter | $\mu$ (mu) | $\sigma$ (sigma) |
| Sample (Roman letters) | Statistic | $\bar{X}$ | SD |

statistics are always intended to rely on data collected from a sample to achieve one of the following three objectives:

1) Estimation – Make inferences about a population parameter (e.g., the mean) that is unknown from a sample statistic which can be measured.
2) Hypothesis testing – Make assumptions about a population in advance, and then use a sample to test them.
3) Prediction – Make predictions about the future state of a population based on the analysis of a sample at present.

Before discussing the three functions of inferential statistics, we need to consider related topics such as shape of a distribution and correlation.

#### *13.1.2.1 Shape of a distribution*

In Section 13.1.1, we describe and present (1) measuring central tendency of a distribution using the mode, median, and mean, and (2) measuring dispersion of a distribution with the measure of qualitative variation, range (including maximum and minimum), IQR, and SD. The third and last aspect about a distribution is its shape, with which we can provide a full illustration of the distribution. Like central tendency and dispersion measures, the shape of a distribution contributes to inferential statistics in its unique ways.

When a distribution is presented graphically or visualized, it shows its shape. We usually look at the number of modes, symmetry, and the like to determine the shape of a distribution. Some distributions only have one mode or unimodal (e.g., number of hours each person sleeps everyday), some have two modes or bimodal (e.g., males and females each having a typical height), and others are multimodal (e.g., a typical temperature in each of the four seasons). In research methods, we stop counting when the number of modes reaches three. We would simply refer to that as multimodal. If a distribution is symmetric, it implies that the mean, median, and mode are all of the same value which splits the distribution into two equal halves. This concept is useful in choosing proper techniques for analyzing the data further and in visualizing the data. One distribution that is always presented in books on quantitative data analysis is the normal distribution or normal curve (see Figure 13.1).

As shown in Figure 13.1, the normal distribution is unimodal and symmetric in addition to possessing other properties. The middle point in the normal distribution represents the mean, median, and mode. We'll not go any further about the shape of a distribution due to the scope of this book. We only need to know that the word "normal" in this context denotes "idealized" so that researchers can use it and other kinds of distributions to do estimation in inferential statistics (Rowntree 1981, 69).

Figure 13.2 is the well-publicized illustration derived from the Centers for Disease Control and Prevention (CDC 2007) in the US during the COVID-19 pandemic regarding why we should help flatten the curve of COVID-19 cases. It shows two unimodal, symmetric curves. The purpose of flattening the curve of COVID-19 cases is to delay the arrival of peak daily number of COVID-19 cases (i.e., the mean/median/mode) as well as to lower the daily number of COVID-19 cases so that our health care systems would be able to handle all the COVID-19 cases within their capacity. Figure 13.2 suggests that we should all do our best to flatten the curve by taking preventive measures such as wearing a mask, washing hands, and keeping a safe distance when interacting with

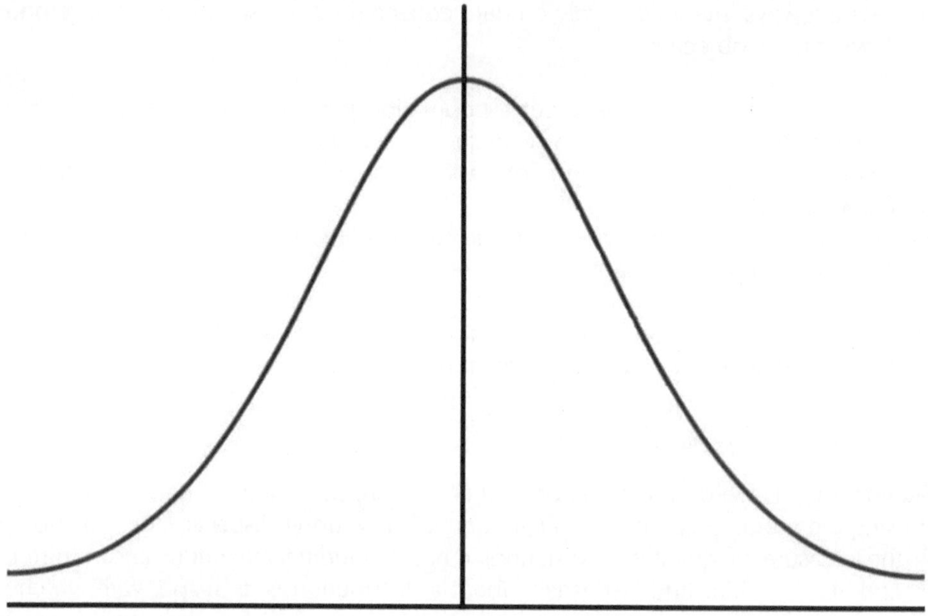

*Figure 13.1* The Normal Distribution[1]

other people. This way, we would prevent our health care systems from overwhelmed so that they could treat all COVID-19 patients as expected and desired.

*13.1.2.2 Correlation*

Correlation, association, or relationship are treated as synonyms and often used interchangeably in research publications. Finding out correlation between two variables or among three and more variables is one objective many researchers strive to achieve in their studies. In such data analysis, researchers want to determine if an independent variable (e.g., age, gender) has any impact on or correlation with a dependent variable (e.g., income, attitude toward climate change). Section 2.3.1 describes and presents the

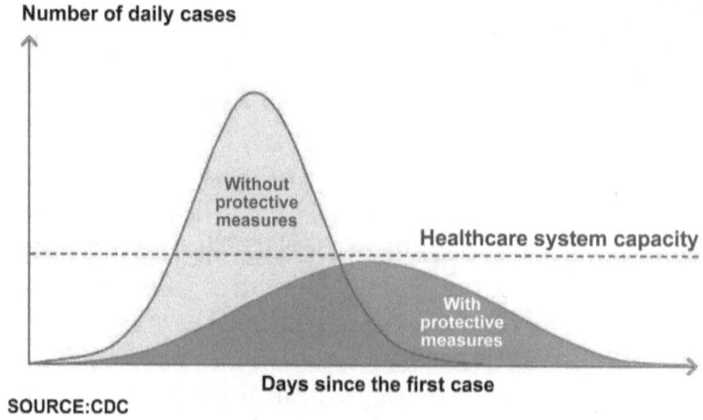


*Figure 13.2* Flattening the Curve of COVID-19 Cases[2]

concepts of independent and dependent variables. There are two kinds of measurements for correlation or relationship: direction and strength.

#### 13.1.2.2.1 DIRECTION OF RELATIONSHIP

The direction of relationship can be positive or negative. Do note that positive or negative in this context only refers to whether a correlation coefficient, a measure for strength of relationship (to be discussed immediately below), is a positive or negative number. A positive relationship means that the values of the independent and dependent variables increase or decrease simultaneously. For example, when more funding (independent variable) is available to a library, the library can offer more services (dependent variable).

A negative relationship implies that when the value of one variable increases, the value of the other variable will decrease, and vice versa. For instance, when the climate becomes warmer, the natural resources available to us would become less due to climate-related disasters. In this example, try to think which variable is independent and which one is dependent.

#### 13.1.2.2.2 STRENGTH OF RELATIONSHIP

The strength of relationship is always measured with a value of correlation coefficient between negative 1 (−1) and positive 1 (1). Figure 13.3 demonstrates how the coefficient value should be interpreted. When a research publication reports, for example, that there is a strong relationship between Variable A and Variable B, the coefficient value would be between 0.7 and 0.89.

There are different kinds of correlation coefficients for measuring the strength of relationship based on levels of measurement. Common correlation coefficients reported in research publications include the following:

- Pearson's r, used for variables measured at the interval and ratio levels;
- Gamma (γ) and Spearman's rho (ρ) for variables measured at the ordinal level;
- Lamda (λ), also known as the Guttman coefficient of predictability, for variables measured at the nominal level.

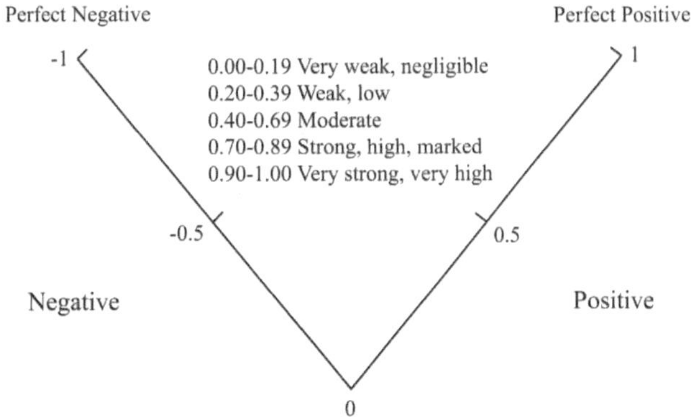


*Figure 13.3* Direction and Strength of Relationship[3]

#### *13.1.2.3 Functions of inferential statistics*

After the data set or distribution is fully described and summarized using measures of central tendency, dispersion, and shape, researchers are ready to perform inferential statistical data analysis. As briefly indicated earlier, inferential statistics are needed because the data gathered in a study often represent only a subset (i.e., sample) from the entire data pool (i.e., population) for time, money, and feasibility reasons. In other words, inferential statistical data analyses are performed only with data from a sample. Estimation, hypothesis testing, and prediction are the three functions of inferential statistics.

##### 13.1.2.3.1 ESTIMATION

When a sample is used to estimate the properties (e.g., the mean) of a population, some estimation error is inevitable. The ideal situation is to minimize the estimation error rate, also known as margin of error. If the margin of error is 3%, for example, researchers using a sample to estimate the properties of population would run a 3% of risk of being wrong in making the estimation. Assuming we would like to learn about subscribers' satisfaction with Netflix – a video streaming service, we surveyed a sample of 2,000 subscribers randomly selected from its over 232,000,000 subscribers (i.e., population) in the first quarter of 2023 (Statista 2023). Eighty-six percent of the 2,000 subscribers surveyed are very satisfied with Netflix services with an estimation error or margin of error ±3 (plus or minus 3). This result indicates that the percentage of all the Netflix subscribers who are satisfied with its service is 86±3 or between 83 (=86−3) and 89 (=86+3). That is how the estimation is done (without the statistical details) and reported. Section 2.5.2.1 briefly discusses the topic of estimation in the context of sample size determination.

You may often hear or read about poll results on certain issues from mass media. For instance, 93% of the public believe that climate warming is a serious issue with a margin of error ±4. This poll result can be interpreted as 89 (=93−4) to 97 (=93+4)% of the public believe that climate warming is a serious issue. In any case, researchers should strive to keep their estimation error as low as possible.

##### 13.1.2.3.2 HYPOTHESIS TESTING

Section 2.3.2 depicts and presents what a hypothesis or research hypothesis is as well as when and how to formulate one. Research hypotheses must be transformed to null hypotheses and then hypotheses in statistical terms before being tested. Of null hypotheses and hypotheses in statistical terms, the latter is not to be further elaborated due to the orientation of this book. A null hypothesis is a statement that nullifies the research hypothesis. For example, if the research hypothesis reads: "Age is a factor that negatively affects the adoption of handheld devices", its corresponding null hypothesis will be: "Age is not a factor that negatively affects the adoption of handheld devices". Do note that hypothesis testing is always performed on the null hypothesis, and it is based on the testing outcome of a null hypothesis that investigators can determine if the research hypothesis is rejected or retained.

While estimation refers to making inferences about a population parameter that is unknown from a sample statistic that can be measured, hypothesis testing denotes making assumptions about a population property (e.g., the mean or proportion) in advance and then using a sample to test them. Hypothesis testing generally serves the following two interrelated purposes:

- Evaluate the accuracy of assumptions. The estimation function of inferential statistics presented above focuses on making an estimation of a population property, whereas hypothesis testing concentrates on assessing the accuracy of such an estimation or assumption.
- Infer relationships between and/or among variables. That is, hypothesis testing in this context is intended to find out whether a relationship exists between and/or among variables as assumed. For instance, a research hypothesis assumes that age is a factor that negatively affects the adoption of handheld devices. This hypothesis testing not only infers a relationship between age and the adoption of handheld devices but also presumes the direction of the relationship is negative (see Section 13.1.2.2).

Detailed procedures for hypothesis testing can be found in many books on research methods (e.g., Frankfort-Nachmias, Nachmias and DeWaard 2015, 403–22).

###### 13.1.2.3.3 PREDICATION

Unlike the afore-discussed other two functions of inferential statistics – estimation and hypothesis testing, prediction involves a time dimension in its analysis. It is for making predictions about the future state of a population based on the analysis of a sample at present. Prediction in inferential statistics is also known as regression analysis. When making prediction, researchers also want to minimize the prediction error, the same as in the case of estimation. Specifics about regression analysis can be found in numerous books on statistics (e.g., Babbie 2016, 456–60; Frankfort-Nachmias, Nachmias and DeWaard 2015, 347–54).

Prediction can be made based on data gathered from a sample about one, two, or more independent variables. In other words, researchers predict how much change the variation in each independent variable will result in the target dependent variable. The life expectancy is often one such dependent variable. Socioeconomic factors such as income, education, employment, community safety, and social support are some of the independent variables that need to be considered when predicting life expectancy. As the world is becoming more populated, researchers often predict what the world population will be, for instance, in 20 years from now or one specific year in the future. This represents another example of prediction. According to Ruiz, Noe-Bustamante, and Saber (2020), the global population is projected to reach 9.5 billion by 2050.

The three functions of inferential statistics can help researchers to choose an appropriate technique for performing inferential statistical data analysis. If the intention is to make an estimate about a population parameter, statistical procedures for estimation should be adopted in data analysis. Likewise, techniques for hypothesis testing or prediction can be chosen accordingly based on the purpose of inferential statistical data analysis.

### *13.1.3 Software for quantitative data analysis*

Software of different types can greatly facilitate quantitative data analysis. Generally speaking, the following three kinds of software are available for quantitative data analysis although this topic will not be explored in detail because of this book's coverage scope and focus.

The first type is spreadsheets such as Microsoft Excel and Google Sheets. Such software usually comes with your word processor in an application suite (e.g., Microsoft 365, Google Workplace). Spreadsheets are more than adequate for basic statistical data analysis and also include some functions for advanced data analysis (e.g., Pearson's r) and data visualization. For extensive research projects with quantitative data, however, dedicated software is needed for data analysis.

The second type refers to dedicated statistical packages such as SPSS (previously Statistical Package for the Social Sciences; currently Statistical Product and Service Solutions) and SAS (previously Statistical Analysis System; currently just SAS). Those software packages can do all kinds of statistical procedures. Many research projects would need such software to perform quantitative data analysis. The price tags for them are nevertheless very hefty. The R language is free and open source but needs programming.

The third type of software for quantitative data analysis includes free websites for computing statistics. Statistics Calculator Online at https://www.easycalculation.com/statistics/statistics.php is one example of such sites (easycalculations.com 2023). However, such sites usually carry a lot of advertisements.

### 13.1.4 More on selecting quantitative data analysis techniques

As seen, the techniques for analyzing quantitative data are so many that it is virtually impossible for any book to provide a comprehensive coverage of them regardless of its size. This book attempts to approach this huge topic by highlighting the functionality of major quantitative data analysis techniques in order to help researchers to select techniques or procedures suitable for their collected data. Considerations for choosing quantitative data analysis techniques include research questions (RQs), number of variables, dependent versus independent variables, levels of measurement, and more. Of all such considerations, levels of measurement specifically serve as an objective and concrete criterion for selecting appropriate statistic procedures or quantitative analysis techniques for the data collected. Loether and McTavish (1980, 20), for instance, synthesize the possible outcomes with regard to choosing statistical procedures based on levels of measurement of data (see Table 13.7).

Table 13.7 reinforces what is discussed above on levels of measurement versus selection of central tendency and dispersion measures. It also recapitulates the general guidelines for choosing proper statistical procedures for data analysis according to levels of measurement. Selecting a statistical procedure for analyzing data measured at the same level (i.e., the diagonal cells in Table 13.7) is always legitimate. However, when choosing a statistical procedure for a lower level of measurement to analyze data measured at a higher level, the information embedding in the data would get lost. For example, if we calculate the mode, a statistic for nominal data, for data measured at the ordinal level, we

*Table 13.7* Levels of Measurement: Data Versus Statistics Used[4]

| *Level of Measurement of Statistic Used* | *Level of Measurement of Data* | | |
|---|---|---|---|
| | *Nominal* | *Ordinal* | *Interval/Ratio* |
| Nominal | Ok | Losing | |
| Ordinal | Logical | Ok | Data |
| Interval/Ratio | | Error | Ok |

would lose the ordering information. On the other hand, we would commit a logical error if we attempt to apply a statistical procedure for a higher level (e.g., the mean) to data measured at a lower level (e.g., ordinal data) because the data do not allow the calculations (e.g., addition, division) needed to perform the statistical analysis.

A website created by Van Eck (2014) based on a book by Andrews et al. (1981) interactively displays which statistical technique(s) could be used according to one's response to a series of prompts (e.g., How many variables does the problem have? How do you want to treat the variables with respect to scale of measurement? Note: scale of measurement is synonymous with level of measurement). This site is of value in guiding one through the process of selecting quantitative data analysis techniques.

## 13.2 Techniques for analyzing qualitative data

Qualitative data consist of texts such as words, phrases, sentences, and passages. Such data can be gathered with any of the data collection techniques presented in this book (see Table 13.1) although some techniques (e.g., interview, observation) are able to yield more qualitative data than others (e.g., questionnaire).

Unlike quantitative data analysis, qualitative data analysis features no fixed, sequential, and standard procedures. As Tesch (1990, 4) fittingly puts it: "When we talk about the analysis of qualitative data, we are not dealing with a monolithic concept like 'statistics'. No one has 'codified' the procedure for qualitative analysis, and it is not likely that anyone ever will". It instead simultaneously uses multiple analytical techniques like abstraction, aggregation, categorization, comparison, conceptualization, induction, inference, interpretation, and summarization in the process. Yet, none of those techniques is singled out and treated as a stand-alone method for analyzing qualitative data because they are presumably imparted in the education process about reasoning, critical thinking, and other similar subjects before the subject of research methods is introduced.

Researchers normally examine and code qualitative data (e.g., interview transcripts, observation notes) to achieve the purpose of data analysis by concurrently applying some or all of the afore-listed analytical techniques. The two prominent methods for analyzing qualitative data are called qualitative coding and content analysis. Both of them are covered in Chapter 9, Content Analysis, from the perspective of data collection. That is, data are collected from sources such as newspaper articles, advertisements, speeches, emails, websites, and social media posts by analyzing their contents using coding and other techniques. In this chapter, qualitative coding and content analysis are to be described and presented from the standpoint of data analysis. It must be pointed out that some principles and practices of these two techniques are shared no matter whether they are used for data collection or for data analysis. For this reason, considerable references are to be made to Chapter 9, Content Analysis, where appropriate in order to avoid unnecessary repetitions.

### *13.2.1 Qualitative coding*

Qualitative coding or coding involves the analytic processes through which data are categorized, conceptualized, and integrated for research purposes. "Coding" is qualified with the term "qualitative" to differentiate it from regular data coding done as one step in data preparation before data analysis (e.g., coding male as 1 and female as 2) or coding in computer programming. For the sake of simplicity, we tend to just use "coding" instead of "qualitative coding" when the context of qualitative data analysis is clear.

When data are categorized in the coding process, researchers may or may not employ a coding scheme which contains all the possible codes or categories needed to perform the coding. Sections 9.3.3 and 9.3.4 already specify when a coding scheme is needed, how to develop or adapt one if needed as well as the general coding process. There are three kinds of coding usually performed in analyzing qualitative data: open coding, axial coding, and selective coding. Each coding type has its functions and features.

#### *13.2.1.1 Open coding*

Open coding is the analytic process through which concepts or categories are identified and their properties (i.e., characteristics that define and describe categories) and dimensions (i.e., variations within properties that give specificity and range to concepts) are discovered in data. It can generate categories and their properties for seeking to determine how categories vary dimensionally. Therefore, open coding can be done without a coding scheme or used for creating a coding scheme. Below is an example of open coding from Strauss and Corbin (1998, 106–7), two renowned authors in qualitative research with a focus on grounded theory (see Section 6.2.1) and extensive coverage of qualitative coding. The open coding in the following example is done on a passage of an interview transcript about teens and drug use without a coding scheme.

- Interviewer: Tell me about teens and drug use.
- Respondent: I think teens use drugs as a release from their parents **["rebellious act"]**. Well, I don't know. I can only talk for myself. For me, it was an experience **["experience"]** [*in vivo* code]. You hear a lot about drugs **["drug talk"]**. You hear they are bad for you **["negative connotation"** to the **"drug talk"]**. There is a lot of them around **["available supply"]**. You just get into them because they're accessible **["easy access"]** and because it's kind of a new thing **["novel experience"]**. It's cool! You know, it's something that is bad for you, taboo, a "no" **["negative connotation"]**. Everyone is against it **["adult negative stance"]**. If you are a teenager, the first thing you are going to do is try them **["challenge the adult negative stance"]**.

All the codes in the above example are in bold, with double quotation marks and square brackets. The codes can be either derived from the data or assigned by the coder. If the former, it is called *in vivo* code. The term "experience" in the above passage is an example of *in vivo* coding. The category or concept identified in the passage is drug use. Rebellious act, easy access, and novel experience represent subcategories of teens' drug use. Do note that in real-life data analysis, coding may not be performed with such intensity as the above example.

#### *13.2.1.2 Axial coding*

Axial coding is the process of relating categories to their subcategories by their properties and dimensions. It is termed "axial" because coding occurs around the axis of a category. Below is an example of axial coding, taken from Strauss and Corbin (1998, 125), that provides an explanation why teens experiment with drugs. The interview transcript passage is the same as the one used in open coding above but the coding type is axial. All the codes in the following example are in bold and *in vivo*.

- Interviewer: Tell me about teens and drug use.
- Respondent: I think teens use drugs as a **release** from their parents. Well, I don't know. I can only talk for myself. For me, it was an **experience**. You **hear** a lot about drugs. You hear they are **bad** for you. There is **a lot of them around**. You just get into them because they're **accessible** and because it's kind of a **new thing**. It's **cool**! You know, it's something that is bad for you, **taboo**, a "no". Everyone is **against** it. If you are a teenager, the **first** thing you are going to do is **try** them.

As seen, the same passage is coded for the second time to discover why teens use drugs. The reasons for using drugs, in those teens' minds, include a "release" from parents, "a lot of them around", and a "new thing". The axial coding in the above example is done around the why property (i.e., an axis of the drug use category). Other possible properties include types of drugs tried and consequences of drug use if the entire interview transcript is coded.

According to Strauss and Corbin (1998, 124), axial coding links categories to subcategories along the lines of their properties (e.g., when, where, why, who, how, with what consequences) and dimensions (e.g., access is "easy", drug talk is "negative", and the experience is "novel" about the why property). This practice, however, is above and beyond the need of many qualitative data analysis tasks. As in the case of open coding, axial coding in the above example is done more intensively than in actual research projects.

#### *13.2.1.3 Selective coding*

Selective coding refers to the process of integrating and refining what is already coded in open coding and axial coding based on the RQ (Strauss and Corbin 1998, 143). In the example above, the coder would pull together all the different reasons why teens use drugs and eventually form a coherent answer to the RQ: Why do teens use drugs? During selective coding, the data that had been broken down and compared during open coding and axial coding are reassembled and refined for addressing the RQ.

Simply put, researchers break apart their data by identifying categories and subcategories in open coding. They also link and integrate those categories and subcategories in axial coding and selective coding before finalizing their qualitative coding results.

These three kinds of coding may appear linear according to the descriptions above. However, they are not necessarily sequential analytic steps. How and when each of them is used all depends on the research need.

### *13.2.2 Content analysis*

As pointed out in Chapter 9, content analysis can be both a data collection technique and a data analysis technique. Chapter 9 already presents content analysis as a data collection technique which is accomplished by analyzing the content of data sources. This section is devoted to content analysis as a data analysis technique. Content analysis in this context is defined as careful, detailed, systematic examination and interpretation of data gathered. Apparently, content analysis plays a dual-role (i.e., for data collection and for data analysis) in the research process. Table 13.8 compares these two functions of content analysis by data source, analysis process, outcome, and other features.

*Table 13.8* Comparison of Content Analysis for Data Collection and for Data Analysis

| *Feature* | *Content Analysis* | |
|---|---|---|
| | *For Data Collection* | *For Data Analysis* |
| Data source | Books, newspapers, videos, photos, emails, websites, ... | Data collected for a study |
| Analysis process | Similar | |
| Outcome | Source data for a study | Results of a study |
| Type | Same (i.e., manifest and latent) | |
| Other | Involving data analysis in order to collect data | Performing only data analysis |

As shown in Table 13.8, content analysis differs in three aspects when it is used for data collection or for data analysis. First, the data sources for content analysis as a data collection technique include books, newspaper or magazine articles, advertisements, speeches, official documents, videos, musical lyrics, photographs, emails, websites, and social media posts. In contrast, what is collected for a study (e.g., focus group discussions, observation notes) becomes the target data for analysis when content analysis is chosen as a data analysis technique. For example, data gathered via interviews are the target data to be analyzed when content analysis is applied as a data analysis technique.

Second, the outcomes are dissimilar if content analysis is adopted for different purposes. When it is utilized as a data collection technique, the outcome is expected to be source data for a research project. By comparison, the outcome would be results of a study when content analysis is employed as a data analysis technique.

The third difference emerging in this comparison lies in that the data analysis component is always present in research even if content analysis is only meant to be a data collection technique. That is because content analysis for data collection still involves or depends on data analysis. Or, it is through analyzing the content of data source that content analysis achieves its data collection purposes.

Table 13.8 displays the two identical features of content analysis regardless of whether it is for data collection or for data analysis. One refers to that in both cases content analysis is categorized as manifest or latent. Section 9.2 illustrates the characteristics of each type of content analysis in detail as well as offers guidelines with regard to how each can be done accordingly. The other similar aspect shared by content analysis in its dual-role is the analysis process depicted in Section 9.3. What is presented in that section in principle is applicable to content analysis when it is used as a data analysis technique. Researchers likewise need to determine units of analysis; decide if a coding scheme is used and other associated implementations; and generate content categories and/or subcategories by coding and simultaneously using such analytical techniques as abstraction, aggregation, constant comparison, induction, interpretation, and summarization. Among all such analytic techniques that are not individually discussed in data analysis, constant comparison is the most frequently adopted.

It is worth repeating that using content analysis to analyze data has no established, sequential, and standard procedures. Content analysis is often done via qualitative coding, and qualitative coding involves content analysis as well because coding can be performed only after the content of the target data is analyzed. Relationships among qualitative coding, content analysis, and other analytical techniques for qualitative data are neither sequential nor mutually exclusive. As shown above, both qualitative coding and content analysis also rely on other analytical techniques (e.g., categorization, comparison, summarization) to be fully functional.

Some authors (e.g., Yin 2015) regard qualitative data analyses as a process of disassembling and reassembling the target data. These two activities in fact recapitulate qualitative data analysis well. Researchers disassemble their data by identifying categories, subcategories, or even their properties and dimensions as Strauss and Corbin (1998) put when analyzing qualitative data. Researchers then reassemble the analyzed data with identified subcategories and categories to discover recurring themes, build models, create theories, or draw conclusions. Qualitative coding, content analysis, and other analytical techniques hence are utilized simultaneously and coherently throughout the process of qualitative data analysis.

### *13.2.3 Quality assurance of qualitative data analysis*

Because of the very nature of qualitative data analysis (e.g., no standard or established analysis procedures), Lincoln and Guba (1985, 289–331), Yin (2015, 85–90), and other authors suggest the following measures for ensuring its quality.

- Encourage data analysis activities as prolonged and persistent engagement in the process. Unlike quantitative data analysis, qualitative data analysis actually is a reiterative process. In few cases can researchers complete qualitative data analysis as one distinctive phase (Bryman and Burgess 2002, 217) or from the beginning to the end in one shot.
- Do peer debriefing to seek their criticisms and comments as two heads are better than one. In addition, those peers can review the data analysis from a fresh and different perspective as outsiders. Their feedback should be honest, valuable, and helpful.
- Do member checking with participants to make sure that their input (e.g., interview data) is interpreted accurately and in context.
- Search for discrepant evidence and negative cases in the data set to ensure that the analysis accounts for all known cases without exception and does not only cover positive cases.
- Archive a portion of the data gathered for later use. After the analysis of the other data is complete, examine the archived data in light of the available analysis results with an eye to inconsistencies and gaps.
- Use some descriptive statistics such as frequency and percentage to provide a quick overview of the analyzed qualitative data.

The afore-suggested quality assurance measures for qualitative data analysis are by no means a panacea for guarding against all the pitfalls and mistakes in the process. Instead, researchers should fully understand the unique features of qualitative data analysis, perform coding and content analysis prudently along with the application of other analytical techniques such as constant comparison. In addition, keep the above suggestions in mind when analyzing qualitative data.

### *13.2.4 Software for qualitative data analysis*

Software has been developed to help researchers analyze qualitative data. An application for qualitative data analysis should ideally have all the functions listed in Table 13.9 (Weitzman 2000, 805–6). Those functions can be placed into three groups. The first group

*Table 13.9* Functions of Qualitative Data Analysis Software

| *By Word Processor* | *By Qualitative Analysis Software* | *By People* |
|---|---|---|
| • Making notes<br>• Writing up or transcribing<br>• Editing<br>• Storage<br>• Search and retrieval<br>• Memoing/commenting<br>• Report writing | • Coding<br>• Data linking<br>• Data display<br>• Graphic mapping | • Content analysis<br>• Conclusion drawing and verification<br>• Theory building |

represents those tasks that a word processor can accomplish; the second group consists of those functions that qualitative data analysis software can fulfill; and the third group contains those functions that only people can perform.

As shown in Table 13.9, a regular word processor can perform half of the 14 functions of qualitative data analysis. Besides, qualitative data analysis software can handle four additional functions. However, it must be pointed out that both the word processor and qualitative data analysis software can only deal with the mechanical part of the work. Specifically, it is the coder who actually selects and determines which term (i.e., word or phrase) can serve as a code, while the software would mark it up. Likewise, for example, it is the researcher who writes up the report, whereas the software can only do word processing. As for the remaining three functions listed in Column 3 of Table 13.9 (i.e., content analysis, conclusion drawing and verification, theory building), they are all intellectual activities that only people, rather than software, can perform. Apparently purchasing a software application for qualitative data analysis could be a serious financial investment. Yet, the dedicated software can merely handle the mechanical part of the four functions enumerated in Column 2 of Table 13.9. By contrast, a word processor, an application utilized by most, if not all, researchers, can readily perform the mechanical part of the functions listed in Table 13.9 without incurring any additional cost. In other words, meaningful qualitative data analysis is still mostly done by people instead of by dedicated software for qualitative data analysis. It is people who do all the analytical thinking, including qualitative coding, content analysis, and the use of other analytic techniques. It is also necessary for us to know that the qualitative data to be analyzed must be imported into the dedicated software before any qualitative data analysis is conducted.

Below are some dedicated software applications for qualitative data analysis. They all offer demos or tutorials, and some may provide trial software with limited capability.

- ATLAS.ti (http://www.atlasti.com) – An application developed in Germany
- Ethnograph (http://www.qualisresearch.com) – An application developed in the United States
- HyperResearch (http://www.researchware.com) – An application developed in Canada
- NVivo at Lumivero (https://lumivero.com/products/nvivo/) – Initially developed in Australia and formerly part of QSR International at http://www.qsrinternational.com
- Provalis Research (http://www.provalisresearch.com) – An application developed in Canada
    - WordStat for content analysis and text mining
    - QDA Miner for qualitative data analysis

In sum, such software is helpful in aggregating, summarizing, and organizing qualitative data (e.g., frequencies, links among points and texts). The actual analysis is, however, still performed by researchers.

## 13.3 Selection of data analysis techniques

In the previous two sections, we already discuss how to select analytical techniques for quantitative and qualitative data separately. Figure 13.4 provides a complete decision chart for selecting data analysis techniques in a study.

The first step in selecting data analysis techniques is to determine whether the data collected for a study are quantitative or qualitative. Once the data type is determined, researchers can continue the selection process by following the proper path.

If the data to be analyzed are quantitative, there are multiple options to be considered in the selection process of quantitative data analysis techniques. One type of options relates to the undertaken research that includes RQs, number of variables, independent and dependent variables, and more. The other kind of options is about levels of measurement that directly determines individual techniques that can be utilized for analyzing quantitative data. As shown in Table 13.7, quantitative data analysis techniques ought to be chosen according to levels of measurement. If the data are measured at the ordinal level, for example, quantitative analysis techniques for that level of measurement should be chosen. Only then can the corresponding analysis not lose data or commit any logical error.

By contrast, selecting qualitative data analysis techniques is a relatively simple and straightforward process because there are not many choice options. Qualitative coding and content analysis are the only two major, interrelated techniques for analyzing qualitative data after RQs for a study are taken into consideration. However, simple selection of qualitative data analysis techniques does not mean easy application of them. To the contrary, analyzing qualitative data is a complicated and involving task as already discussed in Section 13.2.

Needless to say, each study is unique and requires a set of data analysis techniques carefully chosen after all the relevant factors outlined in Figure 13.4 are considered in the selection process. It is for the same reason that only some general guidelines can be offered for selection of data analysis techniques.

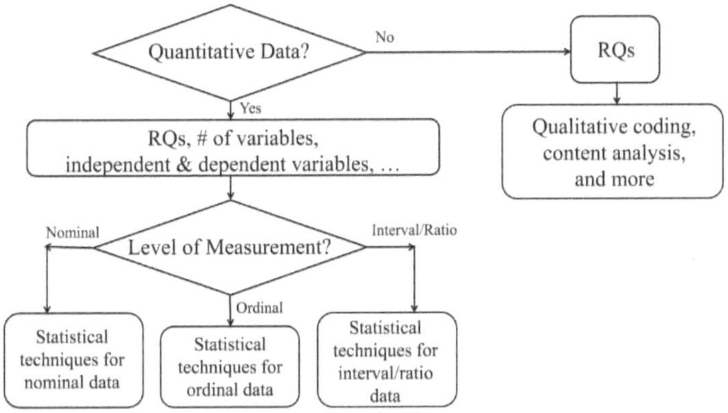


*Figure 13.4* Decision Chart for Selection of Data Analysis Techniques

As pointed out numerous times in this book, no study collects only quantitative or qualitative data. Instead, almost all studies collect both kinds of data although with a focus on one of the two kinds of data. There are two possible scenarios in terms of the quantitative-qualitative division.

1 Research projects gather mainly quantitative data with a small amount of qualitative data. Questionnaire research represents a pertinent example in this category. As known, questionnaire research primarily relies on closed-ended questions to solicit quantitative data, while open-ended questions are used occasionally that yield qualitative data. Consequently, most data from questionnaire research need to be analyzed quantitatively and a small percentage of the collected data should be analyzed qualitatively. Corresponding techniques thus should be selected for analyzing each type of data.
2 Studies collect primarily qualitative data with a small portion of quantitative data. Interview research can serve as a fitting example in this scenario. As discussed in Chapter 4, interview questions are mainly open-ended and solicit qualitative answers. Closed-ended questions account for a small percentage in interview research and typically yield quantitative data. Hence, qualitative techniques are usually selected to analyze interview data while the small amount of quantitative data shall also be analyzed with proper techniques.

In either scenario, researchers need to choose analysis techniques for both quantitative and qualitative data with a focus on one of the two types of data accordingly.

In addition, it is a recommended practice to use some descriptive statistics such as frequency and percentage (e.g., number of identified categories and themes) to provide a quick overview of the analyzed qualitative data. Similarly, it also helps gain a better understanding of the analyzed quantitative data by discussing quantitative results with a qualitative lens. In other words, quantitative and qualitative techniques should be utilized in combination and integrally for data analysis according to data type and other considerations (e.g., RQs), forming the integrated data analysis approach briefly introduced in Section 1.2. Data analysis techniques in no case should be selected and applied in isolation and out of the context of a research project.

**Chapter Summary**

- Quantitative or qualitative techniques constitute the two major kinds of techniques for data analysis. All the data collected for research can be broadly categorized into two kinds: quantitative or qualitative. Quantitative techniques are developed for analyzing numeric or quantitative data while qualitative methods are established to analyze text or qualitative data.
- Quantitative data analysis techniques are also known as statistical data analysis techniques that are placed into two categories: descriptive and inferential. Descriptive statistical data analyses enable researchers to describe, summarize, and gain an overview of what is collected before deciding what advanced statistical procedures can be utilized to further analyze the data on hand. By comparison,

inferential statistical data analyses rely on data collected from a sample to achieve one of the following three objectives: estimation, hypothesis testing, and predication. There are well-established procedures for performing both types of statistical data analyses.

- Qualitative data analysis features no fixed, sequential, and standard procedures. This brings challenges to qualitative data analysis and requires the adoption and implementation of quality control measures (e.g., encouraging prolonged data analysis, doing peer debriefing).
- Qualitative coding and content analysis are the two major techniques for analyzing qualitative data. However, other techniques such as constant comparison should be utilized together with qualitative coding and content analysis during the qualitative data analysis process.
- When the data to be analyzed are quantitative, multiple factors need to be considered: RQs of the study, number of variables, independent and dependent variables as well as levels of measurement in selecting techniques for data analysis. In contrast, the selection of data analysis techniques is relatively straightforward when the data to be analyzed are qualitative because the interrelated qualitative coding, content analysis, and other techniques are usually applied simultaneously in the data analysis process.
- The integrated approach to data analysis implies that quantitative and qualitative techniques should be utilized in combination and integrally for data analysis based on the type of data collected and other considerations (e.g., RQs). As no study collects only quantitative or qualitative data, techniques for analyzing those data shall be selected accordingly. In addition, use some descriptive statistics such as frequency and percentage (e.g., number of identified categories and themes) to provide a quick overview of the analyzed qualitative data. Similarly, present and discuss quantitative results with a qualitative lens to help gain a better understanding of the analyzed quantitative data.

## Notes

1 Figure 13.1 is reprinted with permission from Kazarinoff (2022).
2 Figure 13.2 is derived from Figure 1 in CDC (2007, 18). As CDC is a US government agency, its publications are in the public domain.
3 Figure 13.3 is adapted from Rowntree (1981, 163 and 170).
4 Table 13.7 is adapted from Figure 2.2 in Loether and McTavish (1980, 20).

## References


Andrews, Frank M., Laura Klem, Terrence N. Davidson, Patrick M. O'Malley, and Willard L. Rodgers. 1981. *A Guide for Selecting Statistical Techniques for Analyzing Social Science Data*. 2nd ed.. Ann Arbor: Survey Research Center, Institute for Social Research, the University of Michigan.

Babbie, Earl R. 2016. *The Practice of Social Research*. 14th ed. Boston: Cengage Learning.

Bryman, Alan and Robert G. Burgess., eds. 2002. *Analyzing Qualitative Data*. New York: Routledge.

CDC. 2007. *Interim Pre-Pandemic Planning Guidance: Community Strategy for Pandemic Influenza Mitigation in the United States. Early, Targeted, Layered Use of Nonpharmaceutical Interventions*. Centers for Disease Control and Prevention, U.S. Department of Health and Human Services.

easycalculation.com. 2023. *Statistics Calculator Online*. Accessed June 21, 2023. https://www.easycalculation.com/statistics/statistics.php.

Frankfort-Nachmias, Chava, David Nachmias, and Jack DeWaard. 2015. *Research Methods in the Social Sciences*. 8th ed. New York: Worth Publishers.

Kazarinoff, Peter. 2022. "Calculate the Probability Under a Normal Curve." Accessed September 19, 2023. https://pythonforundergradengineers.com/probability-under-normal-curve-with-python.html.

Loether, Herman J. and Donald G. McTavish. 1980. *Descriptive and Inferential Statistics: An Introduction*. Boston: Allyn and Bacon.

Lincoln, Yvonna S., and Egon G. Guba. 1985. *Naturalistic Inquiry*. Beverly Hills: Sage.

Rowntree, Derek. 1981. *Statistics Without Tears: A Primer for Non-Mathematicians*. Boston: Allyn and Bacon.

Ruiz, Neil, Luis Noe-Bustamante, and Nadya Saber. 2020. "Coming of Age." *Finance and Development* 57 (1):46–47. Accessed June 17, 2023. https://www.imf.org/en/Publications/fandd/issues/2020/03/infographic-global-population-trends-picture.

Statista. 2023. "Number of Netflix Paid Subscribers Worldwide from 1st Quarter 2013 to 1st Quarter 2023." Accessed June 16, 2023. https://www.statista.com/statistics/250934/quarterly-number-of-netflix-streaming-subscribers-worldwide/.

Strauss, L. Anselm, and Juliet M. Corbin. 1998. *Basics of Qualitative Research: Techniques and Procedures for Developing Grounded Theory*. Thousand Oaks: Sage.

Tesch, Renata. 1990. *Qualitative Research: Analysis Types and Software Tools*. New York: The Falmer Press.

Van Eck, Neal. 2014. *The Decision Tree for Statistics*. Accessed June 20, 2023. https://statisticaldecisiontree.microsiris.com/default.htm.

Weitzman, Eben. 2000. "Software and Qualitative Analysis." In *Handbook of Qualitative Research*, edited by Norman K. Denzin and Yvonna S. Lincoln, 803–20. 2nd ed. Thousand Oaks: Sage.

Wikipedia Contributors. 2022. "Qualitative Variation." In *Wikipedia, The Free Encyclopedia*. Last modified November 29, 2022. Accessed June 16, 2023. https://en.wikipedia.org/wiki/Qualitative_variation.

Wikipedia Contributors. 2023. "A List of Countries by GDP (Nominal) Per Capita." In *Wikipedia, The Free Encyclopedia*. Last modified June 8, 2023. Accessed June 16, 2023. https://en.wikipedia.org/wiki/List_of_countries_by_GDP_(nominal)_per_capita.

Wilcox, Allen R. 1967. *Indices of Qualitative Variation*. Oak Ridge: Oak Ridge National Laboratory. Accessed June 14, 2023. https://www.osti.gov/servlets/purl/4167340.

Wilcox, Allen R. 1973. "Indices of Qualitative Variation and Political Measurement." *The Western Political Quarterly* 26 (2): 325–43.

Yin, Robert K. 2015. *Qualitative Research from Start to Finish*. 2nd ed. New York: The Guilford Press.

# 14 Research Design via Method Selection

**Chapter Overview Questions**

- What role does method selection or research methodology development play in research design?
- What are the two major factors to be considered in the selection of data collection techniques for a study?
- How does research method applicability help in the selection process of research methods?
- When should researchers determine what data analysis techniques should be chosen for a study?
- Why should researchers select multiple techniques and apply them integrally for data collection and analysis in a study?

Research design refers to the overall plan of a project starting from identification of a research problem to presentation of research findings. The essence of any research design, however, mainly centers on research methodology development or the selection of research methods, consisting of both data collection and analysis techniques, according to research questions of a study and functions/features of those methods.

## 14.1 Research questions and functions/features of data collection methods

In Chapter 2, we address some major points of research design, including problem identification and statement, literature search and review, from problem statement to research questions, research validity and reliability, and research ethics. Development of research methodology could not be covered in Chapter 2 because its components – data collection techniques and data analysis techniques – were not yet presented at that point. Now that main data collection techniques are covered in Chapters 3–12, and Chapter 13 is devoted to techniques for data analysis and their selection, the time is right for discussing research methodology development, a key step in research design.

Research methodology development involves selection and justification of data collection and analysis techniques based on the research questions formulated for a study and functions/features of related research methods. As pointed out throughout this book, research methods are named after data collection techniques. The general procedure is that researchers look closely at their research questions to find out what data need to be

DOI: 10.4324/9781003021711-18

collected to answer them before deciding which data collection techniques will enable them to accomplish the data collection task primarily based on functions and features of relevant data collection techniques.

Functions and features of each data collection method included in the current book are already described and discussed in detail from Chapters 3 through 12. Table 14.1 recapitulates the functions and features of the major research methods and corresponding chapter number for each is also listed for easy reference. It is those method functions and features that truly help researchers to make decisions in selecting data collection techniques appropriate for their projects.

Readers are encouraged to consult the functions and features of each data collection method presented in each chapter before making a method selection. For example, researchers can choose questionnaire as their data collection technique if a study intends to gather data from a large number of participants who are located in different geographical areas. Similarly, researchers would select focus groups as their data collection technique if their research objective is to find out collective viewpoints on a given topic and interaction among participants in each group. If individuals' viewpoints on a topic are what a research project intends to gather, interview would be an appropriate data collection technique. Such rationale could go on, and Table 14.1 can serve as a guide for selecting data collection techniques for any research project.

In Section 2.3.1, the following research questions are formulated for a study that intends to explore factors affecting the adoption of handheld devices.

- Research Question 1: Is age a factor affecting the adoption of handheld devices?
- Research Question 2: Is gender a factor affecting the adoption of handheld devices?
- Research Question 3: Are years of education a factor affecting the adoption of handheld devices?
- Research Question 4: Is prior ownership of small electronic devices a factor affecting the adoption of handheld devices?
- …
- Research Question n: Are there other factors that affect the adoption of handheld devices?

What data should be collected in order to answer the above list of questions? It can be seen from the first four research questions that data need to be collected about the following variables: age, gender, years of education, ownership of small electronic devices, and adoption of handheld devices. In addition, the adoption of handheld devices variable is optionally defined in Section 2.3.1 as purchasing and owning a handheld device (e.g., a mobile phone) or borrowing one (e.g., a tablet) from a library. Data about this variable should thus be collected accordingly. As for the variable of other factors in Research Question n, it is for researchers to solicit from participants what other factors that should also be considered in the study. This example demonstrates how to determine what data need to be collected by closely examining research questions.

Data about age, gender, years of education, ownership of small electronic devices, and adoption of handheld devices apparently can be gathered using closed-ended questions. Data about the variable of other factors, however, shall be collected via an open-ended question, allowing participants to enumerate factors affecting the adoption of handheld devices but not already specified in other research questions. When data can be mostly gathered with closed-ended questions, questionnaire seems to be an appropriate research

*Table 14.1* Major Data Collection Techniques and Their Functions/Features

| *Research Method* | *Method Function/Feature* | *Chapter* |
|---|---|---|
| Bibliometrics and altmetrics | Collect publication, citation, and social media data unobtrusively for evaluating research, tracing the development of a discipline, understanding the intellectual structure of a field or domain, exploring scholarly exchanges among researchers, gauging the sentiment of social media users, and more. | 7 |
| Content analysis | Collect data by examining and analyzing the content in texts unobtrusively for understanding the manifest and latent messages imbedded in texts. Texts in content analysis can be books, newspaper or magazine articles, advertisements, speeches, official documents, videos, musical lyrics, photographs, emails, websites, and social media posts. | 9 |
| Delphi study | Collect data with a questionnaire or survey from a group of experts to address a research problem in order to reach consensus or make forecasts via several rounds of exchanges. Anonymity, iteration, statistical summary and aggregation of individual participants' responses, and controlled feedback of responses to all participants are its distinctive features. | 12.2 |
| Ethnography/field study | Collect data using multiple individual techniques specifically selected for learning about or understanding the members of a community in their physical setting (i.e., field) through their actions, behaviors, exchanges, interactions, and other activities. Data collection is typically done over a period of time (e.g., months or even years) using interview, observation, and other techniques. | 11 |
| Experiment | Collect data by following a procedure to test or try out what is studied in either a laboratory or a real setting in order to study causal relationships between independent and dependent variables. Multiple techniques are adopted for data collection in experiment research. For instance, use questionnaire or interview for pretest and posttest, perform observation in the experiment process, or employ think-aloud protocols for gathering cognitive data from participants. | 10 |
| Focus groups | Collect data from several groups of participants on their collective viewpoints on a given research topic as well as on interaction among participants in each group through discussion and commenting. | 8.1 |
| Historical method | Collect data for studying problems concerning the past by critically assessing, analyzing, interpreting, and presenting evidence from both published and unpublished materials. | 8.4 |
| Interview | Collect data via a real-time question-and-answer conversation in which the researcher asks each participant questions relating to a research problem and the participant will in turn answer the questions asked. It is especially helpful for researchers to gather in-depth data from a small number of participants. | 4 |
| Observation | Collect data via carefully and attentively watching and systematically making notes on what is being studied. It requires researchers to use all their five senses and draw on their knowledge and common sense. Observation allows researchers to collect data from their own perspective. Researchers can observe and record what occurs in real time without the need for participants to recall from memory as in doing questionnaire and interview research | 5 |

(*Continued*)

*Table 14.1* (Continued)

| *Research Method* | *Method Function/Feature* | *Chapter* |
|---|---|---|
| Questionnaire | Collect data via a list of questions specifically created for a study. The majority of survey questions are closed-ended that can be further divided into factual, opinion/attitude, and filter questions. A small number of open-ended questions may also be asked in a survey for collecting qualitative data. It is particularly useful if researchers plan to get data from a large number of participants dispersed geographically. | 3 |
| Research diary/journal | Collect data of diary entries created and maintained by participants or researchers to record events, activities, observations, thoughts, and questions as they occur at a specific interval for (1) seeking to understand hard-to-reach populations (e.g., victims of domestic violence) or private spheres (e.g., drug addicts) that can pose practical and ethical issues to observe or record; and for (2) asking participants to log the events of their daily lives or providing a space for participants to reflect on experiences in their lives, particularly those linked to health and other personal issues. | 8.3 |
| Theoretical approach | Collect data through conceptual analysis, theoretical examination, or other similar activities. Some theoretical approach studies aim to develop a model, framework, or theory in a respective domain, while others intend to address their research questions by performing theoretical or conceptual analysis of collected data. | 6 |
| Think-aloud protocol | Collect data about participants' cognitive and thinking processes by asking them to verbalize whatever crosses their mind as they perform a specified task (e.g., conducting a search, solving a problem). It is the only method for gathering cognitive data from participants. | 12.1 |
| Transaction log analysis | Collect data automatically captured and recorded by a system (e.g., search engines, e-commerce sites) that include the type, content, and time of transactions or interactions between the system and users of that system (e.g., searchers, online shoppers) for either evaluating the system's performance or users' information behavior unobtrusively. Transaction log data, once permission for data collection is properly obtained, involve no recall, biases, low responses, and incorrect memory from research participants. | 8.2 |

method. This example hence illustrates how one can select a data collection technique according to what data need to be gathered by examining research questions and in consideration of method functions and features.

In sum, research questions of a study determine what data should be gathered. The latter then can be utilized to decide which techniques can be adopted to collect such data. Data analysis techniques are chosen, as discussed in Chapter 13, after research data are collected.

## 14.2 Method selection and applicability

Section 1.5 introduces the topic of research method applicability and establishes four categories of research methods by applicability: super-methods (e.g., content analysis), stand-alone methods (e.g., bibliometrics and other metric-based methods, focus groups, historical method, interview, observation, questionnaire, research diary, theoretical approach, transaction log analysis), meta-methods (e.g., experiment, ethnography/field study), and contingent methods (e.g., Delphi study, think-aloud protocol). This taxonomy of research methods by applicability can help researchers to decide whether a research method can be used alone or otherwise. The presentation of research methods in this book is also organized according to this framework.

From the perspective of research method applicability, each stand-alone research method can be utilized by itself in a study or adopted in conjunction with other methods within its own category or from other categories. Examples of such applications are many. For instance, bibliometrics can be used for gathering publication and citation data about the development of one specialty. Interviews or focus groups can then be utilized with researchers from that specialty to collect data on their assessment of its development.

Super-methods such as content analysis basically are applied alone or with other method in any study explicitly or implicitly. Some studies focus exclusively on analyzing the content of data sources for data collection purpose (e.g., Bie and Tang 2015 – the case study for Chapter 9 Content Analysis). Some research applies content analysis explicitly as one of the data collection techniques (e.g., content analysis of site documents in Athanases and Heath 1995 – Case Study 1 in Chapter 11 Ethnography/Field Study). In contrast, some investigations may perform content analysis implicitly. Take interview as an example, content analysis of interviewees' responses is usually performed during the interview process to determine if additional questions need to be asked in semi-structured and unstructured interviews although content analysis in this example is not listed explicitly as a data collection method.

Meta-methods inherently comprise more than one method but their compositions may vary from one research project to another. For example, observations, interviews, surveys, and content analysis of site documents are among the data collection techniques chosen for Case Study 1 in Chapter 11 Ethnography/Field Study (Athanases and Heath 1995). By comparison, participant observation and semi-structured interviews constitute the methodology for Case Study 2 in the same chapter (Bundgaard et al. 2012). Both are field research projects but the individual methods each study selected are based respectively on their research questions and method functions/features.

The individual data collection techniques in contingency methods are pre-determined. Namely, think-aloud protocols rely on the experiment method (e.g., performing a task, doing a test) in its implementation, whereas Delphi study is built on a series of linked surveys to be functional. However, such integration of data collection techniques does not exclude either think-aloud protocols or Delphi study from selecting other research methods in their application. For example, the case study by Brancheau, Janz, and Wetherbe (1996) for Section 12.2 conducted a follow-up interview with its participants. The researchers, in addition to soliciting the participants' general reaction to the results, asked which forces were driving the top issues, what made the issues so troublesome, and whether they thought the framework had the right balance between management and technology issues. Those interviews are a fine complement to the Delphi method as they can solicit in-depth data from participants that would not be able to be gathered via the Delphi survey.

In sum, method applicability can facilitate the selection of data collection techniques. Knowing the taxonomy of research methods by applicability, researchers would be more at ease in choosing appropriate individual research methods to form a methodology for a given study. Transaction log analysis, for example, is a stand-alone method. Nevertheless, we can also use interviews together with this method in order to gather data about participants' views and opinions that cannot be collected using the transaction log analysis method.

## 14.3 Integrated research method designs

As discussed in Section 14.1, research method selection relies on research questions and method functions/features. Often times a research project requires the use of more than one method in order to gather all the data needed for the study because each method, as explained from Chapters 3 to 12, is only good for collecting certain kind of data. For instance, bibliometrics is suitable for gathering publication and citation data but needs to rely on interview or questionnaire if researchers plan to collect data on reasons for citation.

Mixed methods are likely a term we frequently encounter when considering the utilization of more than one method in a study. However, that term cannot properly reflect the true intention of using multiple methods in a research project in two aspects. First, the term of mixed methods erroneously assumes that there are quantitative methods and qualitative ones. This assumption is false because, as discussed throughout this book and summarized in Table 13.1, each research method can be used to collect both quantitative and qualitative data even if it is more suitable for gathering one of the two types of data. In other words, there are no quantitative or qualitative research methods. Instead, research methods can be used to collect mainly quantitative or qualitative data. Second, choosing more than one research method for a study requires serious consideration of how selected individual methods can complement each other before implementing them integrally in research. This process is obviously far more than a casual mix of different methods. Because of the afore-stated two points, integrated methods rather than mixed methods would be a more appropriate term for research that adopts more than one research method for data collection.

As seen from above, using integrated methods in a study implies that researchers need to think through the functions and features of each method besides considering how to apply chosen methods integrally so that the advantage of one method can help overcome the limitation of the other. For example, questionnaire can gather data from a large number of geographically dispersed participants but offers no opportunity for asking clarification and follow-up questions. If interview is used in conjunction with questionnaire in a study, researchers can ask clarification and follow-up questions while mitigating the limitation of the interview method: It can only involve a small number of participants. Questionnaire and interview therefore constitute one common integrated method design.

Other integrated method design examples are many in the scholarly world as well as in the case studies included in this book. For instance, Logan, Barrett, and Pagotto (2019), the case study for transaction log analysis (see Section 8.2.5), also administered a pre-chat survey as well as an optional exit survey in order to gather additional data on both reference staff and users' behaviors during the chat reference sessions.

In the case study selected for research diaries (see Section 8.3.5), Välimäki, Vehviläinen-Julkunen, and Pietilä (2007) chose research diaries as the primary data collection technique. All the participants were also randomly assigned to one of the two groups. The experiment or intervention group received intensive psychoeducational courses to help family caregivers of people with Alzheimer's disease, while the control group did not. This experiment component should allow researchers to gather data and then compare if the intensive psychoeducational courses are of help to the family caregivers of people with Alzheimer's disease. However, Välimäki, Vehviläinen-Julkunen, and Pietilä (2007) did not report any comparison result given their exclusive focus on research diaries as a data collection technique in the publication.

Experiments represent another set of examples of integrated method designs. In a true experiment, researchers may choose questionnaire or interview for pretest and posttest, perform observation in the experiment process, and employ think-aloud protocols if cognitive data need to be gathered from participants. As seen, each individual technique in this integrated method design serves its specific function and collects a particular kind of data. It is the researchers who decide which and how many data collection techniques shall be selected for a project based on research questions and method functions/features. Take Case study 2 for Chapter 10 Experiment (see Section 10.5.2) as an example, Venkatraman and Zaheer (1990) used content analysis of an insurance carrier's internal records to gather each interfaced agent's performance data rather than data collection techniques like questionnaire or interview for pretest and posttest purposes in the quasi-experiment study. Although the researchers collected pretest and posttest data unconventionally, content analysis is one of the integrated methods used in this quasi-experiment study.

The research Athanases and Heath (1995) reported, Case Study 1 for Chapter 11 Ethnography/Field Study (see Section 11.5.1), sets a pertinent example of applying multiple data collection techniques in a single study. Section 11.3.2.1 in that chapter actually is devoted to the discussion on selecting multiple techniques for data collection. Observations, interviews, surveys, and content analysis of site documents are among the data collection techniques chosen for the study. As detailed in the case study, each of the chosen methods was utilized to collect certain type of data for the study. Specifically, observations were made in the classrooms to gather data about the setting and what was going on at the time of observation. Interviews were conducted with teachers, students, parents, and other school personnel, enabling the collection of data on a variety of topics (e.g., school system curriculum, school organization). Surveys were distributed to students and teachers to solicit written data from them. Content analysis of site documents (e.g., school newspapers, and brochures) was performed to collect data from existing sources. In addition, discussions – a form of focus groups – were organized with students to gather their reactions to the English classes they had two years ago. Although there is room for improvement in the application of those data collection techniques (e.g., discussion of the classes students had two years ago), Athanases and Heath (1995) demonstrate how to select multiple methods and apply them integrally to achieve the purpose of data collection.

There are as numerous different combination and integrations of research methods as we can choose for research methodology development. It is neither possible nor the intention of this book to enumerate all such integrated method designs. However, the principles underlying those designs remain the same: Select techniques that can gather data to address the research questions formulated in a study according to method functions and features.

A total of 15 research methods receive extensive coverage in this book. Readers are therefore encouraged to refer to the corresponding chapters to learn more or review relevant research methods in the process of research methodology development.

## 14.4 Final words on research methods and design

Chapters 3–12 present 15 research methods and can serve as a toolbox in the research design process regarding method selection. Together with what is covered in Chapter 2, The Research Process and Research Design, and the rest of the book, design a study by following all the principles and practice discussed in this book and especially by choosing proper data collection techniques. Only then can researchers determine how the collected data, as discussed in Chapter 13, Techniques for Data Analysis and Their Selection, are to be analyzed to answer the research questions formulated for the study.

Throughout all the 14 chapters, one point worth reemphasizing is that research methods should be categorized and named by data collection technique instead of anything else. Only in doing so, can we have unique and indicative names of research methods such as questionnaire, interview, and observation. Anything that cannot be utilized for data collection is not a research method. It might instead merely refer to the sample size of research (e.g., case study) or temporal dimension of a study (e.g., longitudinal research).

Research methods should never be categorized as quantitative or qualitative simply because almost all research methods can be utilized for collecting both quantitative and qualitative data although each method is more suitable for collecting one of the two types of data. However, data analysis techniques are always labelled as quantitative and qualitative because all the data analysis techniques are naturally divided into these top two groups (see Chapter 13).

Research methods can also be chosen by applicability in addition to other established selection criteria (e.g., research questions of a study and method functions/features). Some methods can be used alone (i.e., stand-alone methods); some can be used in all occasions explicitly or implicitly (i.e., super-methods); some methods need to be used together (i.e., meta-methods), and others have to rely on one method in order to be functional (i.e., contingent methods). This applicability of research methods can additionally help scholars with their selection of data collection techniques.

As each method is designed to collect certain kind of data, researchers are encouraged to choose multiple methods in order to gather all the data a study needs. After the data collection process is completed, selection of data analysis techniques can begin based on the type of data collected, research questions of the study, and other related considerations (e.g., number of variables in the study). This integrated approach is highly recommended so that each technique is chosen carefully and integrated with others to maximize their benefits while mitigating their limitations. The so-called mixed methods, as a phrase, may not accurately reflect researchers' intention to choose and integrate more than one method for data collection and analysis in a study.

Research design through method selection is by no means an easy task. Researchers need to be clear about their research questions and be familiar with the functions/features of each viable research method before making an educated choice. This book provides the principles and practice in research methods and design, hoping to guide readers through this complex process smoothly and successfully.

**Chapter Summary**

- Research method selection or research methodology development plays an essential role in research design. It is through those selected techniques for data collection and analysis that a research design gets implemented.
- Research questions of a study and functions/features of relevant methods are the two major factors to be considered in the selection process of data collection

techniques. We look at the research questions formulated in a study to find out what data need to be collected. Then we choose data collection techniques based on their functions and features to gather what is needed for addressing the research questions of the study.

- Research method applicability can facilitate the method selection process by offering additional guidance regarding which method can be used alone or otherwise.
- Researchers can determine what data analysis techniques should be chosen for a study after its data collection process is complete. At that point, researchers would know for certain if they have quantitative data, qualitative data or both before choosing data analysis techniques accordingly.
- As each data collection technique can only gather certain kind of data, it is common that more than one technique are adopted in order to collect all the data a study needs to address its research questions. Likewise, each data analysis technique has its unique functions and features. Therefore, multiple techniques should be selected and applied integrally for data collection and analysis so that their benefits are maximized and their limitations are mitigated.

## References

Athanases, Steven Z., and Shirley Brice Heath. 1995. "Ethnography in the Study of the Teaching and Learning of English." *Research in the Teaching of English* 29 (3): 263–87. Accessed June 19, 2023. https://www.jstor.org/stable/40171429.

Bie, Bijie, and Lu Tang. 2015. "Representation of Autism in Leading Newspapers in China: A Content Analysis." *Health Communication* 30 (9): 884–93. https://doi.org/10.1080/10410236.2014.889063.

Brancheau, James C., Brian D. Janz, and James C. Wetherbe. 1996. "Key Issues in Information Systems Management: 1994–95 SIM Delphi Results." *MIS Quarterly* 20 (2): 225–42. https://doi.org/10.2307/249479.

Bundgaard, Karin, Karl Brian Nielsen, Charlotte Delmar, and Erik Elgaard Sørensen. 2012. "What to Know and How to Get to Know? A Fieldwork Study Outlining the Understanding of Knowing the Patient in Facilities for Short-Term Stay." *Journal of Advanced Nursing* 68 (10): 2280–88. https://doi.org/10.1111/j.1365-2648.2011.05921.x.

Logan, Judith, Kathryn Barrett, and Sabina Pagotto. 2019. "Dissatisfaction in Chat Reference Users: A Transcript Analysis Study." *College & Research Libraries* 80 (7): 925–44. https://doi.org/10.5860/crl.80.7.925.

Välimäki, Tarja, Katri Vehviläinen-Julkunen, and Anna-Maija Pietilä. 2007. "Diaries as Research Data in a Study on Family Caregivers of People With Alzheimer's Disease: Methodological Issues." *Journal of Advanced Nursing* 59 (1): 68–76. https://doi.org/10.1111/j.1365-2648.2007.04273.x.

Venkatraman, N., and Akbar Zaheer. 1990. "Electronic Integration and Strategic Advantage: A Quasi-Experimental Study in the Insurance Industry." *Information Systems Research* 1 (4): 377–93. https://doi.org/10.1287/isre.1.4.377.

# Appendix

## List of Case Study Publications

A full copy of all the case study publications used in this book is available from Google Scholar by searching the title unless indicated otherwise.

**Chapter 3 Questionnaire**

Case Study 1

National Research Council. 2006. "Assessment of Research Doctorate Programs: Admitted-to-candidacy Doctoral Student Questionnaire." In *A Guide to the Methodology of the National Research Council Assessment of Doctorate Programs*, edited by Jeremiah P. Ostriker, Paul W. Holland, Charlotte V. Kuh, and James A. Voytuk, 141–58. Washington: The National Academies Press. Accessed June 19, 2023. https://gradschool.unc.edu/facultystaff/nrc/NRC-Methodology-Guide.pdf. (Search Google Scholar by *A Guide to the Methodology of the National Research Council Assessment of Doctorate Programs*.)

Case Study 2

Niu, Xi, Bradley M. Hemminger, Cory Lown, Stephanie Adams, Cecelia Brown, Allison Level, Merinda McClure, Audrey Powers, Michele R. Tennant, and Tara Cataldo. 2010. "National Study of Information Seeking Behavior of Academic Researchers in the United States." *Journal of the American Society for Information Science and Technology* 61 (5):869–90. https://doi.org/10.1002/asi.21307.

**Chapter 4 Interview**

Case Study 1

Boyce, Carolyn, and Palena Neale. 2006. *Conducting In-depth Interviews: A Guide for Designing and Conducting In-depth Interviews for Evaluation Input.* Pathfinder International Tool Series, Monitoring and Evaluation – 2. Watertown: Pathfinder International.

Case Study 2

Ng'oma, Mwawi, Samantha Meltzer-Brody, Ellen Chirwa, and Robert C. Stewart. 2019. "'Passing Through Difficult Times': Perceptions of Perinatal Depression and Treatment Needs in Malawi - A Qualitative Study to Inform the Development of a Culturally Sensitive Intervention." *PLoS ONE* 14 (6): e0217102. https://doi.org/10.1371/journal.pone.0217102.

**Chapter 5 Observation**

Case Study 1

Northcote, Jeremy, and Michael Livingston. 2011. "Accuracy of Self-Reported Drinking: Observational Verification of 'Last Occasion' Drink Estimates of Young Adults." *Alcohol and Alcoholism* 46 (6): 709–13. https://doi.org/10.1093/alcalc/agr138.

Case Study 2

Mueller, Susan, Thierry Volery, and Björn von Siemens. 2012. "What Do Entrepreneurs Actually Do? An Observational Study of Entrepreneurs' Everyday Behavior in the Start-Up and Growth Stages." *Entrepreneurship Theory and Practice* 36 (5): 995–1017. https://doi.org/10.1111/j.1540-6520.2012.00538.x.

**Chapter 6 Theoretical approach**

Case Study 1

Lewicki, Roy J., Edward C. Tomlinson, and Nicole Gillespie. 2006. "Models of Interpersonal Trust Development: Theoretical Approaches, Empirical Evidence, and Future Directions." *Journal of Management* 32 (6): 991–1022. https://doi.org/10.1177/0149206306294405.

Case Study 2

Saarijärvi, Hannu, P. K. Kannan, and Hannu Kuusela. 2013. "Value Co-creation: Theoretical Approaches and Practical Implications." *European Business Review* 25 (1): 6–19. https://doi.org/10.1108/09555341311287718. (Inaccessible at Google Scholar on June 19, 2023, but it can be obtained from Research Gate after logging in or signing up for a free account at https://www.researchgate.net).

**Chapter 7 Metrics-based research methods: From bibliometrics to altmetrics**

Case Study for Bibliometrics

Harzing, Anne-Wil. 2013. "A Preliminary Test of Google Scholar as a Source for Citation Data: A Longitudinal Study of Nobel Prize Winners." *Scientometrics* 94 (3): 1057–75. https://doi.org/10.1007/s11192-012-0777-7.

Case Study for Altmetrics

Bornmann, Lutz. 2014b. "Validity of Altmetrics Data for Measuring Societal Impact: A Study Using Data from Altmetric and F1000Prime." *Journal of Informetrics* 8 (4): 935–50. http://dx.doi.org/10.1016/j.joi.2014.09.007.

**Chapter 8 Other stand-alone methods**

Case Study for §8.1 Focus Groups

Gullifer, Judith, and Graham A. Tyson. 2010. "Exploring University Students' Perceptions of Plagiarism: A Focus Group Study." *Studies in Higher Education* 35 (4): 463–81. https://doi.org/10.1080/03075070903096508.

Case Study for §8.2 Transaction Log Analysis

Logan, Judith, Kathryn Barrett, and Sabina Pagotto. 2019. "Dissatisfaction in Chat Reference Users: A Transcript Analysis Study." *College & Research Libraries* 80 (7): 925–44. https://doi.org/10.5860/crl.80.7.925.

Case Study for §8.3 Research Diaries

Välimäki, Tarja, Katri Vehviläinen-Julkunen, and Anna-Maija Pietilä. 2007. "Diaries as Research Data in a Study on Family Caregivers of People with Alzheimer's Disease: Methodological Issues." *Journal of Advanced Nursing* 59 (1): 68–76. https://doi.org/10.1111/j.1365-2648.2007.04273.x.

Case Study for §8.4 Historical Method

Mackintosh, Carolyn. 1997. "A Historical Study of Men in Nursing." *Journal of Advanced Nursing* 26 (2): 232–36. https://doi.org/10.1046/j.1365-2648.1997.1997026232.x.

**Chapter 9 Content analysis: A super-method**

Case Study

Bie, Bijie and Lu Tang. 2015. "Representation of Autism in Leading Newspapers in China: A Content Analysis." *Health Communication* 30 (9): 884–93. https://doi.org/10.1080/10410236.2014.889063.

**Chapter 10 Meta-methods: Experiments**

Case Study 1

Clariana, Roy B. and Doris Lee. 2001. "The Effects of Recognition and Recall Study Tasks with Feedback in a Computer-Based Vocabulary Lesson". *Educational Technology Research and Development* 49 (3): 23–36. https://doi.org/10.1007/bf02504913.

Case Study 2

Venkatraman, N. and Akbar Zaheer. 1990. "Electronic Integration and Strategic Advantage: A Quasi-Experimental Study in the Insurance Industry." *Information Systems Research* 1 (4): 377–93. https://doi.org/10.1287/isre.1.4.377.

**Chapter 11 Meta-methods: Ethnography/field study**

Case Study 1 Ethnography

Athanases. Steven Z., and Shirley Brice Heath. 1995. "Ethnography in the Study of the Teaching and Learning of English." *Research in the Teaching of English* 29 (3): 263–87. Accessed June 19, 2023. https://www.jstor.org/stable/40171429.

Case Study 2 Field Study

Bundgaard, Karin, Karl Brian Nielsen, Charlotte Delmar, and Erik Elgaard Sørensen. 2011. "What to Know and How to Get to Know? A Fieldwork Study Outlining the Understanding of Knowing the Patient in Facilities for Short-term Stay." *Journal of Advanced Nursing* 68 (10): 2280–88. https://doi.org/10.1111/j.1365-2648.2011.05921.x.

**Chapter 12 Continent methods**

Case Study for §12.1 Think-aloud Protocols

Alhadreti, Obead and Pam Mayhew. 2018. "Rethinking Thinking Aloud: A Comparison of Three Think-Aloud Protocols." *CHI 2018* Paper 44. April 21–26, 2018, Montréal, QC, Canada. 1–12. https://doi.org/10.1145/3173574.3173618.

Case Study for §12.2 Delphi Study

Brancheau, James C., Brian D. Janz, and James C. Wetherbe. 1996. "Key Issues in Information Systems Management: 1994–95 SIM Delphi Results." *MIS Quarterly* 20 (2): 225–42. https://doi.org/10.2307/249479.

# Index

*Italicized* and **bold** pages refer to figures and tables respectively.